Lecture Notes in Networks and Systems

Volume 39

Series editor

Janusz Kacprzyk, Polish Academy of Sciences, Warsaw, Poland
e-mail: kacprzyk@ibspan.waw.pl

The series "Lecture Notes in Networks and Systems" publishes the latest developments in Networks and Systems—quickly, informally and with high quality. Original research reported in proceedings and post-proceedings represents the core of LNNS.

Volumes published in LNNS embrace all aspects and subfields of, as well as new challenges in, Networks and Systems.

The series contains proceedings and edited volumes in systems and networks, spanning the areas of Cyber-Physical Systems, Autonomous Systems, Sensor Networks, Control Systems, Energy Systems, Automotive Systems, Biological Systems, Vehicular Networking and Connected Vehicles, Aerospace Systems, Automation, Manufacturing, Smart Grids, Nonlinear Systems, Power Systems, Robotics, Social Systems, Economic Systems and other. Of particular value to both the contributors and the readership are the short publication timeframe and the world-wide distribution and exposure which enable both a wide and rapid dissemination of research output.

The series covers the theory, applications, and perspectives on the state of the art and future developments relevant to systems and networks, decision making, control, complex processes and related areas, as embedded in the fields of interdisciplinary and applied sciences, engineering, computer science, physics, economics, social, and life sciences, as well as the paradigms and methodologies behind them.

Advisory Board

Fernando Gomide, Department of Computer Engineering and Automation—DCA, School of Electrical and Computer Engineering—FEEC, University of Campinas—UNICAMP, São Paulo, Brazil
e-mail: gomide@dca.fee.unicamp.br

Okyay Kaynak, Department of Electrical and Electronic Engineering, Bogazici University, Istanbul, Turkey
e-mail: okyay.kaynak@boun.edu.tr

Derong Liu, Department of Electrical and Computer Engineering, University of Illinois at Chicago, Chicago, USA and Institute of Automation, Chinese Academy of Sciences, Beijing, China
e-mail: derong@uic.edu

Witold Pedrycz, Department of Electrical and Computer Engineering, University of Alberta, Alberta, Canada and Systems Research Institute, Polish Academy of Sciences, Warsaw, Poland
e-mail: wpedrycz@ualberta.ca

Marios M. Polycarpou, KIOS Research Center for Intelligent Systems and Networks, Department of Electrical and Computer Engineering, University of Cyprus, Nicosia, Cyprus
e-mail: mpolycar@ucy.ac.cy

Imre J. Rudas, Óbuda University, Budapest Hungary
e-mail: rudas@uni-obuda.hu

Jun Wang, Department of Computer Science, City University of Hong Kong, Kowloon, Hong Kong
e-mail: jwang.cs@cityu.edu.hk

More information about this series at http://www.springer.com/series/15179

Mohan L. Kolhe · Munesh C. Trivedi
Shailesh Tiwari · Vikash Kumar Singh
Editors

Advances in Data and Information Sciences

Proceedings of ICDIS 2017, Volume 2

Springer

Editors
Mohan L. Kolhe
Smart Grid and Renewable Energy
University of Agder
Kristiansand
Norway

Munesh C. Trivedi
Department of Computer Science
    and Engineering
ABES Engineering College
Ghaziabad, Uttar Pradesh
India

Shailesh Tiwari
Department of Computer Science
    and Engineering
ABES Engineering College
Ghaziabad, Uttar Pradesh
India

Vikash Kumar Singh
Department of Computer Science
Indira Gandhi National Tribal University
Amarkantak, Madhya Pradesh
India

ISSN 2367-3370   ISSN 2367-3389   (electronic)
Lecture Notes in Networks and Systems
ISBN 978-981-13-0276-3   ISBN 978-981-13-0277-0   (eBook)
https://doi.org/10.1007/978-981-13-0277-0

Library of Congress Control Number: 2018933483

Printed on acid-free paper

This Springer imprint is published by the registered company Springer Nature Singapore Pte Ltd.
The registered company address is: 152 Beach Road, #21-01/04 Gateway East, Singapore 189721,
Singapore

Preface

The ICDIS-2017 is a major multidisciplinary conference organized with the objective of bringing together researchers, developers, and practitioners from academia and industry working in all areas of computer and computational sciences. It is organized specifically to help computer industry to derive the advances in next-generation computer and communication technology. Researchers invited to speak will present the latest developments and technical solutions.

Technological developments all over the world are dependent upon the globalization of various research activities. Exchange of information, and innovative ideas is necessary to accelerate the development of technology. Keeping this ideology in preference, the International Conference on Data and Information Sciences (ICDIS-2017) has been organized at Indira Gandhi National Tribal University, Amarkantak, Madhya Pradesh, India, during November 14–15, 2017.

The International Conference on Data and Information Sciences has been organized with a foreseen objective of enhancing the research activities at a large scale. Technical Program Committee and Advisory Board of ICDIS-2017 include eminent academicians, researchers, and practitioners from abroad as well as from all over the nation.

In this book, selected manuscripts have been subdivided into three tracks namely—Web and Informatics, Intelligent Computational Techniques, and Intelligent Communication and Networking. A sincere effort has been made to make it an immense source of knowledge by including 32 manuscripts in this proceedings volume. The selected manuscripts have gone through a rigorous review process and are revised by authors after incorporating the suggestions of the reviewers. These manuscripts have been presented at ICDIS-2017 in different technical sessions.

ICDIS-2017 received around 230 submissions from around 550 authors of different countries such as India, Malaysia, Bangladesh, Sri Lanka. Each submission went through the plagiarism check. On the basis of plagiarism report, each submission was rigorously reviewed by atleast two reviewers with an average of 1.93 per reviewer. Even some submissions had more than two reviews. On the basis of these reviews, 59 high-quality papers were selected for publication in two proceedings volumes, with an acceptance rate of 25.6%.

We are thankful to our keynote speakers, delegates, authors for their participation and interest in ICDIS-2017 as a platform to share their ideas and insights. We are also thankful to the Prof. Dr. Janusz Kacprzyk, Series Editor, AISC, Springer Nature, and Mr. Aninda Bose, Senior Editor, Hard Sciences, Springer Nature, India, for providing continuous guidance and support. Also, we extend our heartfelt gratitude to the reviewers and Technical Program Committee members for showing their concern and efforts in the review process. We are indeed thankful to everyone directly or indirectly associated with the conference-organizing team leading it toward the success.

Although utmost care has been taken in compilation and editing, a few errors may still occur. We request the participants to bear with such errors and lapses (if any). We wish you all the best…

Organizing Committee
ICDIS-2017

Organizing Committee

Patron

Prof. T. V. Kattimani, Vice-Chancellor, Indira Gandhi National Tribal University, Amarkantak

General Chairs

Dr. Mohan L. Kolhe, University of Agder, Norway
Dr. Shekhar Pradhan, Devry University, New York

Program Chair

Dr. K. K. Mishra, Motilal Nehru National Institute of Technology Allahabad, India
Prof. Shailesh Tiwari, ABES Engineering College, Ghaziabad, Uttar Pradesh, India

Conference Chairs

Dr. Vikas Kumar Singh, Indira Gandhi National Tribal University, Amarkantak, Madhya Pradesh, India
Dr. Munesh C. Trivedi, ABES EC, Ghaziabad, Uttar Pradesh, India

TPC Chairs

Dr. Nitin Singh, Motilal Nehru National Institute of Technology Allahabad, India
Dr. B. K. Singh, RBS College, Agra, Uttar Pradesh, India

Publication Chairs

Dr. Deepak Kumar Singh, Sachdeva Institute of Technology, Mathura, India
Dr. Pragya Dwivedi, Motilal Nehru National Institute of Technology Allahabad, India

Publicity Chairs

Dr. Anil Dubey, Government Engineering College, Ajmer, India
Dr. Deepak Kumar, Amity University, Noida, India

Dr. Nitin Rakesh, Amity University, Noida, India
Dr. Ravi Prasad Valluru, Narayana Engineering College, Nellore, AP, India.
Dr. Sushant Upadyaya, MNIT, Jaipur, India
Dr. Akshay Girdhar, GNDEC, Ludhiana, India

Publicity Co-chair

Prof. Vivek Kumar, DCTM, Haryana, India

About the Book

With the advent of the digital world, the information and data science came into existence with a wide scope of innovations and implementations. Both of these play a major role in the making of policies and taking decisions within or outside any organization, institutions, society, etc.

Data science and information science are complementary to each other but distinct. Data science is related to an inference of knowledge and meaningful information from data. However, information science deals with the design and development of strategies, methods, and techniques concerned with the analysis, classification, storage, retrieval, dissemination, and protection of information.

Nowadays, information and data science field entered into a new era of technological advancement, which we call *Smart and Intelligent Information and Data Science*. *Smart and Intelligent Information and Data Science* provides the use of artificial intelligence techniques to solve the complex problems related to policy and decision making. We can say that the main objective of *Ambient Computing and Communication Sciences is* to make software, techniques, computing and communication devices, which can be used effectively and efficiently.

Keeping this ideology in preference, this book includes the insights that reflect the immediate surroundings developments in the field of *Smart and Intelligent Information and Data Science* from upcoming researchers and leading academicians across the globe. It contains the high-quality peer-reviewed papers of *International Conference on Data and Information Sciences (ICDIS-2017)*, held at Indira Gandhi National Tribal University, Amarkantak, MP, India, during November 17–18, 2017. These papers are arranged in the form of chapters. The contents of this book cover three areas: *Web and Informatics, Intelligent Computational Techniques, and Intelligent Communication and Networking*. This book helps the perspective readers' from industry and academia to derive the immediate surroundings developments in the field of data and information sciences and shape them into real-life applications.

Contents

Editors and Contributors

About the Editors

Prof. (Dr.) Mohan L. Kolhe is with the University of Agder, Norway, as Full Professor in Electrical Power Engineering with focus on smart grid and renewable energy in the Faculty of Engineering and Science. He has also received the offer of full professorship in smart grid from the Norwegian University of Science and Technology (NTNU). He has more than 25 years of academic experience in electrical and renewable energy systems at the international level. He is a leading renewable energy technologist and has previously held academic positions at the world's prestigious universities, e.g., University College London, UK/Australia; University of Dundee, UK; University of Jyvaskyla, Finland; Hydrogen Research Institute, QC, Canada.

Prof. (Dr.) Munesh C. Trivedi currently works as a Professor in Computer Science and Engineering Department, ABES Engineering College, Ghaziabad, India. He has published 20 textbooks and 80 research publications in different international journals and proceedings of international conferences of repute. He has received Young Scientist and numerous awards from different national as well as international forums. He has organized several international conferences technically sponsored by IEEE, ACM, and Springer. He is on the review panel of IEEE Computer Society, International Journal of Network Security, Pattern Recognition Letter and Computer & Education (Elsevier's Journal). He is Executive Committee Member of IEEE UP Section, IEEE India Council, and also IEEE Asia Pacific Region 10.

Prof. (Dr.) Shailesh Tiwari currently works as a Professor in Computer Science and Engineering Department, ABES Engineering College, Ghaziabad, India. He is an alumnus of Motilal Nehru National Institute of Technology Allahabad, India. His primary areas of research are software testing, implementation of optimization algorithms, and machine learning techniques in various problems. He has published more than 50 publications in international journals and in proceedings of international conferences of repute. He is editing Scopus, SCI, and E-SCI-indexed journals. He has organized several international conferences under the banner of IEEE and Springer. He is a Senior Member of IEEE, Member of IEEE Computer Society, Fellow of Institution of Engineers (FIE).

Dr. Vikash Kumar Singh is with Indira Gandhi National Tribal University, Amarkantak, MP, India, as Associate Professor in Computer Science with focus on artificial intelligence in the Faculty of Computronics. He has also received UGC-NET/JRF. He has more than 17 years of academic experience. He has completed MCA along with Ph.D. His academic and research work includes more than 250 research papers, and he has attended more than 15 national and international conferences, workshops, and seminars. He has been invited by many national/international organizations for delivering expert lectures/courses/keynote addresses/workshops.

Contributors

B. M. Abir BRAC University, Mohakhali, Dhaka, Bangladesh

Bibhudendra Acharya Department of Electronics and Telecommunication, National Institute of Technology, Raipur, Chhattisgarh, India

Priyanka Agrawal National Institute of Technology, Raipur, India

Naveed Ahmed Faculty of Engineering and Computer Science, National University of Modern Languages (NUML), Islamabad, Pakistan

Himanshu Ahuja Computer Science Department, Delhi Technological University, New Delhi, India

Sachin Ahuja Department of Computer Science and Engineering, Chitkara University Institute of Engineering and Technology, Chitkara University, Chandigarh, India

Deepak Arora Amity Institute of Engineering and Technology, Amity University, Noida, Uttar Pradesh, India

Anjum Mohd Aslam Department of Computer Science & Engineering, NITTTR, Chandigarh, India

Atta-ur-Rahman Department of Computer Science, College of Computer Science and Information Technology, Imam Abdulrahman Bin Faisal University (IAU), Dammam, KSA, Saudi Arabia

Rabindra Kumar Barik School of Computer Application, KIIT University, Bhubaneswar, India

Amitabha Chakrabarty BRAC University, Mohakhali, Dhaka, Bangladesh

Vivek Kumar Chandra C.S.I.T, Durg, India

Stuti Chaturvedi Government Engineering College, Ajmer, India

Sanjay Chaudhary School of Engineering and Applied Science, Ahmedabad University, Ahmedabad, Gujarat, India

Sachin Chavan Mukesh Patel School of Technology Management and Engineering, SVKMS NMIMS, Dhule, Maharashtra, India

Priyansha Chouksey Department of Industrial and Production Engineering, S.G. S.I.T.S, Indore, India

Smita Das National Institute of Technology Agartala, Agartala, Tripura, India

Sujata Dash Department of Computer Science & Application, North Orissa University, Odisha, India

Mrinal Kanti Debbarma National Institute of Technology Agartala, Agartala, Tripura, India

Aparna Deshpande Department of Industrial and Production Engineering, S.G.S. I.T.S, Indore, India

S. B. Dhok Center for VLSI and Nanotechnology, Visvesvaraya National Institute of Technology, Nagpur, India

Veer Sain Dixit Department of Computer Science, ARSD College, University of Delhi, New Delhi, India

Harishchandra Dubey Electrical Engineering, The University of Texas at Dallas, Richardson, USA

Jayesh Deep Dubey Amity Institute of Engineering and Technology, Amity University, Noida, Uttar Pradesh, India

Kartikey Dwivedi Department of Electronics and Communication, Manipal Institute of Technology, Manipal, India

Deeksha Ekka School of Studies in Engineering and Technology, Guru Ghasidas University, Bilaspur, CG, India

Sanjay Garg Institute of Technology, Nirma University, Ahmedabad, Gujarat, India

Ajay K. Gupta Department of Computer Science & Engineering, M.M.M. University of Technology, Gorakhpur, India

Kunal Gupta Mukesh Patel School of Technology Management and Engineering, SVKMS NMIMS, Dhule, Maharashtra, India

Ratneshwer Gupta SC&SS, Jawaharlal Nehru University, New Delhi, India

Z. M. Hanapi Faculty of Computer Science and Information Technology, Department of Communication Technology and Network, Universiti Putra Malaysia, Selangor, Malaysia

Ahsan Hussain Department of Computer Science and Engineering, National Institute of Technology Goa, Farmagudi, Goa, India

Md Saiful Islam BRAC University, Mohakhali, Dhaka, Bangladesh

Shaily Jain Department of Computer Science and Engineering, Chitkara University Institute of Engineering and Technology, Chitkara University, Chandigarh, India

Mala Kalra Department of Computer Science & Engineering, NITTTR, Chandigarh, India

Raksha Kiran Karda Department of Computer Science & Engineering, NITTTR, Chandigarh, India

S. Karthiga Department of Information Technology, Thiagarajar College of Engineering, Madurai, Tamil Nadu, India

Bettahally N. Keshavamurthy Department of Computer Science and Engineering, National Institute of Technology Goa, Farmagudi, Goa, India

Pooja Khanna Amity Institute of Engineering and Technology, Amity University, Noida, Uttar Pradesh, India

Pravin Kumar National Institute of Technology, Raipur, Raipur, India

Sanjay Kumar Department of Information Technology, National Institute of Technology, Raipur, India

Vaibhav Kumar Computer Engineering Department, National Institute of Technology, Kurukshetra, Kurukshetra, Haryana, India

Vinay Kumar Department of ECE, Visvesvaraya National Institute of Technology, Nagpur, India

Manisha Kumari School of Studies in Engineering and Technology, Guru Ghasidas University, Bilaspur, CG, India

Y. L. Kweh Faculty of Computer Science and Information Technology, Department of Communication Technology and Network, Universiti Putra Malaysia, Selangor, Malaysia

Naresh Lodha Mukesh Patel School of Technology Management and Engineering, SVKMS NMIMS, Dhule, Maharashtra, India

Vibha Lohan Department of Computer Science and Engineering, Guru Jambheshwar University of Science and Technology, Hisar, Haryana, India

Narendra D. Londhe National Institute of Technology, Raipur, India

Somania Nur Mahal BRAC University, Mohakhali, Dhaka, Bangladesh

Maqsood Mahmud Department of Management Information System, College of Business Administration, Imam Abdulrahman Bin Faisal University (IAU), Dammam, KSA, Saudi Arabia

Saikat Majumder Department of Electronics and Telecommunication, National Institute of Technology, Raipur, Chhattisgarh, India

Toshanlal Meenpal Department of Electronics and Telecommunication, National Institute of Technology, Raipur, Chhattisgarh, India

Haribansh Mishra DST-CIMS, Banaras Hindu University, Varanasi, India

Nandini Nayar Department of Computer Science and Engineering, Chitkara University Institute of Engineering and Technology, Chitkara University, Chandigarh, India

Mohamed Othman Faculty of Computer Science and Information Technology, Department of Communication Technology and Network, Universiti Putra Malaysia, Selangor, Malaysia

Sudhakar Pandey Department of Information Technology, National Institute of Technology, Raipur, India

Minal Patel Computer Engineering Department, A. D. Patel Institute of Technology, Anand, Gujarat, India

G. Priyanka Department of Information Technology, Thiagarajar College of Engineering, Madurai, Tamil Nadu, India

Anukriti Rautela Mukesh Patel School of Technology Management and Engineering, SVKMS NMIMS, Dhule, Maharashtra, India

Anil Kumar Sahu SSGI (FET), SSTC, Bhilai, India

Satya Prakash Sahu National Institute of Technology, Raipur, India

D. N. Sandeep Department of ECE, Visvesvaraya National Institute of Technology, Nagpur, India

Karan Sanwal Computer Science Department, Delhi Technological University, New Delhi, India

Karthick Seshadri Department of Computer Science and Engineering, Thiagarajar College of Engineering, Madurai, India

Udai Shanker Department of Computer Science & Engineering, M.M.M. University of Technology, Gorakhpur, India

Rika Sharma Department of Computer Applications, National Institute of Technology, Raipur, India

Vishnu P. Sharma Government Engineering College, Ajmer, India

Bikesh Kumar Singh Department of Biomedical Engineering, National Institute of Technology, Raipur, India

Poonam Singh Department of Computer Science and Engineering, Chitkara University Institute of Engineering and Technology, Chitkara University, Chandigarh, India

Rishi Pal Singh Department of Computer Science and Engineering, Guru Jambheshwar University of Science and Technology, Hisar, Haryana, India

Sandeep Kumar Singh Department of Computer Science & Engineering, Jaypee University, Noida, India

Vikash Kumar Singh Department of Computer Science & Engineering, IGNTU, Amarkantak, India

Vikram Singh Computer Engineering Department, National Institute of Technology, Kurukshetra, Kurukshetra, Haryana, India

G. R. Sinha IIIT Bangalore, CMR Technical Campus Hyderabad, Hyderabad, India

Saurabh Kumar Srivastava Department of Computer Science & Engineering, Jaypee University, Noida, India

A. Subash Babu Department of Mechanical Engineering, IIT-Bombay, Mumbai, India

R. Raja Subramanian Department of Computer Science and Engineering, Thiagarajar College of Engineering, Madurai, India

R. Suganya Department of Information Technology, Thiagarajar College of Engineering, Madurai, Tamil Nadu, India

Prernamayee Tripathy Department of Information Technology, National Institute of Technology, Raipur, India

Munesh C. Trivedi Department of Computer Science & Engineering, ABES Engineering College, Ghaziabad, India

Manu Vardhan National Institute of Technology, Raipur, Raipur, India

Anjani Kumar Verma Department of Computer Science, University of Delhi, New Delhi, India

Kesari Verma Department of Computer Applications, National Institute of Technology, Raipur, India

Shrish Verma National Institute of Technology, Raipur, India

R. Vinodhini Department of Information Technology, Thiagarajar College of Engineering, Madurai, Tamil Nadu, India

Seema Wazarkar Department of Computer Science and Engineering, National Institute of Technology Goa, Farmagudi, Goa, India

Nishi Yadav School of Studies in Engineering and Technology, Guru Ghasidas University, Bilaspur, CG, India

Sadanand Yadav Department of ECE, Visvesvaraya National Institute of Technology, Nagpur, India

Sakshi Yadav Department of Computer Science & Engineering, ABES Engineering College, Ghaziabad, India

Nur Arzilawati Md Yunus Faculty of Computer Science and Information Technology, Department of Communication Technology and Network, Universiti Putra Malaysia, Selangor, Malaysia

Part I
Web and Informatics

Aspect-Based Sentiment Analysis of Tweets Using Independent Component Analysis (ICA) and Probabilistic Latent Semantic Analysis (pLSA)

Pravin Kumar and Manu Vardhan

Abstract Twitter is an ocean of diverse topics while sentiment classifier is always limited to a specific domain or topic. Twitter lacks data labeling and a mechanism to acquire sentiment labels. Sentiment words extracted from the twitter are generalized. It is important to find the correct sentiment from a tweet; otherwise, it may generate different sentiment than the desired. Sentiment analysis work done up to now has limitation, i.e., it is based on predefined lexicons. Sentiment of a word based on this lexicon is not generalized. State of art of our work suggests a solution that will make sentiment analysis based on the sentence sentiments and not just only based on predefined lexicons. In this paper, we propose a framework to analyze sentiment from the sentence, by applying independent component analysis (ICA) in coordination with probabilistic latent semantic analysis. We view pLSA as word categorization technique based on some topics, wherein a given corpus is split among different topics. We further utilize these topics for tagging sentiment with the help of ICA, in this way, we are able to assign more accurate sentiment of a sentence than the existing approaches. The proposed work is efficient as its outcome is more precise and accurate than the lexicons-based sentiment analysis. With adequate unsupervised machine learning training, accurate outcomes with a normal precision rate of 77.98% are accomplished.

Keywords Probabilistic latent semantic analysis · Sentiment analysis · Text sieve · Term frequency

P. Kumar (✉) · M. Vardhan (✉)
National Institute of Technology, Raipur, Raipur, India
e-mail: pk.mmmec@gmail.com

M. Vardhan
e-mail: mvardhan.cs@nitrr.ac.in

© Springer Nature Singapore Pte Ltd. 2019
M. L. Kolhe et al. (eds.), *Advances in Data and Information Sciences*, Lecture Notes in Networks and Systems 39, https://doi.org/10.1007/978-981-13-0277-0_1

1 Introduction

Twitter provides a platform for individuals to post their sentiments and suppositions on different issues. The posting of opinion on any issue is not just giving a passionate depiction of the online social media, yet in addition has potential business impacts, money related, and sociological esteems [1]. Nonetheless, confronting the monstrous assumption tweets, it is hard for the individuals to get a general impression without programmed opinion grouping and investigation. Along these lines, there are raising numerous leaning order works demonstrating interests in tweets sentiment analysis [2], and [3]. Sentiment analysis is a critical issue in tweets mining. This twitter information need marked information for utilizing them in the analysis. Subjects on twitter are more assorted while assessment classifiers have constantly committed themselves to a particular domain or area. Along with the domain, sentiment analysis is the required task to depict sentiment, versatile to various subjects without adequate domain information.

Many Researchers have talked about in the course of events and proposed different sentiment analysis methods for on Twitter data. A sentiment classifier determination is constantly committed to its preparation period, to the domain named information highlighted. For example, "read eBook" might be positive as far as for the environment protection views, yet might be negative regarding eye health perspective. It is exceptionally hard to determine sentiments of tweets for a wide range of subjects that will be talked about later on. Different domain and cross-subject adjustment with regards to conclusion examination are contemplated and considered in [2, 4].

Sentiment analysis means to find the polarity of the opinion presented in a written sentence; which also shows its semantic introduction. For example, a sentence in a particular tweet or review, or a survey about a specific item or issue, can be broken down to decide if it is positive, negative, or neutral, together with the extreme word of the conclusion. This extreme word itself can be shown completely as positive, negative, or neutral.

Two primary methodologies in sentiment analysis are characterized in writing: a first methodology is called lexicon-based [5] and the second methodology utilized the supervised learning algorithms [5]. The lexicon approach determines the semantic introduction of given content by the polarities of essential words or expressions [5], for example, the SentiWordNet [6]. In lexicon-based approach, distinctive components of the content might be extricated through word polarities [7], for example, usual word polarity, or quantity of skewed words, yet there is no administered learning. However, content is frequently regarded as bag-of-words, as it were, changed time to time from constituent words with monitoring the area of those words by appropriate authorities.

Complementary options to the bag-of-word methodology are likewise feasible, where polarities of the primary sentence and so forth are ascertained independently [7]. Moreover, as words might have distinctive essences in various areas (e.g., "small" may have a positive undertone in mobile phone review domain, apart from this, it might have negative undertone in lodging space in car), one can utilize domain-specific vocabulary at whatever point accessible. The generally utilized SentiWord-Net [6] and SenticNet [8] are two generally known area autonomous dictionaries.

Supervised learning based methodologies utilize machine learning procedures to set up a polarity from a directed corpus of surveys with related marks. For example in [9], researchers utilize the Naive Bayes classifier based machine learning algorithm to isolate positive sentiments from negative sentences by taking in the restrictive likelihood appropriations of the considered corpus in two classes, i.e., positive or negative. Note that in machine learning based sentiment analysis approaches, an extremity dictionary may at present be utilized to extricate components from content, for example, normal word polarity what's more, the quantity of positive words and so forth, that are later utilized as a part of a learning calculation.

On the other hand, in some regulated methodologies, the dictionary is not required. For example in the latent Dirichlet allocation (LDA) [10] approach, a preparation corpus is utilized to learn the likelihood disseminations of point and word events in the distinctive classifications (e.g., positive or negative) and another content is characterized by its probability of originating from these distinctive conveyances [11]. While administered sentiment analysis using domain adapted and sentence-based analysis approaches are more effective than dictionary-based ones, gathering a substantial measure of sentiment from domain information can be an issue.

In this work, we introduce an unsupervised learning based sentiment analysis approach. To start with, we apply a basic word weight frequency based approach to make a sieve from these bag of words proposed in [1] to form a vocabulary for a partic-ular domain. Next, we propose a sentence-based investigation of the sentiment word categorization, utilizing the refreshed sentiment dictionary for highlight extraction. While word-level polarities give a basic, yet viable strategy for evaluating a sentence polarity, the whole from word-level polarities to survey extremity is too enormous. The rest of this paper contains the following sections. Section 2 gives an outline of related work. Section 3 proposes the word weight mythology; Sect. 4 outlines the probabilistic latent semantic analysis procedure of a domain-independent vocabu-lary. Section 5 portrays our problem formulation. Section 6 depicts the mathematical formulation presented in this proposed work. Section 7 presents the experimental details and Sect. 8 reports results outcomes. At last, in Sect. 9 we finish up with conclusions with our thoughts for future work.

2 Background Details and Related Work

Sentiment analysis has been a dramatic course of interest in recent 10–15 years, with expanding scholarly and business enthusiasm in this field. A detailed review of the previous work done in sentiment analysis has been discussed in [1], while we just outline essential research points here. Document is commonly seen as bag of words and document's polarity is evaluated from the extremity of the words within the document [4, 12]. Since taking few words of the entire archive just as a bag of words is extremely oversimplified, later research effort concentrated on examination of expressions and sentences. Some research work centered on polarity determination of expressions to make utilize this data while deciding the polarity of the record. Wiebe [13] found subjective modifiers from corpus. At that point, Wiebe [13] finds the impacts of adjective course and tradability on sentence polarity. The objective of this approach was to decide regardless of whether a sentence is positive or not, by looking at the descriptive words in that sentence. In this manner, a few investigations concentrated on sentence-level polarity determination in various documents. Some current research work additionally inspected disambiguation and subjectivity as the sentiment relation among the sentences, so as to separate adequate data for a more precise sentiment analysis [12]. Wiebe et al., present a wide study of sentiment analysis utilizing different elements and features in sentences [14].

Deciding the polarity estimation of a document, despite basically characterizing it as positive or negative, is a relapse issue that is addressed utilizing marginally extraordinary administered learning strategies. In a relapse issue, the assignment is to take in the mapping $y = f(x)$, where $x \in Rd$ and $y \in R$. It can be thought that relapse issue is more troublesome than the grouping issue, as the last can be expert once sentiment polarities are evaluated. In the event that one considers three notion classifications (negative, neutral and positive), at that point treating the issue as a relapse issue as opposed to a grouping issue might be the more fitting methodology since class marks are ordinal.

Polarity-based lexicon demonstrates the estimated polarity of words. SentiWordnet [6] and SenticNet [8] are most normally utilized polarity lexicons used in sentiment analysis. In [1], researchers talk about the three fundamental methodologies for sentiment vocabulary building: manual approach, lexicon-based approach, and corpus-based approach. The real weakness of the manual approach is the cost (time and exertion) to manually choose words to fabricate such a dictionary. There is additionally the likelihood of missing essential words that could be caught with programmed strategies.

Word lexicon based methodologies (e.g., [4, 6, and 14]) work by extending a little arrangement of seed sentiment words, by utilizing lexical property, for example, the WordNet [15]. Considering these methodologies, the subsequent polarities are domain independent.

Corpus-based methodologies could be utilized to be trained domain dependent lexicons utilizing a domain corpus of marked surveys. Wilson et al. stretch the significance of logical polarity to separate by the earlier polarity of a word [16]. They extricate relevant polarities by characterizing a few logical components. In [16], a twofold engendering strategy is utilized to separate both lean words and elements, consolidated with a polarity determination strategy beginning with a seed set of words. Researchers utilize straight programming to refresh the polarity of words in view of determined hard or delicate limitations.

Another use of corpus-based methodologies shows up in [11] to take in an opinion dictionary which is domain specific as well as viewpoint subordinate. Another current work grows a given lexicon of words with known polarities by first delivering a new arrangement of equivalent words with polarities and utilizing these to additionally find the polarities of different words [16]. A basic corpus-based sentiment analysis strategy proposed by Demiroz et al. is utilized as a part of our framework [1]. Researchers consider Naïve Bayes classification Using Domain-based Sentence-Based Analysis the TF_IDF [17] polarity of each word in positive or negative value, and adjust word polarities as per this distinction.

Sentence-level sentiment analysis is a well-known field of sentiment analysis. A few analysts drew nearer the issue by first analysis subjective sentences in a corpus, with the expectation of killing insignificant sentences that would create clamor as far as extremity estimation [7].

However, an additional approach is to use the composition in sentences, as opposed to considering a document as a bag of words [14–16]. For example in [14], conjunctions were investigated to acquire polarities of the words that are associated with the conjunct. What's more, Wilson et al. [18] bring up the issue of getting proviso level supposition quality as a training step for sentence-level sentiment analysis. In [5], researchers focused on sentence-based polarities dependently, again to get sentence polarities all the more effectively, with the objective of enhancing audit extremity thus. The principal line extremity has moreover been utilized as an element by [7].

Our work is inspired by our perception that the seed words of a sentence are frequently exceptionally demonstrative of the actual polarity. Beginning from this straightforward perception, we planned more advanced elements for sentence-level estimation examination. In order to do that, we played out an inside and out an investigation of various sentence sorts.

3 Weight of the Word in Topic

Text document is generally represented as vector space model [19]. In this vector space model, words are represented as vectors. A collection of document can be represented as the vector document matrix.

$$D = (x_{ij}) \tag{1}$$

Here, x_{ij} is the weight of word I in document d, and this weight of word is calculated by using following formula:

$$x_{ij} = \frac{\log(1 + f_{ij}) \log\left(\frac{N}{n_i}\right)}{\sqrt{\sum_{k=1}^{n} \left(\log(1 + f_{kj}) \log\left(\frac{N}{n_k}\right)\right)^2}} \tag{2}$$

Here, n = total number of words, N = total number of document, f_{ij} = number of times the word appears in document 'd'. Utilizing the logarithmic estimation of the word frequency to figure the weight specifically decreases the effect on the weight ascertained because of the excessively high word recurrence. Standardization maintains a strategic distance from the effect on the weights figured because of the diverse lengths of writings. This is a regularly utilized weight computation technique for documents, and researchers have demonstrated the absolute best outcomes. Since each word does not typically show up in each record, the document 'd' is generally inadequate. The quantity of lines, n, of the document relates to the quantity of words in the lexicon. Lines n can be huge. Subsequently, dimensionality lessening is important.

4 Probabilistic Latent Semantic Analysis

Probabilistic latent semantic analysis (pLSA) is a strategy for the classification of topic models. Its fundamental objective is to find co-occurrence data under a probabilistic system with a specific end goal to find the fundamental semantic structure of the information. It was introduced in 1999 by Hofmann [20]. It was first utilized for text-based applications, (for example, ordering, recovery, and grouping); however, its utilization right away spread in different fields. The objective of pLSA is to utilize this co-occurrence matrix to extricate the alleged "Topics" and clarify the archives as a blend of them.

According to PLSA, data can be categorized into three sets of variables:

- Documents: $d \in D = \{d1, \ldots d_n\}$ d is observed variables. N is the number defined from the size of our given corpus.
- Words: $w \in W = \{w1, \ldots w_m\}$ w is observed variables. M is the number of distinct words from the corpus.
- Topics: $z \in Z = \{z1, \ldots z_k\}$ z is latent variables. Their number K is to be specified a priori.

There are some imperative presumptions made by the exhibited show:

- Bag-of-words. Naturally, each record is viewed as an unordered accumulation of words. More accurately, this means that the joint variable(d, w) is represented as

$$P(D, W) = \prod_{(d,w)} (d, w) \tag{3}$$

- Conditional autonomy. This implies that words and records are restrictively autonomous given the topic:

$$P(w, d|z) = P(w|z)P(d|z) \tag{4}$$

(This can be effortlessly demonstrated by utilizing d-division into our graphical model: the way from d to w is hindered by z.) The model can be totally characterized by indicating the joint appropriation. We get P (d, w) by utilising the data set items.

$$P(d, w) = P(d)P(w|d)$$

$$P(w|d) = \sum_{z \in Z} P(w, z|d)$$

$$= \sum_{z \in Z} P(w|d, z) P(z|d) \tag{5}$$

Using the conditional independence, we obtain the formulae as follows:

$$P(w|d) = \sum_{z \in Z} P(w|z, z|d)$$

$$P(w|d) = \sum_{z \in Z} P(z) P(d|z)P(w|z) \tag{6}$$

5 Problem Formulation

Twitter sentiment analysis can be seen as a text analysis because we analyze collection of tweets as documents. The principle work of topic modeling is to make topics with a predefined set of classes. Sentiment analysis has been connected in different types of documents, for example, report ordering, archive separating, word sense disambiguation, and so forth. As studied in Sebastiani [6], one of the focal issues in sentiment analysis is the means by which to speak to the substance of a document so as to find the sentiment in it. From investigation in data recovery frameworks, a standout among the most prominent and fruitful strategies is to speak to a document by the gathering of terms that shows up in it. The similitude between documents is characterized by utilizing the term frequency inverse document frequency (TFIDF) measure [19]. In this approach, the terms or elements used to speak to a document are controlled by taking the union of all terms that show up in the accumulation of writings used to determine the classifier. This, as a rule, brings about a substantial number of components. In this manner, dimensionality diminishment is a related issue that should be tended too. This is an issue regular to tweet examination when a list of capabilities comprises of every extraordinary word, called tokens, of the arrangement of gathering tweets. Also, each tweet is a short instant message up to 140 characters gathered from Twitter. Along these lines, the setting data of a tweet is significantly more restricted in contrast with a content report. The problem here is the means by which to find a viable approach to extract sentiment from a tweet's phrase according to the domain, in which user has opined his/her view.

6 Proposed Approach

From the weight of the word, we create a bag of sentiment words. These sentiment words are similar to the training sample sentiment words. These sentiments words will be used for sieving purpose for all the corpuses. We can make it as standard sentiment words, which can be verified and updated after review by experts. We term this bag of words as a sieve. After forming this sieve, find all the sentiment words from the corpus. And using a standard independent component algorithm (ICA), extract these sentiment words from the sieve. After filtering from the sieve, the extracted sentiments contain the polarity according to our sieve. The advantage of this approach is that it allows the corpus to gain real polarity in the domain of topics.

Let D refers to our corpus, we represent it as $D(+, -, *)$. Here, '+' refers to positive tweets, '−' refers to negative tweets, and '*' refers to neutral tweets. Our sieve is represented as S. sieve such that $S \subset D$. The extracted sentiment word is termed as approximate space D_s here $D_s = (U, T)$ and $S(+, -, *)$. Here, $D_s = (U_k, k)$ k is a finite set of sentiment words. D_s is the set of tweets in U represented by keywords in k.

In sieve approach, knowledge or information of a subject is induced from an approximation space $S = (U, k)$ accuracy of approximation is defined as the ratio of the extensions of the lower approximation to the upper approximation.

$$D_s = \{e \in U : [e] \subseteq X\}$$
$$T = \{e \in U : [e] \cap x \neq \phi\} \qquad (7)$$

Using this approximation space

$$\sum_{i=1}^{n} D_s = \sum_{i=1}^{n} S - \sum_{i=1}^{n} D_k \qquad (8)$$

Final sieve words D_s are the words with domain aspect sentiment. Polarities of these words are more accurate than the lexicon-based sentiment analysis. Next, we utilize these set of words D_s for sentiment analysis of our twitter corpus.

7 Experiments

Data observation and motivation: To get an efficient observation, we collect the tweets of Google hashtag. With the help of Twitter OAuth API, we collected more than 10,000 tweets and apply the basic preprocessing to form a corpus. This corpus's sentiment labels on various topics. This data set is a random sample of streaming tweets unlike data collected by Google queries. The size of our handheld data allows

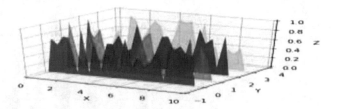

Fig. 1 Polarity graph

us to perform cross-validation experiments and check the performance of the classifier across fields.

We acquire approximately 10,000 manually annotated twitter data (tweets) restriction like language, location, etc., which were barred during the steaming process. In fact, we also collect tweets in foreign language, then we use Google Translate for converting foreign language to English language. We remove the junk tweets label means; these tweets are ambiguous in meaning. Manual analyses of a random sample of tweets labeled as junk suggested that many of these were not translatable well using Google Translate. We use the standard tokenizer to tokenize the tweets. We use a stop words dictionary to identify stop words, that is found in wordnet and counted as English words.

8 Results

To display our finding, we create a graph as shown in Fig. 1. In this graph, red color represents the negative words, green color shows the positive, blue shows the neutral, and yellow color shows the junk tweets. X-axis here represents the size of corpus taken. Y-axis shows the number of sentiment classes, here we considered four classes: negative, positive, neutral, and junk. By using our approach, we decide the sentiment of negative and positive class. For neutral and junk class, we used a lexicon to determine the polarity of the corpus. For verifying our finding, we compare our approach from some existing standard algorithms. From the comparison as shown in Table 1 and Fig. 2, it is clear that we have achieved some increase in accuracy and F-score. F-score and accuracy are calculated with the recall and precision.

Table 1 Accuracy and F-score of different approaches

	NB	DT	MSVM	RF	MS3VM	COMS3VM	ACOMS3VM	Proposed approach
Accuracy	0.778	0.8152	0.8086	0.7364	0.8166	0.8058	0.8252	0.83
F-score	0.722	0.732	0.4175	0.732	0.3784	0.4614	0.5164	0.4252

Fig. 2 Accuracy and
F-score of different
approaches

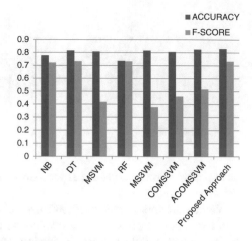

9 Conclusions

Sentiment analysis is an emerging research topic, in which numerous issues need to be tackled. In the proposed work, finding the aspect-based polarity of twitter corpus is done consequently. Aspect-based data extrication from unstructured data is a great degree troublesome for a machine and assigning the polarity accurately according to the twitter data. Apart from preprocessing, many other tasks like creating a runtime sieve, extraction, etc., are quite time consuming. Creating a runtime sieve is a difficult issue to be viewed and to resolve this issue practically, few arrangements have been made. In our work, we utilized the probabilistic latent semantic analysis (pLSA) and independent component algorithm (ICA) to create a sieve. The proposed work is efficient as its outcome is more precise and accurate than the lexicons-based sentiment analysis. With adequate unsupervised machine learning training, accurate outcomes with a normal precision rate of 77.98% are accomplished.

Some other issues like mnemonics sentiments, etc., are beyond this work, so in the current work done, such cases are not considered. This work is left for future research.

References

1. Pang B, Lee L (2008) Opinion mining and sentiment analysis. Found Trends Inf Retr 2(1–2):1–135
2. Spencer J, Uchyigit G (2012) Sentimentor: sentiment analysis of twitter data. CEUR Workshop Proc 917:56–66. https://doi.org/10.1007/978-3-642-35176-1_32
3. Mejova YA (2012) Sentiment analysis within and across social media streams, vol 190
4. Abbasi A, Chen H, Salem A (2008) Sentiment analysis in multiple languages: feature selection for opinion classification in web forums. ACM Trans Inf Syst 26(3):1–34. https://doi.org/10.1145/1361684.1361685

5. Medhat W, Hassan A, Korashy H (2014) Sentiment analysis algorithms and applications: a survey. Ain Shams Eng J 5(4):1093–1113. https://doi.org/10.1016/j.asej.2014.04.011
6. Esuli A, Sebastiani F (2006) SENTIWORDNET: a publicly available lexical resource for opinion mining. In: Proceedings of the 5th conference on language resources and evaluation (LREC'06), Genova, IT, pp 417–422
7. Düsterhöft A, Thalheim B (2003) Natural language processing and information systems. Natural Lang Process Inf Syst (June 2013) 220–233. https://doi.org/10.1007/978-3-319-41754-7
8. Cambria E et al (2010) SenticNet: a publicly available semantic resource for opinion mining. In: AAAI fall symposium: commonsense knowledge, vol 10, no 0
9. Tan, S et al (2009) Adapting naive bayes to domain adaptation for sentiment analysis. Adv Inf Retr 337–349
10. Blei, DM, Ng AY, Jordan MI (2003) Latent dirichlet allocation. J Mach Learn Res 993–1022
11. Zafar MB, Bhattacharya P, Ganguly N, Gummadi KP, Ghosh S (2015) Sampling content from online social networks. ACM Trans Web 9(3):1–33. https://doi.org/10.1145/2743023
12. Najaflou Y, Jedari B, Xia F, Member S, Yang LT, Obaidat MS (2013) Mobile social networks. IEEE Syst J 9(3):1–21
13. Li X, Li J, Wu Y (2015) A global optimization approach to multi-polarity sentiment analysis. PLoS ONE 10(4):1–18. https://doi.org/10.1371/journal.pone.0124672
14. Varghese R, Jayasree M (2013) Aspect based sentiment analysis using support vector machine classifier. In: 2013 international conference on advances in computing, communications and informatics (ICACCI), 22–25 Aug 2013, pp 1581–1586. https://doi.org/10.1109/ICACCI.2013.6637416
15. Mejova Y, Srinivasan P (2011) Exploring feature definition and selection for sentiment classifiers. In: Fifth international AAAI conference on weblogs and social media, pp 546–549
16. Prager J (2006) Open-domain question-answering. foundations and trends®. Inf Retr 1(2):91–231. https://doi.org/10.1561/1500000001
17. Martineau Justin, Finin Tim (2009) Delta TFIDF: an improved feature space for sentiment analysis. Icwsm 9:106
18. Wilson T, Wiebe J, Hwa R (2004) Just how mad are you? Finding strong and weak opinion clauses. In: Proceedings of AAAI, pp 761–769 (Extended version in Comput Intel 22(2):73–99, 2006)
19. Salton Gerard, Buckley Christopher (1988) Term-weighting approaches in automatic text retrieval. Inf Process Manag 24(5):513–523
20. Hofmann, T (1999) Probabilistic latent semantic analysis. In: Proceedings of the fifteenth conference on uncertainty in artificial intelligence. Morgan Kaufmann Publishers Inc

Bio-inspired Threshold Based VM Migration for Green Cloud

Raksha Kiran Karda and Mala Kalra

Abstract Cloud data centers are always in demand of energy resources. There are very limited numbers of nonrenewable energy resources but data centers have large energy consumption as well as carbon footprints. It is a big challenge to reduce the energy consumption. VM migration techniques are used for server consolidation in order to reduce the power consumption. In general Power consumption varies directly proportional to CPU utilization thus three threshold energy saving algorithm (TESA) having a larger impact on the performance of the system with respect to CPU utilization. But with existing TESA too many migrations leads to performance degradation. The proposed ant colony optimization (ACO) is the algorithm which is applied for the VM placement to select the appropriate host, the host which has the least chances of overutilization and requires minimum migrations is selected as the best machine for task migration. The performance of the proposed algorithm is tested in cloudsim in terms of energy efficiency, number of migrations, and SLA violation.

Keywords ACO · Green cloud computing · MIMT · TESA

1 Introduction

Now it's the time we live in and around the world of data. Every aspect of our lives is almost controlled by data. Data is growing at a tremendous rate due to increase in the use of Internet-connected devices. We are moving toward the era of the Internet of things where instead of devices, things are connected to the Internet so the data is growing rapidly. In order to store, maintain, and compute the big data, there is always a need to establish quality resources with huge processing power. To meet

R. K. Karda (✉) · M. Kalra
Department of Computer Science & Engineering, NITTTR, Chandigarh, India
e-mail: rakshakiran75@gmail.com

M. Kalra
e-mail: malakalra2004@gmail.com

© Springer Nature Singapore Pte Ltd. 2019
M. L. Kolhe et al. (eds.), *Advances in Data and Information Sciences*, Lecture Notes in Networks and Systems 39, https://doi.org/10.1007/978-981-13-0277-0_2

15

Table 1 Power consumption in data centers of US [4]

Year	End use energy (BKWh)	Elec. bills (The US, $B)	Power plants (500 MW)	CO_2 (US Million MT)
2013	91	$9.0	34	97
2020	139	$13.7	51	147
Increase (%)	53	52	50	52

the infrastructural and services, most of the users and business enterprises are opting the cloud computing which offers on-demand resource provisioning, multi-tenancy, elasticity, resource pooling, and effective utilization of high-priced infrastructure resources with minimum cost. These resources can be accessed as well as released with much ease and thus are helpful in providing services as pay per use to multiple users [1]. Cloud computing allows IT industries to focus on doing what they actually want without spending money on infrastructure and wasting time in managing them.

Cloud can be seen as three different service models: infrastructure as a service (IaaS), platform as a service (PaaS), and software as a service (SaaS) [2]. As a large number of organizations are increasingly moving toward cloud computing, it led to the need for more data centers which consume a huge amount of energy and carbon footprints. Enormous amounts of electrical power used in computing, data center infrastructure, electrical equipment and networking equipment lead to high operational costs and also reduction in the profit margin of cloud service providers.

Green cloud is now a stressing need for cloud service providers because it decreases the profit of cloud service providers due to the large investment in data centers required for power provisioning and cooling infrastructure. Gartner estimates that information and communication technology (ICT) is responsible for 2% of worldwide CO_2 emission which is due to the PCs, servers, telephony, local area network, and printers [3]. Table 1 shows the values of different parameters related to power consumption.

One of the appropriate ways to minimize power consumption in cloud data centers is virtualization technology which plays a significant role in the efficient management of resources at data centers. Virtualization allows resources of a single physical host to be shared by many individually separated machines called virtual machines (VMs). Hypervisors provide an abstraction layer so that each VM works independently. Thus, minimizing the cost of hardware investment and other operational costs which are related to power and cooling, etc. VM migration technology allows us to migrate the VMs from one host to another so that idle hosts can be switched off in order to achieve optimized power consumption. VM migration is done based on the host CPU utilization, i.e., overutilization and underutilization.

The problem of energy efficient resource utilization in cloud computing is NP-hard problem as there exists no algorithm which may produce the optimal result in polynomial time. Many of the previous studies focused on nature-inspired techniques [5]. ACO, PSO, and GA are some of the popular metaheuristics used in this domain. ACO is a multi-agent approach used to find the solution for combinatorial NP-hard problem of energy efficiency [6]. It gives us the optimized solution when there are a number of solutions that exist. ACO has gained huge popularity in the past years due to its efficiency and effectiveness to solve large and complex problems.

In this proposed work, the aim is to minimizing energy consumption with less number of migrations. For this, a VM placement algorithm is presented which is based on ACO metaheuristic algorithm while considering the host's CPU utilization thresholds. ACO algorithm is implemented with cloudsim package.

The remainder of this paper is organized as follows. Section 2 discusses related work and Sect. 3 provides an overview of techniques related to proposed work. Section 4 describes the simulation model whereas experimental setup and results are shown in Sect. 5. Finally, conclusions and future directions are given in Sect. 6.

2 Related Work

Energy consumption in data centers has drawn a significant attention in past few years due to negative impacts on the environment and increase in the operational costs. This section discusses the various solutions proposed to decrease the power consumption.

Yin and Wang [7] applied ACO to allocation problem where the nonlinear type of resources is available. ACO seeks to search an optimal allocation of a finite amount of resources to a large number of tasks to optimize their nonlinear objective function.

Van Ast et al. [8] described the basic ACO algorithm and its variants have success-fully been applied to various optimization problems, such as the traveling salesman problem, optimal path planning for mobile robots, telecommunication routing, job shop scheduling, and load balancing.

Beloglazov and Buyya [9] also analyzed that compared to non-power-aware algo-rithms (NPA), dynamic voltage frequency scaling (DVFS) works better with respect to energy saving. Single threshold technique produces almost two times better results than DVFS. But double threshold with minimum of migrations (MM) has a signif-icant reduction in energy consumption with a minimum number of migrations but little increase in SLA violation.

In cloud data centers, the main factor behind energy consumption is over-provisioning of virtual machines which consumes more energy. Copil et al. [10] developed a technique for negotiation in GSLA based on particle swarm optimization which will promote the user and service provider to decrease the resource allocation and helped to achieve the optimized energy consumption.

Huang et al. [11] proposed the novel heuristics for placement (best fit host (BFH) and best fit VM (BFV)) which find the best fit combinations of VM from the over-utilized host for migrations based on 0/1 knapsack dynamic programming for less number of migrations and to optimize energy consumption. It is considered that combining one or more virtual machine into the host (knapsack) so that the predicted usage of the host become less than 1 or maximize whereas the sum of MIPS of all the selected virtual machines does not surpass the effectual capacity of the host in order to reduce migration.

Xiong et al. [12] concentrated on three options to minimize energy consumption in the data center. First is either the quality of service can be compromised for the limited extent to reduce energy consumption, or fix some tradeoff between energy and performance up to the definite amount, or manage both parameters in such an efficient way so that neither energy nor performance compromised. The third criterion is chosen by the author and performed better with the help of virtualization technology. The energy is minimized but the algorithm is restricted to only two resources (CPU and Disk) but ignoring the others such as network resources, memory, etc., is not taken into account.

The authors Zhou et al. [13] proposed a deployment algorithm for virtual machines called three-threshold energy-saving algorithm (TESA), keeping in mind the end goal of increasing the energy productivity of data centers. The results demonstrated that Minimization of migration policy based on TESA (MIMT) performed best among all the proposed policies for selection of VMs. MIMT also proved to be one of the best methods to improve the energy proficiency in data centers.

Farahnakian et al. [14] describe two widely used techniques used to reduce energy consumption, i.e., dynamic server provisioning and virtual machine consolidation. The author proposed the ant colony based VM consolidation approach to finding the optimized solution for NP-hard problems. Ant colony algorithm is best suited for dynamic workloads. The system architecture uses local agents and global agents to monitor the current CPU utilization for consolidation. Proposed approach reduces the energy consumption with less SLA violation as well as maintaining the QOS.

Mustafa et al. [15] explained that modified best fit decreasing (MBFD) is the modified algorithm of best fit decreasing (BFD) algorithm which is based on the bin packing problem. In MBFD, the VMs are sorted in descending order based on their CPU utilization. After sorting, all VMs are deployed to the hosts based on the power model. According to which it is checked that how much change is observed in energy consumption of the servers after the placement of VM. VM is placed on a server which shows minimal change in energy consumption.

To enhance the performance and to reduce migration cost of data center Reguri et al. [4] suggested a new scheme of traffic vector in which number of VMs need migration is maintained as a VM cluster for migration thus reduces migration cost. Virtual machine clusters are formed according to highly correlating services and the traffic exchange between them. By clustering, the overall energy saving is 23% with a very less increase in the SLA violation. It is considered that after 50% of CPU utilization performance to power ratio increases sub-linearly.

The increase in CPU utilization adversely affects the energy consumption of data center. Choi et al. [16] proposed task classification based energy-aware consolidation algorithm which used a double threshold to select overutilized hosts and then for migrating the VM from that host divided the tasks into two classes computation intensive tasks and data-intensive tasks so that with this task classification information VM could appropriately assign to the suitable host.

Selim et al. [17] proposed the CPU utilization variance (CUV) technique to select the VM for migration from overutilized servers and selected a host to place the VM which had the minimum variance of CPU utilization among all the hosts. This proposed technique significantly decreased the energy consumption and reduced the number of migrations.

Singh and Chana et al. [18] concluded that allocation of resources improves the resource utilization if it considers the type of workload, i.e., homogeneous or heterogeneous. For a better resource allocation, automatic resource provisioning system with consideration of all the parameters of QoS (quality of service) like execution time, availability, security, and energy consumption will improve the performance of data center.

From the literature review, it has been observed that there is the scope for improvement in the deployment policies and placement policies of VMs. In this paper, we propose a novel approach which considers the CPU utilization factor and finds the best destination host for placement through ACO. The proposed algorithm provides less number of migrations by reducing the overall power consumption in the data center.

3 Methodologies Used for Proposed Work

In this section, the various methodologies used in the proposed work are discussed.

3.1 Virtualization

VMs refer to an instance of an operating system alongside at least one application is run in an isolated partition inside the computers. Multiple VMs can exist on a single physical server. If a machine gets overloaded or becomes faulty, VMs from that machine need to be transferred to another physical machine. The interruption to the users should be less in such situations. VM migration is known as the process of shifting virtual machine from one physical machine to another physical machine in the form of image file [19].

Additionally, migration of VMs permits the transparent transfer of task from one physical host to another physical host. This is required for dynamic changing user requirements and varying application demands. However, this flexibility also introduces new management demands since large pools of VMs must be provisioned and managed. The VM allocation is categorized into two parts, the first part deals with when a new request enters and accepted for VM provisioning and deployment on physical servers and the second part considering the optimization of the VMs which are previously deployed in the data center as shown in Fig. 1.

With virtualization the problem of energy consumption is solved by four approaches i.e., host underutilization, host overutilization, VM selection, and VM

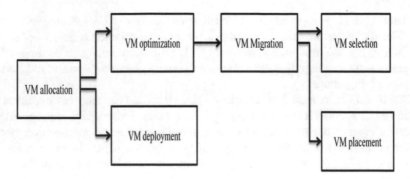

Fig. 1 General issues associated with VM allocation

placement. All hosts which are underutilized in terms of CPU utilization are consolidated so that all VMs are migrated to those hosts which are properly utilized and source hosts are turned off in order to reduce the power consumption. VMs from overutilized hosts are also following migration in order to achieve normal utilization of resources. It is also a challenging task to decide that which VM should be migrated and where it should be placed in order to find the optimized solution for energy efficiency and to ensure less SLA violation.

3.2 Ant Colony Optimization(ACO)

ACO is inspired by the behavior of ants. Ants are not having vision capability but they are very much capable to find the shortest path from their nest to a source of food. They have stigmergic communication, i.e., interacting through local environment. Initially, the ants move randomly to find the food. When they move they deposit some chemical onto their path called pheromone. The ant reaches at food source by shortest path and returns earlier to the nest than the other ants. Thus, pheromone intensity increases onto that path [20]. This intense pheromone allures other ants to trace exactly the same path and they led to deposit pheromones on their way. This behavior of ants, in turn, leads to the uprising of paths that can be shown to be optimal or near optimal. ACO is used as metaheuristic optimization technique that uses these concepts to solve combinatorial optimization problems (COPs). When a number of solutions are available, ACO helps to find the best solution [21]. Basically, ant colony optimization algorithm works through following phases:

i. **Edge selection**

When any host gets overutilized then some of the VMs need to migrate from that host to some other host. Then it is considered that ant consisting in itself VM moves from the host I to host j. In positive traversing strategy, more pheromone is on the way from I to j then there is more possibility to choose that path. To calculate the probability, the equation applied as follows:

$$\rho_{i,j} = \frac{\left(\tau_{i,j}^{\alpha}\right)\left(\eta_{i,j}^{\beta}\right)}{\sum\left(\tau_{i,j}^{\alpha}\right)\left(\eta_{i,j}^{\beta}\right)} \tag{1}$$

where

$\tau_{i,j}$ is the pheromone deposited when ant goes from edge i to j
α is a parameter which controls the relative influence of $\tau_{i,j}$
$\eta_{i,j}$ is the heuristic value of path i, j which is resources available on to that path
β is a parameter which controls the relative influence of $\eta_{i,j}$

ii. **Update pheromone**

In this phase, the machine which has the least probability of overutilization is selected as the best machine for the task migration. To select the best machine pheromone is updated by the equation below:

$$\tau_{i,j} = (1 - \rho)\,\tau_{i,j} + \Delta\tau_{i,j} \tag{2}$$

where

$\tau_{i,j}$ is the pheromone amount
ρ is the evaporation rate of pheromone, i.e., CPU capacity allocated
$\Delta\tau_{i,j}$ is the amount of pheromone deposited, typically given by

$$\Delta\tau_{i,j}^{\kappa} = \begin{cases} 1/L_k & \text{if ant k travels on edge } i,\ j \\ 0 & \text{Otherwise} \end{cases} \tag{3}$$

where

L_k is the cost of the kth ant's tour (typically distance).

3.3 *System Utilization Based Power Consumption Model*

The major portion of power consumption in data centers is contributed by virtual machines and physical servers. Also among the different hardware components, CPU is the largest power consumer in a server. Thus, CPU utilization is the main metric for calculating the energy consumption of the entire system. Hence, the energy consumption of any system is having a linear relationship with CPU utilization. When we assume that the server is not working or in the idle state even then it consumes 70% of its peak power. Power consumed by server when it is in idle state is called static power consumption and dynamic power consumption depends on the CPU utilization when server is in working mode. The power model p(u) can be described as [22].

$$p(u) = p_{idle} + u * (P_{max} - p_{idle}) \tag{4}$$

$$p(u) = Staticpower + dynamicpower * utilization$$

Where

p_{idle}	Power consumed by the physical server when in idle state = static power
P_{max}	Power consumption in the data center 100% maximum
$(P_{max} - p_{idle}) * u$	dynamic power
$p(u)$	Total power consumed by a data center
u	CPU utilization with respect to workload on the server

As the workload in data centers is dynamic in nature, the CPU utilization varies from time to time. Energy, a function of time can be computed as an integration of power consumed in cloud data center over some period of time. Thus, energy calculations can be done as

$$E = \int P(u(t)) \, dt$$

3.4 SLA Violation Model

It is very important to meet the quality of service (QoS) in cloud computing data centers with energy-saving strategy. In fact, QoS requirements are commonly framed as SLA which is an agreement between user and cloud service provider who are bound to satisfy the norms. The SLA violation is defined as follows:

$$SLA = U_{requested} - U_{allocated}/U_{requested} \tag{5}$$

where $U_{requested}$ is the number of instructions that must be executed or requested by VM on any host and $U_{allocate}$ is the instructions which are actually allocated to VM after migration. Instruction allocated may not be according to the requested number of instructions. Thus, SLA violation here represents the percentage of the CPU performance that has not been allocated relatively to the total demand.

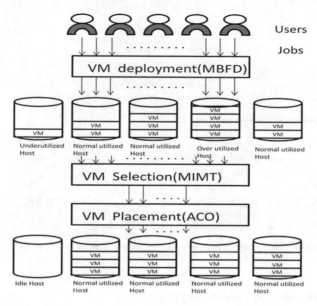

Fig. 2 Proposed energy efficient model

4   Simulation Model

The proposed approach comprises of three main steps. The first step is power efficient deployment of VMs and second is the selection of VMs among all the VMs from overutilized hosts and the lastly the placement of those selected VM on a host to minimize the number of migrations and to reduce energy consumption as shown in Fig. 2.

4.1   VM Deployment (MBFD)

Whenever a new request enters into a cloud data center, it is first allocated to any VM according to some VM provisioning method and that VM belongs to a host. Traditional best fit decreasing (BFD) algorithm allocates sorted VMs on the basis of maximum remaining CPU capacity of hosts or minimum CPU capacity of hosts. It does not consider the energy parameter [23]. But modification of best fit decreasing (MBFD) algorithm is one of the best VM allocations method in which VMs are allocated to hosts according to their power consumption and CPU utilization. In the presented approach, VMs are allocated in such a way that they consume less power after allocation and the CPU utilization should remain in normal threshold values, i.e., greater than the lower threshold and less than the upper threshold value.

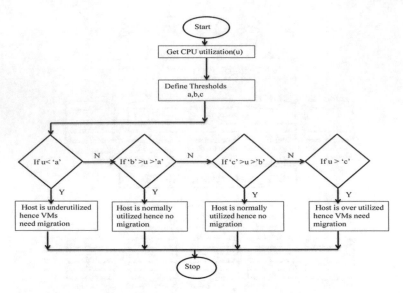

Fig. 3 Three-threshold energy-saving policy [12]

4.2 VM Selection (MIMT)

The proposed approach is based on to improve the performance of data center by selection of VMs with TESA (Three-threshold energy-saving algorithm) which will reduce the energy consumption of the data centers by migrating the VM to maintain a threshold of resource utilization. The TESA drives the relationship which is based on values a, b, and c as shown in Fig. 3. This relationship divides the load of the host in light load, proper load, middle host, and heavy load. The hosts having CPU utilization less than "an" are called underutilized hosts. Minimization of migrations policy based on TESA (MIMT), maximization of migrations policy based on TESA (MAMT), highest potential growth policy based on TESA (HPGT), lowest potential growth policy based on TESA (LPGT), and random choice policy based on TESA (RCT)) are presented by author in [13].

The MIMT performed best out of all these policies which are proven through experimental analysis. It selects the best VMs for migration from the overutilized hosts among all the VMs working on that host. The selection is based on two conditions. First condition is that the utilization of VM should have greater than the difference between the overall utilization of host and "c" threshold and secondly if the VM is migrated from a host, the difference between the new utilization and "c" threshold is minimum across the values of all the VMs.

4.3   VM Placement (ACO)

In this work, the energy efficiency of the data center can be improved by selecting the appropriate destination host for placement of selected VMs for migration. The appropriate host is selected by applying the bio-inspired ant colony optimization technique. In order to traverse the path, ant may choose two strategies, i.e., positive traversing strategy in which there are more chances that ant will traverse through the host with high pheromone level and the other is negative traversing strategy in which there are more chances that ant will traverse through the host with less pheromone [24]. In this proposed work, positive traversing is applied on overutilized hosts to find the host which has the maximum utilization. Then negative traversing is applied on normally utilized hosts to find the host having less utilization so that it can accommodate more VMs. Also, negative traversing is applied on the underutilized host for VM consolidation or placement of VMs in case there is no host available for placement under the pool of normally utilized hosts. During iterations, ant builds a solution by moving from one host to another until they complete a tour. The fitness function is applied on utilization modal to find the probability through edge selection phenomenon which is calculated by Eq. (1) on the basis of pheromone intensity, required resources, and the distance traveled by ant so that hosts should remain in normal threshold values, i.e., fewer chances of overutilization. Then the pheromone is updated for the best-constructed path. This leads to a reduction in a number of migrations and increase in energy efficiency of the data center.

Pseudo Code for ACO

Step 1: Begin
Step 2:   Apply MBFD for deployment of VMs.
Step 3:   Apply thresholds a, b, c for CPU utilization (TESA).
Step 4:   Get a list of over utilized host, underutilized host, and normal utilized host.
Step 5:   Select VMs to migrate from overutilized host by MIMT
Step 6:   Apply Positive traversing on overutilized hosts.
Step 7:   For all VMs that needs to migrate from Overutilized hosts.
Step 8:   Find Probability of Ants by Constructing Ant Solutions.
Step 9:   Apply Negative traversing on Normal Utilized Hosts.
Step 10:   While Normal_Host_Utilization < threshold 'c'.
Step 11:   Update amount of pheromone.
Step12:   Host_Utilization=Host_Utilization + VM_Utilization
Step 13:   End while
Step 14:   Update Pheromone Solution.
Step 15:   Apply Negative traversing on underutilized Utilized hosts.
Step 16:   Repeat steps from 5 to 15.
Step 17:   End For
Step 18:   Migrate VMs that are updated in Pheromone Solution.
Step 19: End

Table 2 Parameters used for simulation

Number of hosts	200
Number of VMs	600
VM CPU capacity	250, 500, 750, 1000
Host CPU capacity	1000, 2000, 3000
VMM	XEN
Number of data center	1
Memory capacity of host (MIPS)	1000
Memory capacity of VM (MIPS)	150

Fig. 4 Comparison based on energy consumption

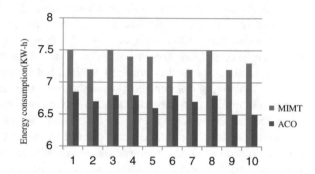

5 Experimental Setup and Results

To analyze the efficiency of proposed approach, the algorithm is compared with MIMT [13]. Cloudsim is used for simulating cloud computing environment because it is easy as compared to experiment on real cloud data centers. Power management or energy consumption depends on the values of thresholds for CPU utilization and probabilistic edge selection criteria. The performance is evaluated by considering the metrics of energy consumption, a number of migrations and SLA violation. It is assumed that the each VM is executing at least one cloudlet. The algorithm is implemented by considering the input parameters described in Table 2.

By considering the above-shown input parameters, i.e., 200 hosts and 600 virtual machines, the experiment is conducted 10 times to find the accurate results. Figure 4 shows that there is 8% less energy consumption as compared to previous work. It is important to maintain QoS, so considering SLA as a parameter our algorithm successfully lowers the percentage of violation as shown in Fig. 5 and as it was one of the main objectives to reduce the number of migrations which also reduced by 31% revealed by Fig. 6.

MIMT will only select some VMs for migration but in the absence of a novel approach for VM placement, it tends to increase the number of migrations. The performance of the proposed algorithm is also evaluated by varying the number of hosts and number of VMs. As shown in Fig. 8, ACO approach can better decrease

Fig. 5 Comparison based on
SLA violation

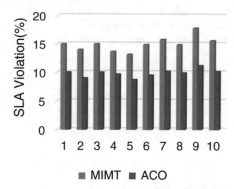

Fig. 6 Number of
migrations comparison

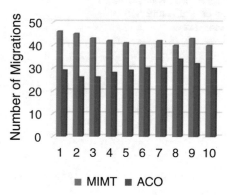

Fig. 7 Energy consumption
based on different no. of
hosts and VMs

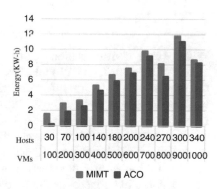

the power consumption when compared with the MIMT. Energy consumption saving
varies from to 6 to 86.84% with number of VMs varying from 1000 to 100 (Fig. 7).
The average energy saving with ACO approach comes out to be 22%. Also numbers
of migrations as shown in Fig. 8 are reduced with less SLA violation shown in Fig. 9.

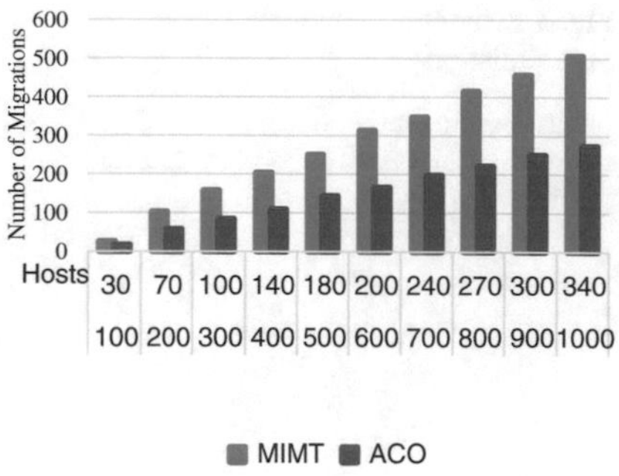

Fig. 8 Number of
migrations based on different
no. of hosts and VMs

Fig. 9 SLA violation based
on different number of hosts
and VMs

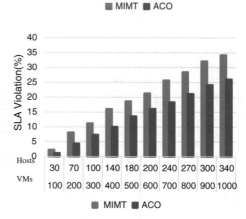

6 Conclusion

With the help of virtualization technique, cloud data centers can improve energy efficiency utilizing different VM allocation and optimization techniques. In this work, the applied ACO algorithm for VM placement reduces the number of migrations by selecting the appropriate host. TESA with ACO is used for proper utilization of resources in order to achieve the objective of optimization of energy consumption. The proposed approach has minimized the energy consumption with less number of migrations. It is being observed that the proposed method also met the QoS by reducing the SLA violation. Future lies in implementing the work by applying metaheuristic techniques like genetic and firefly optimization. It can further be implemented by the hybrid approach of ACO with classifiers like a neural network. Future work may also include dynamic three-threshold mechanism to lower the energy consumption in the dynamic environment of cloud computing.

References

1. Buyya R, Yeo CS, Venugopal S, Broberg J, Brandic I (2009) Cloud computing and emerging IT platforms: vision, hype, and reality for delivering computing as the 5th utility. Future Gener Comput Syst 25(6):599–616
2. Zhang Q, Cheng L, Boutaba R (2010) Cloud computing: state-of-the-art and research challenges. J Internet Serv Appl 7–18
3. Pettey C (2007) Gartner estimates ICT industry accounts for 2 percent of global CO_2 emissions. Dostupno na: https://www.gartner.com/newsroom/id/503867, 2013, vol 14
4. Reguri VR, Kogatam S, Moh M (2016) Energy efficient traffic-aware virtual machine migration in green cloud data centers. In: Big data security on cloud (Big Data Security), IEEE international conference on high performance and smart computing (HPSC), and IEEE international conference on intelligent data and security (IDS), pp 268–273
5. Kalra M, Singh S (2015) A review of meta-heuristic scheduling techniques in cloud computing. Egypt Inf J Elsevier 16(3):275–295
6. Blum C (2005) Ant colony optimization: introduction and recent trends. Phys Life Rev 2(4):353–73
7. Yin PY, Wang JY (2006) Ant colony optimization for the nonlinear resource allocation problem. Appl Math Comput 174(2):1438–1453
8. Van Ast J, Babu R, De Schutter B (2009) Fuzzy ant colony optimization for optimal control, pp 1003–1008
9. Beloglazov A, Buyya R (2010) Energy efficient allocation of virtual machines in cloud data centers. In: Cluster cloud and grid computing, pp 577–578
10. Copil G, Moldovan D, Salomie I, Cioara T, Anghel I, Borza D (2012) Cloud SLA negotiation for energy saving—a particle swarm optimization approach. In: Intelligent computer communication and processing (ICCP), pp 289–296
11. Huang J, Wu K, Moh M (2014) Dynamic virtual machine migration algorithms using enhanced energy consumption model for green cloud data centers. In: High performance computing and simulation (HPCS), pp 902–910
12. Xiong AP, Xu CX (2014) Energy efficient multiresource allocation of virtual machine based on PSO in cloud data center. Math Prob Eng
13. Zhou Z, Hu ZG, Song T, Yu JY (2015) A novel virtual machine deployment algorithm with energy efficiency in cloud computing. J Central South Univ 22(3):974–83
14. Farahnakian F, Pahikkala T, Liljeberg P, Plosila J, Tenhunen H (2015) Utilization prediction aware VM consolidation approach for green cloud computing. In: Cloud Computing, pp 381–388
15. Mustafa S, Nazir B, Hayat A, Madani SA (2015) Resource management in cloud computing: taxonomy, prospects, and challenges. Comput Electr Eng 186–203
16. Choi H, Lim J, Yu H, Lee E (2016) Task classification based energy-aware consolidation in clouds. In: Scientific programming
17. Selim, GEI, El-Rashidy MA, El-Fishawy NA (2016) An efficient resource utilization technique for consolidation of virtual machines in cloud computing environments. Radio Sci Conf (NRSC) 316–324
18. Singh S, Chana I (2016) A survey on resource scheduling in cloud computing: Issues and challenges. J Grid Comput 14(2):217–264
19. Acharya S, Mello DAD (2013) A taxonomy of live virtual machine (VM) migration mechanisms in cloud computing environment. In: Green computing, communication and conservation of energy (ICGCE), pp 809–815
20. Marino MA (2005) Ant colony optimization algorithm (ACO): a new heuristic approach for engineering optimization. WSEAS Trans Inf Sci Appl 606–610
21. Gao C, Wang H, Zhai L, Gao Y, Yi S (2016) An energy-aware ant colony algorithm for network-aware virtual machine placement in cloud computing. In: Parallel Distributed Computing System (ICPADS), pp 669–676

22. Dayarathna M, Wen Y, Member S, Fan R (2016) Data center energy consumption modeling : a survey. IEEE Commun Surv Tutor 732–794
23. Singh K, Kaushal S (2016) Energy effi cient resource provisioning through power stability algorithm in cloud computing. In: Proceedings of the international congress on information and communication technology, pp 255–263
24. Wen W, Wang C, Wu D, Xie Y (2015) An ACO-based scheduling strategy on load balancing in cloud computing environment. Front Comput Sci Technol (FCST) 364–369

Multi-label Classification of Twitter Data Using Modified ML-KNN

Saurabh Kumar Srivastava and Sandeep Kumar Singh

Abstract Social media has become a very rich source of information. Labeling unstructured social media text is a critical task as features belong to multiple labels. Without appropriate labels, raw data does not make any sense. So it is mandatory to provide appropriate labels. In this work, we have proposed a modified multilabel K nearest neighbor (Modified ML-KNN) for generating multiple labels of tweets which when configured with a certain distance measure and number of nearest neighbors gives better performance than conventional ML-KNN. To validate the proposed approach, we have used two different twitter data sets, one Disease related tweets set prepared by us using five different disease keywords and an other benchmark Seattle data set consisting of incident-related tweets. The modified ML-KNN is able to improve the performance of conventional ML-KNN with a minimum of 5% in both the datasets.

Keywords Twitter · Multi-label classification · Disease dataset
Seattle dataset

1 Introduction and Related Work

Social media is a place where people use a lot of text postings. Social media text classification systems retrieve such posts to user's interest and views in the form of summaries. Textual data over social media belongs to either the unstructured or semi-structured category. Due to the emergence of web 3.0 especially online information is growing enormously. Thus, we require some automatic tools for analyzing such large collection of textual data. In this regard, work in [1] proposed architecture to track real-time disease-related posting for early disease outbreaks prediction. Support vector machine (SVM) used for classifying postings, achieved up to 88%

S. K. Srivastava (✉) · S. K. Singh
Department of Computer Science & Engineering, Jaypee University, 62 Noida, India
e-mail: phd.jiit@gmail.com

S. K. Singh
e-mail: sandeepk.singh@jiit.ac.in

© Springer Nature Singapore Pte Ltd. 2019
M. L. Kolhe et al. (eds.), *Advances in Data and Information Sciences*, Lecture Notes in Networks and Systems 39, https://doi.org/10.1007/978-981-13-0277-0_3

accuracy in terms of performance. It is a very challenging task to determine tweets in multi-labels. The increasing volume of data demands classification into one or more concrete category in automated or a mutually exclusive manner. It is found that unstructured text has multiple labels. Due to overlapping terms, it is a very challenging task to determine tweets in multiple labels. Multi-label classification is reported by many researchers using Twitter data. Health-related discussions are very common on social media. People frequently share their experiences related to disease and use process of diagnosis which can be used to capture health-related insights from the social media. The authors in [2] have used semi-structured data for multi-label classification, problem transformation and algorithm adaptation methods are used in reported literature. Experimentation concluded that Binary Relevance (BR) is better over Label Powerset (LP) and ML-KNN both. The author in [3] proposed an annotation tool to collect and annotate twitter messages related to diseases. The tool automatically makes feature set for relevance filtering. The author in [4] proposed a methodology to identify incident related information using Twitter. This one has identified that assigning a single label to the text may lose important situational information for decision-making. In the paper problem transformation algorithms BR, LP, and Classifier Chain (CC) are used with Support Vector Machine (SVM) as a base classifier. Results are compared using precision and recall values. The above work illustrates that text data is highly sparse and nowadays research tries to utilize this real-time data for preparing an expert system that can use tweets/postings for surveillance over social media. Twitter can be a source of real-time surveillance using spatial and temporal text mining. In the context of real-time text mining, we are doing a very initial level task which can be further utilized by the surveillance module for better insights. In our work, we have introduced modified ML-KNN for health-related surveillance over social media.

2 Algorithms Used

In multi-label classification, reported work has been done in two well-known categories of algorithms.

2.1 Problem Transformation Methods

Problem transformation method [5] are multi-label learning algorithms that transform learning problem into one or more single-label classification. The problem transformation methods are binary relevance, label powerset, and classifier chains method.

2.2 Algorithm Adaptation Methods

Algorithm adaptation methods [5] adapted machine learning algorithms for the task of multi-label classification. Following popular machine learning algorithms have been adapted in the literature like boosting, k-nearest neighbors, decision trees, and neural networks. The adapted methods can directly handle multi-label data. Here, in this research work, we have presented a modified multi-label k-nearest neighbor method that upgrades the nearest neighbor family using appropriate similarity measures and number of nearest neighbors.

3 Conventional Versus Modified ML-KNN

3.1 Conventional ML-KNN

ML-KNN is derived from the popular k-nearest neighbor (KNN) algorithm [6]. It works in two different phases. First, k-nearest neighbors of each test instance in the training set is identified. Then, according to the number of neighboring instances belonging to each possible class, maximum a posteriori (MAP) principle is utilized to determine the label set for the test instance. The original ML-KNN uses Euclidean similarity measure with default 8 nearest neighbors. In our work, the effectiveness of ML-KNN is evaluated based on four similarity measures of Minkowski family mentioned in [7] and their variations with number of nearest neighbors.

3.2 Modified ML-KNN

In modified ML-KNN, we have used four types of similarity measures in which Manhattan, Euclidean, Minkowski, and Chebyshev are used with different nearest neighbors parameter (5, 8, 11, 14) which is used for the evaluation of ML-KNN. The experiment shows that the performance of ML-KNN can be improved by selecting some well-experimented similarity measures and appropriate number of nearest instances belonging to each possible class.

4 Architecture Used for Result Evaluation

Real-time information filtering of relevant postings with their unique labels is an important task on social media. Informative postings can be further used for effective surveillance. The filtering task can improve the performance of system as it contains unique and noise-free data. Generally, we think that postings will belong to only one

Fig. 1 Framework for
empirical analysis

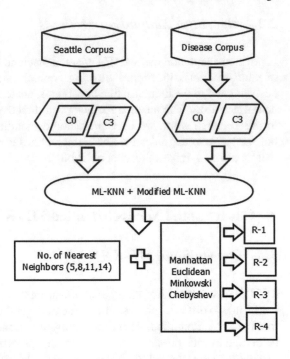

category but in real-world scenario, each tweet is associated with multiple labels. Following architecture shows our methodology for efficient result evaluation.

We have considered two different configurations of dataset. First when it belongs to raw category (C0). Here, raw category is defined by removing link, special symbols, and duplicate tweets from the corpus. Second, when stop words are removed and all the text data is stemmed means processed category (C3). For both the dataset, we have identified the appropriate similarity measure and number of nearest neighbors (NN) which can give better performance. We have used few configurations in ML-KNN for improving the multi-label algorithm. We have used MULAN Library [8] for result evaluation.

5 Data Sets Description and ML-KNN

In our research work, we have created our own disease corpus and found some motivating examples that belong to multiple categories of diseases. We have prepared tweets dataset manually annotated with the help of medical domain expert and the prepared corpus is used for result evaluation. Some of motivating examples which belong to multiple categories are as follows (Fig. 1, Table 1).

Table 1 Tweet belongs to multiple disease category

Tweets	Label sets
Close youre mouth when youre sneezing and coughing its not that difficult its actually pretty simple	Cold, Cough, Congestion
My stomach hurts so much	Stomachache, Abdominal Cramps
Knocked out at bc my stomach hurt so bad woke up rn and Im about to go back to sleep	Stomachache, Abdominal Cramps
Dear Mr brain eye headaches each day is not fun Its tough to look around nn yours truly a salty irritated person	Conjunctivitis, Headache
Keen yes that one K eyes are watery, inflammation	Conjunctivitis, Inflammation

We have used two different datasets 1. Disease corpus 2. Seattle dataset, both the dataset are based on Twitter data. Seattle is a standard dataset mentioned in paper [9]. We prepared our synthetic dataset based on disease keywords suggested in [10]. The disease data preparation phases are as follows.

5.1 Data Collection Phase

In data collection phase, raw tweets are collected to build a corpus. Twitter is the source of information capturing where we used the disease keyword for capturing relevant disease tweets from social media. Disease corpus is built by collecting tweets for five (D-1 to D-5) different diseases—Abdominal pain, conjunctivitis, cough, diarrhea, and nausea. The keywords to search tweets related to these diseases are taken from one of the classical work [10]. We have used Tweepy streaming API [11] for tweet collection. We collected only textual content of tweets in five different categories. All tweets were processed to remove duplicate tweets as well as other URLs. A total of 2009 unique disease tweets of five different disease categories were used in the final disease corpus.

5.2 Data Cleaning Phase

In cleaning phase, raw tweets are first cleaned before they are subjected to different preprocessing configurations. Cleaning process, generally taken in an effort to reduce noise that improves quality of training model. The idea behind these steps is to remove the noise from the dataset as special symbols, special syntaxes, duplicates, and stop words are viewed as noise and will not be beneficial for the input in any models.

6 Measures Used for Result Evaluation

Following measures are used for performance evaluation of Modified ML-KNN and ML-KNN.

6.1 Subset Accuracy

Subset accuracy [5] evaluates the fraction of correctly classified samples based on their ground truth label set. It is a multi-label counterpart of the traditional accuracy metric.

6.2 Hamming Loss

The hamming loss evaluates the fraction of misclassified instance-label pairs, it is calculated when a relevant label is missed or an irrelevant is predicted. Note that when each example in S is associated with only one label, hloss S(h) will be 2/q times of the traditional misclassification rate.

Zhang and Zhou [5] hloss (h) $= 1/p \sum_{i=1}^{p} 1/q |h(x_i) \Delta y_i|$ Here, Δ stands for the symmetric difference between two sets.

6.3 Example-Based Precision

Example-based precision is defined as—Precision (h) $= 1/N \sum_{i=1}^{N} |h(x_i) \cap y_i|/|y_i|$

6.4 Example-Based Recall

Example-based recall is defined as—Recall (h) $= 1/N \sum_{i=1}^{N} |h(x_i) \cap y_i|/|x_i|$

6.5 Example-Based F Measure

F measure score is the harmonic mean between precision and recall and is defined as F1 $= 1/N \sum_{i=1}^{N} 2^*|h(x_i) \cap y_i|/|h(x_i)|+|y_i|$. F measure score is an example-based metric and its value is an average overall example in the dataset. F measure score reaches its best value at 1 and worst score at 0.

6.6 Micro-Averaged Precision

Micro-precision (precision averaged over all the example/label pairs) is defined as—Micro-precision $= \sum_{j=1}^{Q} tp_j / \sum_{j=1}^{Q} tp_j + \sum_{j=1}^{Q} fp_j$ where tp_j, fp_j are defined as macro-precision.

6.7 Micro-Averaged Recall

Micro-recall (recall averaged over all the example/label pairs) is defined as—Micro-recall $= \sum_{j=1}^{Q} tp_j / \sum_{j=1}^{Q} tp_j + \sum_{j=1}^{Q} fn_j$ where tp_j, fn_j are defined as for macro-recall.

6.8 Micro-Averaged F Measure

Micro-averaged F measure is the harmonic mean between micro-precision and micro-recall. Micro-F is defined as micro-averaged F measure $= 2 \times$ micro-precision x micro-recall/micro-precision + micro-recall.

7 Result Evaluation and Discussion

7.1 Result Discussion

With the above experimentation, it is evident that performance varies in both C0 and C3. We can easily depict that configurations C0 and C3 have different results on ML-KNN when variations in distance measures and number of neighbors are applied over original ML-KNN. For both the datasets, C3 configuration (When stop words are removed and terms are stemmed) gives best subset accuracy and minimum hamming loss.

7.1.1 C0 Configuration

When we use C0 configuration Tables 2 and 4, it is clearly visible that the Euclidean and Minkowski distance measures along with eight neighbors perform best among all with the value of 84.72% subset accuracy. In case of Seattle dataset, Euclidean and Minkowski perform the best when configured with nearest neighbor value 5. The subset accuracy, in this case, is 48.49%. The Chebyshev distance measure is having poor performance among all the considered distance measures for both the datasets.

Table 2 Disease dataset with C0 configuration

Algorithm	Subset accuracy	Subset hamming loss	Configuration	Distance measure	NN−value
M – ML KNN	83.33	4.24	C0	Manhattan	5
M – ML KNN	83.33	4.24	C0	Manhattan	8
M – ML KNN	83.73	4.16	C0	Manhattan	11
M – ML KN N	82.98	4.39	C0	Manhattan	14
M – ML KNN	84.72	4.48	C0	Euclidean	5
M – ML KNN	83.33	4.24	C0	Euclidean	8
M – ML KNN	83.73	4.16	C0	Euclidean	11
M – ML KNN	82.98	4.39	C0	Euclidean	14
M – ML KNN	84.72	4.48	C0	Minkowski	5
M – ML KN N	83.33	4.24	C0	Minkowski	8
M – ML KNN	83.73	4.16	C0	Minkowski	11
M – ML KNN	82.98	4.39	C0	Minkowski	14
M – ML KNN	5.08	19.02	C0	Chebyshev	5
M – ML KNN	5.03	19.02	C0	Chebyshev	8
M – ML KNN	5.13	19.03	C0	Chebyshev	11
M – ML KN N	4.33	19.18	C0	Chebyshev	14

Table 3 Disease dataset with C3 configuration

Algorithm	Subset accuracy	Subset hamming loss	Configuration	Distance measure	NN−value
M – ML KNN	90.74	2.51	C3	Manhattan	5
M – ML KNN	91.44	2.34	C3	Manhattan	8
M – ML KNN	89.50	2.54	C3	Manhattan	11
M – ML KN N	89.15	2.53	C3	Manhattan	14
M – ML KNN	90.74	2.51	C3	Euclidean	5
M – ML KNN	91.44	2.34	C3	Euclidean	8
M – ML KNN	89.50	2.54	C3	Euclidean	11
M – ML KNN	89.15	2.53	C3	Euclidean	14
M – ML KNN	90.74	2.51	C3	Minkowski	5
M – ML KN N	91.44	2.34	C3	Minkowski	8
M – ML KNN	89.50	2.54	C3	Minkowski	11
M – ML KNN	89.15	2.53	C3	Minkowski	14
M – ML KNN	11.45	17.93	C3	Chebyshev	5
M – ML KNN	9.86	18.15	C3	Chebyshev	8
M – ML KNN	9.90	18.14	C3	Chebyshev	11
M – ML KN N	9.51	18.14	C3	Chebyshev	14

Table 4 Seattle dataset with C0 configuration

Algorithm	Subset accuracy	Subset hamming loss	Configuration	Distance measure	NN−value
M – ML KNN	2.25	35.15	$C0$	Manhattan	5
M – ML KNN	2.25	35.19	$C0$	Manhattan	8
M – ML KNN	1.70	35.31	$C0$	Manhattan	11
M – ML KN N	1.98	35.23	$C0$	Manhattan	14
M – ML KNN	48.49	26.85	$C0$	Euclidean	5
M – ML KNN	45.09	26.78	$C0$	Euclidean	8
M – ML KNN	47.45	26.07	$C0$	Euclidean	11
M – ML KNN	45.80	26.37	$C0$	Euclidean	14
M – ML KNN	48.49	26.85	$C0$	Minkowski	5
M – ML KN N	45.09	26.78	$C0$	Minkowski	8
M – ML KNN	47.45	26.07	$C0$	Minkowski	11
M – ML KNN	45.80	26.37	$C0$	Minkowski	14
M – ML KNN	2.25	35.15	$C0$	Chebyshev	5
M – ML KNN	2.25	35.19	$C0$	Chebyshev	8
M – ML KNN	1.70	35.31	$C0$	Chebyshev	11
M – ML KN N	1.98	35.23	$C0$	Chebyshev	14

7.1.2 C3 Configuration

When we use C3 configuration Tables 3 and 5, it means we use concrete feature set for the classification task. We found Manhattan, Euclidean, and Minkowski with eight nearest neighbor performs best among all with 91.44% overall subset accuracy in case of Disease dataset. For the Seattle dataset, we found Manhattan, Euclidean, and Minkowski with 14 nearest neighbor performs best among all with 53.15% overall subset accuracy.

It is clearly visible with the experimentation that there is around 7% more accuracy in case of Disease data set and 5% more accuracy in case of Seattle dataset. This stands that concrete features play important role in classification task irrespective of their belongingness to single, multi-class, or multi-label classification.

Table 5 Seattle dataset with C3 configuration

Algorithm	Subset Accuracy	Subset hamming loss	Configuration	Distance measure	NN−value
M – ML KNN	52.72	26.00	C3	Manhattan	5
M – ML KNN	52.83	25.33	C3	Manhattan	8
M – M L KNN	52.77	25.34	C3	Manhattan	11
M – ML KN N	53.15	25.27	C3	Manhattan	14
M – ML KNN	52.72	26.00	C3	Euclidean	5
M – ML KNN	52.83	25.33	C3	Euclidean	8
M – ML KNN	52.77	25.34	C3	Euclidean	11
M – ML KNN	53.15	25.27	C3	Euclidean	14
M – M L KNN	52.72	26.00	C3	Minkowski	5
M – ML KN N	52.83	25.33	C3	Minkowski	8
M – ML KNN	52.77	25.34	C3	Minkowski	11
M – ML KNN	53.15	25.27	C3	Minkowski	14
M – ML KNN	3.84	34.74	C3	Chebyshev	5
M – ML KNN	3.73	34.81	C3	Chebyshev	8
M – M L KNN	3.46	34.82	C3	Chebyshev	11
M – ML KN N	3.62	34.78	C3	Chebyshev	14

8   Conclusion

In this paper, the performance of the conventional ML-KNN algorithm is validated by changing appropriate similarity measure and number of nearest neighbors. Based on nearest neighboring instances information and distance measures between the feature of test instances, modified ML-KNN utilizes maximum a posteriori principle to determine the label set for the unseen instances. Experiments on two real-world multi-label datasets showed that performance of Modified ML-KNN is improved on the basis of their distance measures and number of nearest neighbors. Manhattan, Euclidean, and Minkowski show that modified ML-KNN outperforms with C3 configuration and there is around 5–7% hike in subset accuracy. In this paper, the distance between instances is simply measured by four distance metric Manhattan, Euclidean, Minkowski, and Chebyshev. Experiment shows Chebyshev distance metric has the worst performance among all.

9 Future Work

Complex statistical information other than the membership counting statistics can facilitate the usage of maximum a posteriori principle. This can be an interesting issue for future work.

References

1. Sofean M, Smith M (2012) A real-time disease surveillance architecture using social networks. Stud Health Technol Inf 180:823–827
2. Guo J, Zhang P, Guo L (2012) Mining hot topics from twitter streams. Procedia Comput Sci 9:2008–2011
3. Rui W, Xing K, Jia Y (2016) BOWL: Bag of word clusters text representation using word embeddings. In: International conference on knowledge science, engineering and management. Springer International Publishing
4. Ding W et al (2008) LRLW-LSI: an improved latent semantic indexing (LSI) text classifier. Lect Note Comput Sci 5009:483
5. Zhang ML, Zhou ZH (2014) A review on multi-label learning algorithms. IEEE Trans Knowl Data Eng 26(8):1819–1837
6. Aha DW (1991) Incremental constructive induction: an instance-based approach. In: Proceedings of the eighth international workshop on machine learning
7. Cha SH (2007) Comprehensive survey on distance/similarity measures between probability density functions. City 1(2):1
8. Tsoumakas G et al (2011) Mulan: a java library for multi-label learning. J Mach Learn Res, 2411–2414
9. Schulz A et al (2014) Evaluating multi-label classification of incident-related tweets. In: Making Sense of Microposts (Microposts2014), vol 7
10. Velardi P et al (2014) Twitter mining for fine-grained syndromic surveillance. Artif Intell Med 61(3):153–163
11. Roesslein J (2009) Tweepy documentation. http://tweepy.readthedocs.io/en/v3.5

A Comparative Evaluation of Profile Injection Attacks

Anjani Kumar Verma and Veer Sain Dixit

Abstract In recent years, the research on shilling attacks has been greatly improved. However, some serious problem in hand such as attack model dependency and high computational cost. Such recommender system also provides an impressive way to overcome information overload problem. In order to preserve the trust of recommender system, it is required to identify and remove the fictitious profiles from the system. Here, we have used machine learning classifiers to detect the attacker's profiles. A new model is proposed that outperforms in most of the cases.

Keywords Recommender systems · Collaborative filtering · Shilling attacks
MAE · RMSE

1 Introduction

In recent years, the effective profile injection or shilling attacks are more emphasized toward the insertion of bogus user profiles into the system database in order to manipulate the recommendation output, which is actually used to promote or demote the predicted ratings for a particular product. In many e-commerce websites, the recommender systems are widely deployed to provide user-purchasing suggestion. With rapid change in technology, most recommender systems adopted collaborative filtering. However, with the open nature of collaborative filtering recommender systems, it suffers from significant vulnerabilities from being attacked by malicious raters, who inject profiles consisting of biased ratings. Hence, they may have an effective impact on produced predictions. The basic function of a recommender system is to sense the consumer's feedback (which can be present in many forms like implicit, explicit, etc.) and understand user's interests to benefit consumers and e-business owners.

A. K. Verma (✉)
Department of Computer Science, University of Delhi, New Delhi, India
e-mail: anjaniverma29@gmail.com

V. S. Dixit
Department of Computer Science, ARSD College, University of Delhi, New Delhi, India
e-mail: veersaindixit@rediffmail.com

© Springer Nature Singapore Pte Ltd. 2019
M. L. Kolhe et al. (eds.), *Advances in Data and Information Sciences*, Lecture Notes in Networks and Systems 39, https://doi.org/10.1007/978-981-13-0277-0_4

Here, recommender system predicts the consumer's interests and ranks the products, which is based on the attributes like relevance, novelty, serendipity, and recommendation diversity [1]. Applications of recommendation system have expanded beyond the traditional domain and traditional items like books, movies, etc., have spread to news, advertisements, and social networks. We have studied collaborative models of recommendation system, as this is popular personalized recommendation technique. It can filter information which computers are unable to automatically analyze, such as music, text, and so on.

A collaborative filtering model uses the collaborative power of the ratings provided by multiple users to make recommendations. There are two types of collaborative filtering method: Memory-based methods—This is the earliest collaborative filtering algorithm, in which the ratings of user–item combinations are predicted on the basis of their neighborhoods such as user-based and item-based. Model-based methods—In this model, machine learning and data mining methods are used in the context of predictive models. We have studied the different papers and have identified the issues which need to be addressed in recommendation systems such as cold start problem, scalability of the approach, recommending the items in the long tail, accuracy of the prediction, novelty and diversity of recommendation, sparse, missing, erroneous and malicious data, conflict resolution while using ensemble/hybrid approaches, ranking of the recommendations, impact of context awareness, impact of mobility and pervasiveness, big data, and privacy concerns [2]. As we know in collaborative recommendation, a new user problem, new item problem, impact of power users lead to power user attack, and anomalous in ratings are the big issues. Hence, we have tackled the situation of profile attacks or shilling attacks and overcome the problem of injection of fake profile through our proposed model.

2 Background Details and Related Work

People make fake profiles to affect the recommendation image of an item, but they do not only rate target items but other items too not to stand out from genuine users, because rating only target item is suspicious. There are different shilling attack models that can be used for mounting the push attack namely, random attack, average attack, bandwagon attack, and segment attack.

There are some types of shilling attacks on the basis of how fake profilers rate other items such as random attack—when fake profilers give intended values to target item and ratings they give to other items are random in range, average attack—when fake profilers give ratings according to average ratings given by genuine users, for this attack they should have some internal knowledge of recommender system and attack in groups—when fake profilers want to affect some specific group of people to change their view. Some parameters are calculated to detect shilling attacks. In this case, we are considering collaborative recommender system only. Now recommender systems have become important tools to recommend products and services online, some people, with malicious intentions to tamper with original ratings to promote or demote a product or a type of product for their benefits, make fake profiles and give maximum or minimum rating to target product according to their plan, which causes shilling attacks. Recommender systems and related business are very vulnerable to shilling attacks which cause loss of credibility and profits.

There are three categories of attack detection techniques: supervised, unsupervised, and semi-supervised. In supervised or classification techniques, profiles are labeled under genuine and fake profiles. They need a large number of labeled users to enhance accuracy and balance should be maintained between fake and real profiles for the proper functioning of classifiers. But these techniques ignore the combined effect of attackers and only look at individual attackers. Some examples are SVM- and KNN-based approaches. In unsupervised techniques, there are no labels, so it involves less number of computations. Here, dynamic learning of data is made possible and has better accuracy. The techniques calculate similarity and dissimilarity in data; some examples are clustering approach and statistical rules. In semi-supervised techniques, a hybrid of supervised and unsupervised technique is used [3].

Zhou et al. [4] proposed an unsupervised technique for identifying group attack profiles, which uses an improved metric based on degree of similarity with top neighbors (DegSim) and rating deviation from mean agreement (RDMA). Their experimental results have shown a good detection performance. However, calculating the DegSim seemingly consumed a lot of time. In addition, Zhou et al. [5, 6] developed detection approaches based on analyzing rating patterns of target items to capture anomalous profiles.

The rest of the paper has been organized as follows: in Sect. 3, the proposed approach is elicited, in Sect. 4, Experiments and results will be explained. In Sect. 5, we draw conclusions based on the experimental results.

3 Proposed Approach

Collaborative filtering approach is highly prone to shilling attacks because it depends on the feedback of the customers. The fictitious users who are indistinguishable from genuine users enter the system. They create the unscrupulous profiles using various attack models and inject them into the database of the system. As a result, the items that are irrelevant to the current user might get recommended to him and may degrade the performance of the system [7, 8]. Particularly, we are interested in attacks that can be inserted into the system's database with less knowledge of the rating distribution. The main objective of this paper is to deal with the shilling attack. The issue needs to be addressed that is testing the performance of the classifiers used to detect the fictitious profiles and to improve the accuracy of the predictive models used to detect the attack profiles. Classification and clustering are two important techniques that are used to find hidden pattern in a dataset. Classification approach is used when the class is already defined in the dataset. Clustering technique is used when the class to which instances belong is not defined. K-means is the most commonly used partition method. Various classification techniques are Naive Bayes, random forest, SGD,

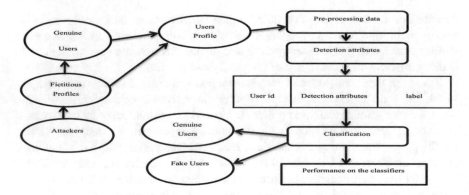

Fig. 1 Attacker's profile classification framework

SMO, decision tree, ZeroR, and many more. The classification techniques perform differently for different datasets. See Fig. 1 that describes a better classification of attacker's profile.

We have proposed here two models namely Model 1 and Model 2 respectively. In Model 1, there are four different classifiers that have been chosen and then their individual performance has been computed over the data. These classifiers are:

```
ZeroR, Logistic, Naïve Bayes (NB), Random Forest (RF)
```

In Model 2, we have integrated two classifiers and make an improvement over the different cases has been taken. In this case, the better improvement has been seen in the integration of

```
Naïve Bayes (NB), Random Forest (RF)
```

Then, we have implemented the algorithm that makes a classification between the fake raters and genuine raters:

```
Algorithm:
Read all set of users who rated on the same movie
do
extract those users who rated either low (nuke) or
high (push)->fake profile
for each movie
do for each user
do
classification using different classifiers
rating pattern from the set obtained->fake rater
```

4 Experimental Setup and Results

We have taken standard data from MovieLens 1M dataset, which is generally used in testing related to recommender systems. MovieLens 1M dataset has been used in the experiments which contain 1,00,000 ratings rated by 943 viewers on 1682 movies [9]. The integer value of ratings varied from 1 to 5 where 1 is assigned to most disliked items and 5 to most liked items. The dataset contains those users who rated at least 20 movies. We have used file namely "ratings.csv" containing ratings of movies. The file consists of four columns of different attributes such as UserId, MovieId, Ratings, and Timestamp. The file has 1,00,005 rows and 4 columns then we have applied some classifiers on our data to understand the distinction. For training and testing, dataset comprises with the instances taken from MovieLens with 15,000 ratings and tests the instances on that and after which it analyzes the accuracy of the performance of different classifiers [10].

There are several metrics that can be used to measure the performance and accuracy of the RS. These metrics can be categorized as follows:

1. Predictive accuracy metrics, for example, mean absolute error (MAE) [11]. It calculates up to what extent a system can predict the ratings of users.
2. Classification accuracy metrics, such as precision, recall, F-measure. These metrics are used to measure how well a system is able to distinguish the items accurately.

Mean Absolute Error (MAE)

$$MAE = (\sum_{[a,q]} |p_{a,q} - r_{a,q}|)/\#r$$

where, $p_{a,q}$ denotes the predicted rating of item q for user a, $r_{a,q}$ is the actual rating and #r represents the total count of the ratings. Lower the value of MAE better is the prediction.

Root Mean Squared Error (RMSE)

$$RMSE = (\sqrt{\sum_{[a,q]} (p_{a,q} - r_{a,q})^2})\#r$$

Precision, Recall, F-Measure

These metrics are used to measure the performance of the classifiers. Precision measures the fraction of relevant items retrieved out of all retrieved items [12]. For example, the percentage of recommended movies is actually good.

$$Precision = TP/(TP + FP)$$
$$= |good\ movies\ recommended|/|all\ recommendations|$$

where false positive (FP) means the count of genuine profiles that are incorrectly classified, true positive (TP) means the fake profiles identified accurately.

Recall determines the fraction of relevant items retrieved out of all relevant items. For example, the percentage of all good movies recommended.

$$\text{Recall} = TP/(TP + FN)$$
$$= |\text{good movies recommended}| / |\text{all good movies}|$$

where a false negative (FN) is the fake profiles that are not correctly identified.

The F-measure combines precision and recall and gives equal weightage to each of them.

$$\text{F-measure} = (2 * \text{Precision} * \text{Recall})/(\text{Precision} + \text{Recall})$$

Results

We have used WEKA tool to analyze the performance of the classifier whose result is shown. To perform an experiment, we have taken 475 instances and integrated the random forest and Naïve Bayes classifier and it gives the following result (see Table 1). Here, we have taken different filler [13] sizes such as 50 and 25% with average attack as a type means we want to make a profile with some attack profile so that in testing phase it can be clearly visible that how much in percent data is giving the result with good accuracy.

Correct classify instances: 474 (99.7895%)

Incorrect classify instances: 1 (0.2105%)

MAE: 0.0033

RMSE: 0.0407

Accuracy:

Similarly, the same data is applied to different classifiers and hence the result has depicted, see Tables 2 and 3 with the different filler size.

The advantage of the integrated model is that it will give good performance in almost all cases. It has been seen in Fig. 2 that shows the graphical view under the different measures with the different filler size attacks [14, 15]. However, the average attack is very similar, but the rating for each filler item is computed based on more specific knowledge of the individual mean for each item [16]. Hence, the average attack is by far the more effective, but it may be impractical to mount, given the degree of system-specific knowledge of the ratings distribution that it requires.

Table 1 Result of integrated (NB + RF) model

	TP rate	FP rate	Precision	Recall	F-mcasnre	Class
Weighted avg.	0.997	0.000	1.000	0.997	0.998	Genuine
	1.000	0.003	0.994	1.000	0.997	Fake
	0.998	0.001	0.998	0.998	0.998	

Table 2 Performance analysis of models for average attack at 50% filler size

Attack size	5%			10%			15%			20%		
Models	PR	Recall	Fm	PR	Recall	Fm	PR	Recall	Fm	PR	Recall	Fm
ZeroR	0.375	0.552	0.393	0.713	0.508	0.713	0.662	0.789	0.696	0.694	0.833	0.757
Logistic	0.861	0.858	0.857	0.894	0.895	0.892	0.922	0.923	0.919	0.953	0.950	0.946
NB	0.971	0.970	0.970	0.973	0.971	0.972	0.980	0.979	0.979	0.992	0.992	0.992
RF	0.972	0.970	0.970	0.981	0.980	0.980	0.993	0.991	0.992	0.994	0.994	0.994

Table 3 Performance analysis of models for average attack at 25% filler size

Attack size Models	5%			10%			15%			20%		
	PR	Recall	Fm	PR	Recall	Fm	PR	Recall	Fm	PR	Recall	Fm
ZeroR	0.375	0.612	0.465	0.316	0.562	0.404	0.434	0.659	0.524	0.518	0.720	0.602
Logistic	0.828	0.827	0.827	0.885	0.883	0.882	0.914	0.909	0.906	0.926	0.921	0.917
NB	0.980	0.980	0.979	0.986	0.985	0.985	0.970	0.970	0.971	0.991	0.991	0.990
RF	0.981	0.980	0.980	0.986	0.985	0.985	0.989	0.989	0.989	0.990	0.989	0.990
Integrated (RF + NB)	0.993	0.993	0.994	0.993	0.994	0.994	0.977	0.977	0.978	0.991	0.991	0.992

Fig. 2 k-fold cross validation at 25% attack size and 50% filler size

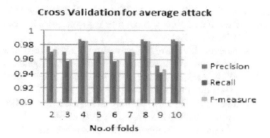

5 Conclusions

In this paper, the main focus is on various components in the field of attacks against recommender systems based on collaborative filtering. It has been investigated that the attack profiles can be generated and injected into the database even with the limited knowledge of the system. It has also been examined that the average attack has the highest impact on the recommender system. But, it is less effective against item-based algorithm and also requires more knowledge about the system. The segment attack has also been studied. Tested classifying techniques and their performance are analyzed using precision, recall, and F-measure and it is identified that Naïve Bayes and random forest models are best performers. More than two models can be integrated for more accurate results and their accuracy can be improved using different methods. Our further work will be to extend this method for much higher dimensionality.

References

1. Aggarwal CC (2016) An introduction to recommender systems recommender systems: the textbook, vol XXI, 1st edn. Springer International Publishing. ISBN: 978-3-319-29657-9, 498 P
2. Lakshmi SS, Lakshmi TA (2014) Recommendation systems: issues and challenges, vol 5(4)
3. Bilge A, Ozdemir Z, Polat H (2014) A novel shilling attack detection method. Proced Comput Sci 31(2014):165–174
4. Zhou W, Koh YS, Wen JH, Burki S, Dobbie G (2014) Detection of abnormal profiles on group attacks in recommender systems. In: Proceedings of the 37th international ACM SIGIR conference on research on development in information retrieval, pp 955–958
5. Zhou W, Wen J, Koh Y, Xiong Q, Gao M, Dobbie G, Alam S (2015) Shilling attacks detection in recommender systems based on target item analysis. PLoS ONE 10(7):e0130968
6. Zhou W, Wen J, Koh YS, Alam S, Dobbie G (2014) Attack detection in recommender systems based on target item analysis. In: International joint conference on neural networks (IJCNN)
7. Chirita PA, Nejdl W, Zamfir C (2005) Preventing shilling attacks in online recommender systems WIDM'05, Bremen, Germany (2005)
8. Lam S, Riedl J (2004) Shilling recommender systems for fun and profit. In: Proceedings of the 13th international conference on world wide web. ACM, pp 393–402

9. Mobasher B, Burke R, Bhaumik R, Williams C (2005) Effective attack models for shilling item-based collaborative filtering systems. In: Proceedings of the 2005 Web KDD workshop, Chicago, Illinois
10. Mahony M, Hurley N, Silvestre C (2005) Recommender systems: attack types and strategies. In: American association for artificial intelligence, pp 334–339
11. Williams C, Mobasher B, Burke R (2007) Defending recommender systems: detection of profile injection attacks. SOCA 1(3):157–170
12. Lee J, Zhu D (2012) Shilling attack detection—a new approach for a trustworthy recommender system. INFORMS J Comput 24(1):117–131
13. Dhimmar J, Chauhan R (2015) An accuracy improvement of detection of profile-injection attacks in recommender systems using outlier analysis. Int J Comput Appl 122(10):22–27
14. Zhang F (2012) A meta-learning-based approach for detecting profile injection attacks in collaborative recommender systems. J Comput 7(1):226–234
15. Gao M, Ling B, Yuan Q, Xiong Q, Yang L (2014) A robust collaborative filtering approach based on user relationships for recommendation systems. Math Probl Eng 2014:1–8
16. Zhang F (2009) Average shilling attack against trust-based recommender 2009. In: International conference on information management, systems, innovation management and industrial engineering, pp 588–591

Part II
Intelligent Computational Techniques

Literature Survey on DNA Sequence by Using Machine Learning Algorithms and Image Registration Technique

R. Vinodhini, R. Suganya, S. Karthiga and G. Priyanka

Abstract The DNA sequence is significantly utilized as a part of a field in medicinal information investigation. It comprehends the inward structure of qualities in the DNA. It comprehends which arrangement codes for what sort of proteins. Analysis of DNA sequences is important in preventing the evolution of viruses, bacteria, and also used to diagnose disease during an early stage. This paper focused on a literature survey on DNA sequence data and structured data examination and furthermore with image registration technique. The 3D graphic portrayal of DNA sequence likewise stays away from the misfortunes of data and furthermore distinguishes the typical and mutant DNA. From the existing work, the authors absorb some clustering algorithm and data analytics techniques like K-mean, k-mer, KNN, SVM, random forest correlation coefficient, and eigenvalue vector are used for predicting neurological disease. From the existing survey, it identifies the following issues: it is difficult to analyze joint DNA sequences, but the combination of two individual sequences is possible in analysis. To address the above issue, we propose the system to predict diseases from DNA sequence by using neural network algorithm through GC content in sequence, and further medical image registration technique is applied to monitor its growth. Medical image registration is an image handling strategy used to adjust different modality into a solitary coordinated image and it is also used to diagnose diseases in the different stages.

Keywords DNA sequence · Neural network · Image registration · Machine learning algorithm

R. Vinodhini (✉) · R. Suganya (✉) · S. Karthiga (✉) · G. Priyanka (✉)
Department of Information Technology, Thiagarajar College of Engineering,
Madurai, Tamil Nadu, India
e-mail: vinodhinicse95@gmail.com

R. Suganya
e-mail: rsuganya@tce.edu

S. Karthiga
e-mail: skait@tce.edu

G. Priyanka
e-mail: priyacse9495@gmail.com

© Springer Nature Singapore Pte Ltd. 2019
M. L. Kolhe et al. (eds.), *Advances in Data and Information Sciences*, Lecture Notes in Networks and Systems 39, https://doi.org/10.1007/978-981-13-0277-0_5

1   Introduction

In the emerging medical domain, there are several diseases are causing critical issues. Diagnosing and preventing in the earlier stages became complex process. The DNA sequence is significantly utilized as a part of a field in medicinal information investigation. It understands the inward structure of qualities in the DNA. In order to analyze the DNA sequence and its structure in the earlier stage helps to prevent the various diseases affecting virus, bacteria. In DNA, four parameter are used to determine individuals behavior, they are A, G, C, T in large amount of the sequence. It is a more challenging task to compute and analyze the large amount of this sequence. To simplify the process and also to predict disease through GC content, we propose a neural network algorithm. The visualization of the DNA sequence is all done by using R software. R is the data analytics software used to do data cleaning, analyzing, and visualizing DNA sequence, for visualization some of the additional packages are used. Image registration is an image processing technique used to align multiple scenes into a single integrated image. It helps to overcome issues such as image rotation, scale, and skew that is common when overlaying images. Image registration is often used in medical and satellite imagery to align images from different sources. Here, we use image registration technique for monitoring a DNA analysis. 3D graphical representation of DNA sequence has been derived from DNA sequence. The 3D graphical representation also avoids losses of information. Here to identify the normal and mutant DNA by analyzing DNA sequences and represented in form of graph. Our main interest is focused on the analysis of structural information and also sequential information of DNA. By using R software package, it generates a visual representation of DNA sequence from that we can predict the disease using machine learning algorithm and further image registration technique is used for my proposed work.

2   Proposed Approach

DNA sequence contains four parameters: A, G, C, T among A and T pair has two bonds but G, C has triple bond that is highly bonded sequence and it is most reliable bond content known as GC content. To diagnose the diseases present in a genome, GC content of the sequence is commonly used. The main objective of this research paper is to focus on the sequential information of DNA sequence. It is analyzed and predicts whether the sequence is affected from the monogenic disorder or not. Thus, the implementation of the sequence is done with four stages which is shown in Fig. 1: (1) sequence visualization representation (2) preprocessing phase, the DNA sequence is converted into numerical data, (3) analyzing phase, applying machine learning algorithm to predict the changes in DNA sequence, (4) image registration technique to monitor the changes in DNA sequence. This paper is adopted neural network algorithm for the prediction of disease from the graphical representation

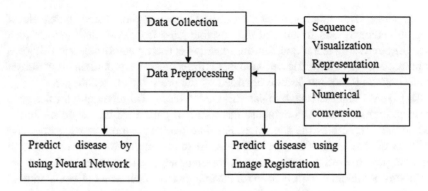

Fig. 1 Block diagram for the proposed methodology

of sequence. Finally, apply the image registration technique to monitor the changes. Image registration is used to analyze disease and find whether it is a normal or mutant one.

3 Literature Survey

In recent years, several authors are interested in doing research in DNA sequences. Yu and Huang [1] describe DNA sequence in both graphical representation and similarity analysis and also represent each DNA sequence with their neighboring nucleotide matrix. Approximate point diagonalization theory transforms each primary DNA sequence into their corresponding eigenvalue vector(EVV) and results from EVV are characterized in form of numeric value. The graphical representation of genome sequence is shown in the form of 2-D phase. Sequences are clustered into subclasses by using K-means algorithm and visualize the sequences in form of dendrogram. The authors computed for the large set of the sequences and analyzed its information, it analyzed all the involved sequence information in jointly (e.g., AA, AC, AT, AG) it was unable to analyze the individual sequence (e.g., A, G, T, C) which is the major drawback in this work.

Dozono and Som [2] used an artificial neural network called self-organized map (SOM) with the frequencies of large amount of tuples (N-Tuples) to categorize the set of DNA sequence with unsupervised learning process. SOM converts nonlinear and statistical data (high dimensional data) into image with a simple geometric relation. For large amount of species among nucleotide, the author computed correlation coefficient. It is the most efficient preprocessing method for DNA sequences. The input data are taken in the form of DNA sequence as the input vector (A, G, T, C). Each DNA sequence is converted into 4-bit binary sequence represents the occurrence of every nucleotide. Likewise in the calculation of combination of 'CC' are scans 16 times of DNA sequence, it takes more time and high computational cost for huge sequence. The existing work described that the author tested thousands of data simultaneously and found that the possible gene from collected data. In data mining preprocessing technique, feature selection is used. Feature selection chooses the optimal subset and performed analysis within that subset contains that subset contains only an important feature obtained by feature selection technique.

Saka [3] says that the image registration is the biological technique for the gene expression is mainly used to compare the common pattern of the sequence and is based on the structure of the sequence. For that, the image must be registered and coordinate of that images has to be found. In image registration, three common methodologies are used they are: manual, correlation based and mutual information. Correlation coefficient is mainly used for the images in the same modality and mutual information is used for multimodality images.

Author [4] describes that DNA in software package called shape R. It is a software package that can be implemented using R software. By using this package, the author analyzes the DNA sequence and visualizes them. Here the input is given in form of DNA sequence or else with genome coordinates and it generated the various graphical representation and visualization. It generated results in form of matrices, they are used for computing machine learning algorithm. The major use for used DNA ShapeR to predict DNA sequence and also used to find the importance of protein binding in DNA structure. Prediction and visualization DNA ShapeR are also used to generated feature vector and also performed clustering by using k-mer (1mer-A, 2mer-AA, 3mer-ACG), this author analyzes the sequence level by level.

Zhou et al. describe [5] web server is used to predict DNA structural and sequential data with high throughput (HT) manner of huge amount of data. The author integrates both sequential and structural or shape analysis using sliding window approach, this approach is used to mine DNA structural information and Monte Carlo method is used for simulation of DNA structure. Input data are taken in the form of nucleotide sequence and also predict multiple structure of DNA as (MGW, Roll, ProT, HelT). The outcomes from the predicted structure of DNA are solved using crystallography and NM and spectroscopy, statistical analysis, and cross validation are done for strong confidence. Input of web server is in the form of sequence, it is in the range between 5 and 10^6 bp in length they could be in the form of FASTA format. Outcome of that previous work predicts structural feature organized along with the plots of graph are also outcomes. In previous work, structural feature could not be analyzed, here they are solved by using HT manner. Calculating binding affinity with various sequence is depending on MGW in central region of its binding sit of five central nucleotide correlated witimportant to find predictionh a logarithm of binding affinity $R^2 - 0.65$. If binding affinity is high complexity, binding affinity calculated for protein element in DNA predicted how much protein DNA content.

Chiu et al. [6] proposed many regulatory mechanisms to measure high degree mechanism for protein DNA binding and they are associated with some high order effects, cooperatively and cofactor are described are determined with some transcription factor such as MGW, helix, propeller twist of all organism in GB framework and it can visualize DNA shape and compute quantitative value of DNA shape feature their function and position of nucleotide solution, various species or organisms DNA sequences are taken, so they have taken DNA from worms, fly, human are compute and found the similarities between them [6]. Input genome is collected in FASTA format, in GB browser there is some default species format and they add extra species gene within GB Shape framework within their assemblies and specific genome to their database. The result of this work is that they annotated and predicted genome with high throughput and their outcomes are converted into big wing data form and display them into genome browser finally they computed clustering of sequences in C++.

Gupta et al. [7] describe that extraction of data from huge amount of data is the biggest challenge. Micro-array technology provides better result in analysis of thousands of gene and monitoring their levels. They used various clustering techniques for microarrdata. Gene are visualized in 2D data also matrix of gene expression of, "n number of samples and the expression level are measured in form of ith gene winder with jth condition hidden functions are identified in own individual gene and associated with another gene". Author describes that about cluster are by hierarchical data structure and also with non-hierarchal clustering with k-means, SOM. They perform cluster analysis process, feature selection and clustering algorithm like fuzzy c-mean, fuzzy adaptive resonance theory, validates reliability of sequence and do clustering indices are help from validating dataset.

Chen et al. [8] describe the difference between the cell-free DNA and the normal DNA. The cell-free DNA (cfDNA) has the certain fragmentation pattern and they are in nonrandom with different in the structural and the sequential feature of the normal DNA. The author used many classification models used to classify the normal and cfDNA, such as k nearest network (KNN), support vector machine (SVM), random forest to obtain the accuracy above 98%. Finally, the author concluded that the random forest is the highest in accuracy compared with the all other algorithms. The data that are used and validated are extracted cfDNA and gDNA(genome) from the same blood samples, cfDNA are taken from the plasma of the blood cells. The author describes the data are classified into two sets they are primary set and validate set are done by the random sampling. Finally for the prediction and classification python language is used and predicted cfDNA are different from the normal DNA.

Wong [9] describes the importance of deciphering and the binding of protein in DNA and also describes the deep understanding of gene regulation. But that is very expensive, time consumption is high and labor intensive. The author describes some of the limitations such as they use tight support and confidence threshold based on the support count. Also describes the association rule mining incorporate the measures and statistical methods to overcome the above the limitations. Finally, author describes the many of the approaches to improve the effectiveness.

Strobelt et al. [10] describe the splicing method in the same DNA that uses different proteins which are called protein isoform. This paper identified the normal and the disease affected genes from the dataset. The author analyzed tumor DNA with the normal DNA of the same organism, the data are in the form of the categorical and continuous data. The author also identified the biological relevant data and correctness of the data is checked to establish trust. The author used three views they are abundance view, junction view, and expression view based on the geometrical sequence. The author describes web server based visualization with java-script as the frontend and python as backend.

Quang et al. [11] describe the structuring and modeling the DNA sequence is the easiest process, but it is difficult for the huge amount of the genome and that is in form of the non-coding DNA that the non-coding DNA are complex for computing when compared with the coding DNA. In human genome, 98% of them are the non-coded genome through that 93% of disease are spread, for this issue this author describes they are overcome in the cloud-based framework with the neural network algorithm. Convolution neural network is used with the Deep-SEA framework are used for the prediction of the DNA.

Zhou et al. [12] describe that prediction of the DNA genome in large sequence and interaction between DNA protein are complex processes. Author describes many methods for the motif feature, it is defined as that the subsequence of the DNA sequence, it is very important to find prediction of DNA-protein binding, this author proposed initially CNN (Convolution Neural Network) captures the motif feature among the sequence and it combines the captured CNN classification motif feature and evolutionary binding feature in a sequence. Author used two datasets PDNA-62 and PDNA-224 to evaluate the performance by using CNN. It is evaluated with the some of the other common metrics: sensitivity (SN), strength (ST), accuracy (ACC) and Mathews correction coefficient (MCC). MCC at the range of 4.3%.

Mathkour et al. [13] describe the increase in the genomic data structuring protein present in it and predict their structure become complex process, so the author propose an integral approach of prediction of tri-nucleotide base pattern in DNA strands. The hidden key of the sequence is also wants to found that in neural network. ANN are computed by three stages, each stage contains one input layer, one output layer and hidden layers with two techniques—feed forward and backpropogation. The four stages present here are preprocessing of data, data segmentation, central working engine, postprocessing output. In order to preprocess the data the two filters are used, in 1st filter they refines the sequence and the pure nucleotide only present in it. In the second filter the sequences are converted into numerical data and their outcome given as the input to the central engine. For data segmentation stage the neural network algorithm are act as the central engine. They are splitted into some samples each of the sample has the input layer output layer, hidden layer for the computation. By using neural network algorithm, optimal results are found with the help of the hidden layer.

Wang [14] describes that the increase in protein DNA binding is the specific identification and straightforward action in DNA for the motif feature. But random forest is the improved version of the decision tree using that author improved the accuracy in the predicting the sequence compared to the all other machine learning algorithm. The data sequence is taken from the website and PDNA62 dataset is used for the protein binding in the DNA structure. Specificity, sensitivity, strength, and accuracy performance are evaluated as same as that with [12].

Can et al. [15] say that registering two common images analyses of H&E modalities are a difficult process because of optical and chemical effects of H&E dyes. The author introduced sequential image registration technique for the solution of

Table 1 Comparison of literature survey on DNA sequence by using machine learning algorithms and image registration technique along with pros and cons

Author	Methodology	Pros	Cons
H. J. Yu and D. S. Haung	EVV, K-means, dendrogram	Justified similarity analysis of sequence	It doesn't obtain sequential information separately
H. Dozono and A. Som	SOM	Categorize DNA sequence	Impossible to calculate 'cc' pair in genome
E. Saka	MATLAB	Compare the common pattern of the sequence and is based on the structure of the sequence	Impossible to done with the different modality
T. P. Chiu, F. Comoglio, T. Zhou, L. Yang, R. Paro, and R. Rohs	DNAShapeR K-mer	Predict disease By machine learning algorithm	Complexity while compute with large sequence
T. Zhou	Sliding window	Predicting DNA structural feature in high throughput manner	Affinity factor different for different organism
Tsupeichiu	GBShape framework, C++	Compute quantitative value of DNA shape feature their function and position of nucleotide solution	Difficult to identify nucleotide position
Shelly Gupta	Fuzzy c-mean, fuzzy	Extraction of data from biggest macro-array	Handling macro-array are complex process
S. Chen	Random forest	Determine the sequence is cfDNA or not	Computation of cfDNA is complex process
M. H. Wong, H. Y. A. Sze-To, L. Y. P. Lo, T. M. C. Chan, and K. S. Leung	Association rule mining	Absorbed by various confidence level of sequence	It does not suite for low confidence level
H. Strobelt	Visual analysis tool	Identify patterns of isoform abundance in groups of samples	Evaluation of quality of data is complex process.
D. Quang and X. Xie	Deep-SEA, Neural network	Prediction of disease by using non-coded DNA	Collision occurs while using large data set
J. Zhou, Q. Lu, R. Xu, L. Gui, and H. Wang	CNN	Prediction of protein binding in sequence	Complex to binding many organism's sequence
H. Mathkour and M. Ahmad	Neural network	Integral approach for predict tri-nucleotide base pattern in DNA strands	It does not mean for double strands
L. Wang	Random forest	It improves accuracy in prediction while compared with other algorithms	Complex to perform and manage this algorithm
A. Can	Hematroxylin_Exoin	It attains 99.8% of the tissue in micro-array, an digitally merge the different imaging modalities into single modality	Difficult to perform macro-array
Y. Ma and J. Tian	Mutual information	Similarity between two different modalities is computed	Difficult to perform large dataset

that issue. This registration algorithm is successfully achieved 99.8% of the tissue in micro-array, author digitally merges the different imaging modalities in single modalities.

Ma and Tian [16] describe the medical image registration with rapid algorithm, in medical image registration there is low speed and huge calculations are computed with the mutual information, normalized mutual information is achieved with the formula NMI(A, B) = (H(A) + H(B)/H(A, B)). Finally, author identified the similarity between the two images A and B are computed with mutual image registration. The comparison of various above literatures along with pros and cons is shown in Table 1.

4 Conclusions

This paper survived with the various data mining algorithms and tools used for prediction of the DNA genome, with image registration technique and for structural analysis of gene. Prediction of the disease by using DNA sequence is mainly used to prevent the people from disease at the early stage without spreading virus. Many machine learning algorithms are used for the prediction of the DNA sequence. Hence, we concluded that in our proposed work, neural network is used to predict the disease in genome and with the help of mutual information in the image registration which is much efficient for the image in the same modality of DNA structure analysis.

References

1. Yu HJ, Huang DS (2013) Graphical representation for DNA sequences via joint diagonalization of matrix pencil. IEEE J Biomed Heal Inform 17(3):503–511
2. Dozono H, Som A Visualization of the sets of DNA sequences using self organizing maps based on correlation coefficients
3. Saka E Image registration and visualization of in situ gene expression images, Aug 2011
4. Chiu TP, Comoglio F, Zhou T, Yang L, Paro R, Rohs R (2016) DNAshapeR: an R/Bioconductor package for DNA shape prediction and feature encoding. Bioinformatics 32(8):1211–1213
5. Zhou T et al (2013) DNAshape: a method for the high-throughput prediction of DNA structural features on a genomic scale. Nucleic Acids Res 41(Web Server issue): 56–62
6. Chiu TP et al (2015) GBshape: a genome browser database for DNA shape annotations. Nucleic Acids Res 43(D1):D103–D109
7. Gupta S, Singh SN, Kumar D (2016) Clustering methods applied for gene expression data: a study. In: Proceedings of the second international conference on computer and communication technologies, pp 724–728
8. Chen S et al (2017) A study of cell-free DNA fragmentation pattern and its application in DNA sample type classification. IEEE/ACM Trans Comput Biol Bioinform 5963(c):1–1
9. Wong MH, Sze-To HYA, Lo LYP, Chan TMC, Leung KS (2015) Discovering binding cores in protein-DNA binding using association rule mining with statistical measures. IEEE/ACM Trans Comput Biol Bioinform 12(1):142–154
10. Strobelt H et al (2016) Vials: visualizing alternative splicing of genes. IEEE Trans Vis Comput Graph 22(1):399–408

11. Quang D, Xie X (2016) DanQ: A hybrid convolutional and recurrent deep neural network for quantifying the function of DNA sequences. Nucl Acids Res 44(11):1–6
12. Zhou J, Lu Q, Xu R, Gui L, Wang H (2016) CNNsite : prediction of DNA-binding residues in proteins using convolutional neural network with sequence features, pp 78–85
13. Mathkour H, Ahmad M (2010) An integrated approach for protein structure prediction using artificial neural network. In: Second international conference computer engineering applications, pp 484–488
14. Wang L (2008) Random forests for prediction of DNA-binding residues in protein sequences using evolutionary information. In: Proceedings of 2nd international conference future generation communication networking, FGCN 2008 BSBT 2008, 2008 international conference bio-science bio-technology, vol 3, pp 24–29
15. Can et al (2008) Multi-modal imaging of histological tissue sections. In: Proceedings of 5th IEEE international symposium biomedical imaging from nano to macro, ISBI, vol 668, pp 288–291
16. Ma Y, Tian J (2010) The algorithm of rapid medical image registration by using mutual information. vol 2, no 1, pp 1–4

What's on Your Mind: Automatic Intent Modeling for Data Exploration

Vaibhav Kumar and Vikram Singh

Abstract With the increased generation and availability of big data in different application domains, there is a need for systems that "understand user's search intents" and "serves data objects" in the background. Search intent is significant objects/topics represents, abstraction of user's information needs, which a user often fails to, specify in information seeking. Thus, the fundamental search is shifting focus on *"Find What You Need* and *Understand What You Find"*. User intention predictions and personalized recommendations are two such methods, to formalize user's intents. We investigated various information factors, which affect user's search intentions and how these intents assist in an exploratory search. In particular, we proposed an algorithm for automatic intent modeling (AIM), based on user's reviews and confidence assessments as search representatives of user's search intents. In this process, we have revisited various related research efforts and highlighted inherent design issues.

Keywords Exploratory search · Exemplar query · Intent mining

1 Introduction

With the exponential growth of information, we often fail to find the exact information that we are seeking for, even with the "powerful" search engines. Studies on two billion daily web searches show that approximately 28% of the queries are modifications of a previous query. In addition, when users attempt to fill a single information need, 52% of them modified their queries [1]. Among those modifications, 35.2% totally changed the query, while 7.1% only added terms. This indicates that users' needs are sometimes too vague to be stated clearly [2, 3]. The major evolution of the

V. Kumar · V. Singh (✉)
Computer Engineering Department, National Institute of Technology,
Kurukshetra, Kurukshetra 136119, Haryana, India
e-mail: viks@nitkkr.ac.in

V. Kumar
e-mail: vaibhav2050@live.com

© Springer Nature Singapore Pte Ltd. 2019
M. L. Kolhe et al. (eds.), *Advances in Data and Information Sciences*, Lecture Notes
in Networks and Systems 39, https://doi.org/10.1007/978-981-13-0277-0_6

search engine is to understand the user's query and user's intent. Indeed, research on search engines is trying today to bring results of research adapted to the intent of the user. Understanding the search intentions of Internet user and dominants on to the search engines. By understanding the user's intention to search, we can define the type of content to be produced in order to maximize your chances of positioning yourself. Search engines are continuously employing advanced techniques that aim to capture user intentions and provide results that go beyond the data that simply satisfy the query conditions. Examples include the personalized results, related searches, similarity search, popular, and relaxed queries [4, 5].

In the past, traditional data retrieval techniques assumed that the user is well informed of the semantics structures of interest and can partially formulate the query [4, 6]. We believe that there are various real-life scenarios in which this is not the case. We are interested in these typical scenarios, where a user is least familiar (knows one single element), and we would like to study possible ways to gather the relevant elements from the probable result set. In other words, the user's initial query works as an element of interests, which are expected to be returned. These queries referred as exemplar queries. Exemplar queries are important practical applications in information searching [7, 8]. This mechanism is particularly suitable for naïve user-initiated search (a student, a curious citizen, an investigator, a lawyer or a reporter), as the user has as a starting point an element from the desired result set. In this process, the result set can be used to retrieve the new relevant data objects or query suggestions by incorporating user's real search intent (in form of feedback or review). This intent-driven approach of data exploration (DE) is traditionally different from existing approach, as it required huge user interaction and system support [7, 9].

Example Exploratory system computes and provides results, which although not in the result set of initial queries. This allows the user to explore additional information and lead toward his interest. For example, user has searched for the movies that are directed by "M. Scorsese" (Q_i). There is a high probability that user is also interested in a movie with similar characteristics such as genre and production year. Generally, users intend carving for a wider query/data spectrum that fetches additional results for queries. Several kinds of adjustments like addition/dropping of predicate terms are required for generating query variations. The various kinds of variations (Q_{i+1}) could be achieved by generating variants of the initial query.

In order to have successful search outcomes, a certain degree of system support to the user/information seeker is required [10]. Search systems can be proactive, hence to provide automatic support via the variable amount of assistance. Modern search systems have the capability to provide real-time support to the system as well as progressively analyzing user logs to provide efficient search strategies for future iterations. We observed that, in a typical information seeking process, three level of system user support is provided, as shown in Fig. 1. At each level of the system, support is intended to assist a naïve user and makes him more informed about his search domains. User interface support is to guide a naïve user in enhancing his knowledge about search space and thus to the reduced cognitive effort in the search. Similarly, knowledge base support is to enrich the log/history or user access pattern in

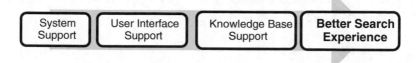

Fig. 1 Three level of user support in IDE

a search session and enrich the exploration of data space. Integrating these dimensions ensures a seamless interactive search experience for a user.

The problem of inferring a user's intentions [11] in search process has been the key research issue for providing personalized experiences and services [1, 12, 13]. A typical search system utilizes these search intents and incorporates into searching process [11]. These systems try to achieve significant accuracy in predicting user's intention, which is close to the precision (92%) of human prediction. To achieve, this traditional fundamental search is shifting focus on *"Find What You Need* and *Understand What You Find"*, moving toward intent-driven data exploration instead data-centric approach. Assisting user in *"Find What You Need"* explores data space for the retrieval relevance data objects, and *"Understand What You Find"* explains these results which result in enhanced user's certainty on his information needs. User intention predictions and personalized recommendations are two such methods to formalize user's intents [14]. We investigated various information factors, which affect user's search intentions and how these intents assist in an exploratory search. In particular, we proposed an algorithm for automatic intent modeling (AIM). In this two factors, user's reviews/feedback and his confidence assessments on query results are pivotal and act as representatives of user's search intents. The model can be trained incrementally and used to improve the overall accuracy of search prediction of user's information.

1.1 Contribution and Outline

The key contribution of the paper is a framework for data exploration based on an automatic intent modeling (AIM) algorithm. We anticipate that framework guides on an effective exploration over various voluminous databases, such as scientific DB, Medical DB, DNA DB, social DB, etc. Another contribution of proposed work is an algorithm of search intent modeling, which is inherently supported by user feedback. Further, various existing intent mining techniques are revisited to identify the possible sources and how these could be used in query modeling.

In Sect. 2, various related research effort and prospects are listed. The proposed approach is discussed in Sect. 3, in which conceptual scheme is depicted in a schematic diagram and algorithm. Section 4 describes various design issues and intrinsic implementation complexity in proposed approach and the analysis of implementations. Lastly, the conclusion is presented.

2   Related Work and Motivation

2.1   Understanding User Intention

Prediction of the user intent is one of the pivotal steps in order to produce most relevant results [1]. Understanding user intention [15] can be very helpful in a different internet-based application like e-commerce, online entertainment, etc. [16]. In order to achieve it, we need to classify Internet queries based on features such as contextual information, keywords, and their semantic relationship [17]. Many present search systems offer an interface to visualize current search related information and capture user intention via personalized tools or operations, such as in [18–20] systems. Most of the interface driven exploratory search systems are using various facets of location, color, movement, and spatial distance of results items in GUI to represent the semantic relationship between current query and relevant data objects or previously retrieved objects. The semantic relationship is pivotal elements in navigational search approach and plays an important role in explorations of the data space. These visual spaces provide a platform to the user for observing the relevance between result objects and query with the respect to their interest, context information, or concern [21]. Users' subjects of interest can be dynamically changed, redefined, added, or deleted from the visual space.

2.2   Understanding Query Intentions

Identifying query intention helps to provide better user experience by adapting the results to the goal. The main aspect of the query intention is to identify the underlying goal of the user when submitting one particular query [8, 22]. The search intentions behind queries are classified as *navigational, informational, and transactional*. Different intents require different answers from the search engine. Thus, automatically identifying or generation of such query intents always was open directions [1, 13, 23, 48]. In traditional search system, to determine the query intent, two different data sources are considered: click-through data, and anchor texts [9]. Using click-through data, the click distribution of each query can be computed [24, 25].When such distribution is highly oriented toward few domains, it can be assumed that query is navigational and if click distribution is relatively flat, it can be considered as informational intent [26]. In [6, 27, 28], it is suggested four different techniques to find whether a search query is navigational or informational. These methods can be used alone but when combined together they give best results (91.7% precision and 61.5% recall according to the authors) [27]. Some other researchers also tried machine learning techniques. The approach of query reformulations discussed in [10, 29–34] is meant for data exploration. Each new query reformulation is expected to capture some intent space. Many researchers believe query as a first formal representation of user interest [21, 35], hence each modified query captures some user search intent.[1]

2.3 Understanding Results Intentions

Over the years, the information retrieval techniques of the search engines have been improved but still, most of the results generated from a query are displayed in form the lists [36, 37]. Since these list-based representations are very helpful in the evaluation of single documents but these do not help user much in a task like manipulating search results, comparing documents and finding a set of relevant documents. Different visualization-based techniques have been developed to tackle these problems. Many systems have been developed to represent search results in a useful manner, both for Web search and for traditional information retrieval systems, such as in [16, 34, 38]. These can be categorized based on whether they provide an entirely textual representation, a visual representation based on the entire document contents, or a visual representation based on an abstraction of the document [39].

Modeling above-discussed limitation for a search task plays a pivotal role in achieving accurate retrieval. We realized the importance of user's search intents in the traditional paradigm for information seeking. These intentions are defined in two levels: action intention and semantic intention. Action intentions are mainly user-initiated, and required interaction. Whereas, semantic intentions are derived from the semantics of query/data stored in the database. In this process, we differentiated user's intentions from user's preferences and two linguistic features (initial user query and review/confidence assessment) are extracted for intention modeling. It is strongly realized that interactive modeling of intents will be the main focus point in future development in search or exploratory systems [35, 36, 40].

3 Automatic Intent Modeling for Data Exploration

With huge data sets generation on daily basis, users do not always have a clear view on the data characteristics or what they are looking for. It seems prudent to propose that the database system itself can provide starting points and guide. Traditionally, exploring information with interactive intent modeling is based on two principles: *visualizing current search intent*, and *balancing exploration and exploitation* of user feedback [7]. The user's cognitive effort is thus reduced, as it is easier to recognize

---

[1]Identify applicable sponsor/s here (*sponsors*).

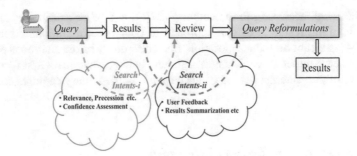

Fig. 2 Modeling user's search intents in exploration

data objects instead of having remembered them during reformulations. The basis for the proposed model is *"Query–Result"* paradigm and information behavior model [41]. A typical user's search intents play a pivotal role in overall information seeking (Fig. 2).

A flexible query answering system aims to exploit data collections in a richer way than traditional systems [37, 47]. In these systems, flexible criteria are used to reflect user preferences; expressing query satisfaction becomes a matter of degree. Nowadays, it becomes more and more common that data originating from different sources and different data providers are involved in the processing of a single query [42]. Also, data sets can be huge or data store can be trusted to the same extent and consequently the results in a query answer can neither be trusted to the same extent. For this reason, data quality assessment becomes an important aspect of query processing [43]. Indeed, to correctly inform users, it is in our opinion essential to communicate not only the satisfaction degrees in a query answer but also the confidence about these satisfaction degrees as can be derived from data quality. As an illustration, a hierarchical approach for search intent modeling and to support the computation of satisfaction degree [44]. Providing confidence information adds an extra dimension to query processing and leads to more soundly query answers.

3.1 Proposed Framework

The block diagram of the proposed approach is shown in Fig. 3. The proposed algorithm starts with presenting data samples to a naive user, these data samples are extracted based on existing user-related statistics (access pattern, profile-related data). The naive user reviews each data sample and categorizes data objects into relevant and nonrelevant. This reviewing process is the first step to apply user's search interest on the result set, in the subsequent exploratory iteration, the data objects labeled as relevant objects only are considered. Further, a data retrieval strategy will be modeled in form of a query. For relevant data items, multiple semantically

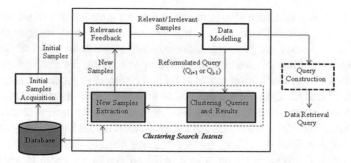

Fig. 3 Proposed architecture of searching intents in DE

equivalent queries are possible. Each query can be considered as query variant for a user interest.

The user feedback on objects of relevance is captured and used in a subsequent iteration. These samples are typically marked as relevant or irrelevant. A sample driven data exploration is based on the iterative and interactive association of the user with the system [45]. Therefore, these relevant samples are retained for the next iteration, due to the probability of the highest utility data objects are in the neighborhood of current relevant objects. The user relevance feedback on the data sample also strengthen the data model defined in the data exploration systems, this data model is a strategy for the final data retrieval. In data exploration systems generally, users stop the exploration process once he satisfies with the quality of the data objects. The data model at the end of the exploration process is the final strategy for result retrieval and thus it is converted into a query.

3.2 Methodology

The proposed work starts with the exploitation of samples. These samples are presented to the user and hence to be exemplified as relevant or not to his initial search. This feedback is the first form of user intention, informally imposed in the search system. We presume a binary, non-noisy, feedback system "for relevance feedback" and starts an iterative steering process [7, 46]. This categorization will be important in the following iterations, as each labeled data is used to characterize the initial data model. The data model may use any subset of attribute or predicate to define the user interest. However, domain experts could restrict the attribute set on which the exploration is performed. A similar scenario is in evidence-based medicine, as in clinical trials, sample tuples extracted from the database are shown to medical assistance and further they are asked to review their facets and label each sample trial as interesting or not. Each iteration of proposed model will add samples and presented to the user. The sample acquisition is purely intent-driven, as data items labeled as relevant are primarily used to extract new sample and train the data model.

The user feedback and trained data model acts as a representative of the user search intents. Automatic intent modeling (AIM) algorithm leverages the current data model to create and identify the possible variants or reformulations, and further, retrieve the data samples. New labeled objects are incorporated into the already labeled sample set and a new classification model is built. The steering process continues, until user realized that the data objects meet his search interest or a sufficient numbers of samples are observed, and accordingly user can terminate the current search process. Finally, AIM algorithm "translates" the classification model into a query expression (Data Extraction Query). This query will retrieve objects characterized as relevant by the user model (Query Formulation). AIM strives to converge to a model that captures the user interest, i.e., eliminating irrelevant objects while identifying and consider relevant objects. Each round refines the user model by exploring further data space (Fig. 4).

Automatic Intent Mining (AIM) Algorithm

Input: User Query Q_i and Initialized data space D_s
Output: Data Retrieval Query (Q_{i+1})

Step1:User *submits* initial Query Q_i on database D and *retrieves* data samples D_s

Step 2:User *review* the data samples D_s and label as relevant/ non-relevant, like D_{Srel}/D_{Snrel}

Step 3:Evaluate degree of search intent incorporated, and
    For each D_{srel} object in relevant sample,
    {*Initialize*, each data point on d-dimension data
    space as Cluster, further Neighborhood is *explored*,
    {if the density of neighboring Cell is greater than
    threshold τ,
    Then *Merge* that Cell and *Form* a Cluster at lower level } }

Step 4: For each higher dimension $k \leftarrow 1$ to d
    {If, overlapping cluster is dense at $k-1$ dimension
    and density is greater than threshold τ,
    Then Intersection again *saved/merge* as cluster}
    Merge_cluster $(c_1, c_2, ..c_n)$,
    { If intersection of clusters has greater then
    threshold τ,
    Then *Form_new_cluster* $(c_{1,2,..n})$,
    Remove lower dimension clusters from cluster set C }

Step 5: Define New Data retrieval model, based on Intents
    {labelled Samples, Old Data model variants)

Step 6:*Extract* new samples D_{snew}, for each reformulated
    query Q_{i+1}

Fig. 4 Automatic intent modeling (AIM) algorithm for DE

4 Design Issues

In the proposed approach, we are trying to incorporate the user's search intents (in query and query's result) into a generation of final data retrieval query and relevant data objects. Multiple design issues are involved in the conceptualized approach, as follows:

Neighborhood Region Creation—Defining the boundary of relevant data object's neighborhood is a key challenge and partially addressed in various research efforts [1]. In our proposed approach, defining a nonoverlapping boundary based on the relevance values is pivotal. As data objects are suggested from each neighborhood regions.

Query Construction/Reformulations—Over the multiple exploratory iterations, the user often meets his information needs and thus terminates the exploratory process. The data model at that iteration can be modeled into a query. In high dimension database, multiple semantically equivalent queries can be constructed and suggested to the user. Selection of appropriate query or reformulations is one key issue; we faced in our proposed work.

Evaluation of data object's relevance—The data object to be retrieved based on a relevance measure with user previous query and previous result as well. The relevance of each data objects is evaluated and used to define the confidence on the user's query. Identification of statistical information to measure the relevance of each data object is key aspects, as they influence overall system's accuracy and measure the effect of intents.

Visualization of retrieved data objects with additional Information—It is not feasible to visualize the entire result set retrieved therefore data summarization technique should be implemented. For example, relevant keywords from the neighborhood dataset are made available to the user in a selective manner, so the user can use them easily in a subsequent iteration.

5 Conclusion

Interactive intent modeling enhances human information exploration through computational modeling, and via helping users search and explore via user interfaces that are highly functional but not cluttered or distracting. In this paper, an automatic data exploration framework is proposed, which assists the user in discovering new interesting data objects/patterns and reduce search efforts. The model is based on the seamless integration of clustered search intents and data modeling that strive to accurately learn user's interests. User's feedback/review on the data objects is act as a representative of his search intent.

In this process, we observed various design challenges such as retrieval of relevant data objects, generation of query variants, and visualization of newly retrieved data samples. We defined two levels of intentions (action intention and semantic intention) and differentiated user's intentions from user's preferences. Two linguistic features (initial user query and review/confidence assessment) are extracted for intention modeling.

References

1. White RW (2011) Interactions with search systems. Cambridge University Press
2. Case DO (2014) Looking for information: a survey of research on information seeking, needs and behavior, 4th edn. Emerald Group Publishing
3. White RW, Roth RA (2009) Exploratory search: beyond the query-response paradigm, 3rd edn. Morgan and Claypool Publishers
4. Dubois D, Prade H (1997) Using fuzzy sets in flexible querying: why and how. In: Flexible query answering systems. Kluwer Academic Publishers
5. Cucerzan S, Brill E (2005) Extracting semantically related queries by exploiting user session information. Technical report, Microsoft Research
6. Jain K et al (1999) Data clustering: a review. ACM Comput Surv 31(3)
7. Boldi P et al (2011) Query reformulation mining: models, patterns, and applications. IR 14(3):257–289
8. Drosou M et al (2013) Ymaldb: exploring relational databases via result-driven recommendations. VLDB J 22(6):849–874
9. Baeza R et al (2004) Query recommendation using query logs in search engines. In: International workshop on clustering information over the web, in EGBT-2014, pp 588–596
10. Chien S, Immorlica N (2005) Semantic similarity between search engine queries using temporal correlation. In: International world wide web conference committee (IW3C2), WWW, pp 2–11
11. Dimitriadou K et al (2014) Explore-by-Example: an automatic query steering framework for interactive data exploration. In: International conference on ACM SIGMOD record, pp 517–528
12. Marchionini G (1989) Information-seeking strategies of novices using a full-text electronic encyclopedia. J Am Soc Inf Sci 40(1):54–66
13. Speretta M, Gauch S (2005) Personalized search based on user search histories. In: Proceedings of international conference on web intelligence (WI'05), pp 622–628
14. Li Y, Belkin NJ (2010) An exploration of the relationships between work task and interactive information search behavior. J Am Soc Inf Sci Tech 61(9):1771–1789
15. Brenes DJ et al (2009) Survey and evaluation of query intent detection methods. In: Proceedings of WSCD, pp 1–7
16. Callahan SP (2006) VisTrails: visualization meets data management. In: International conference on SIGMOD, pp 745–747
17. Mitra M, Singhal A, Buckley C (1998) Improving automatic query expansion. In: Proceedings of 21st international SIGIR conference on research and development in information retrieval, pp 206–214
18. Spoerri A (2004) How visual query tools can support users searching the internet. In: Proceedings of ICIV'04, pp 220–232
19. Chatzopoulou G, Eirinaki M, Polyzotis N (2009) Query recommendations for interactive database exploration. In: International conference on SSDBM, pp 329–334
20. Mishra C, Koudas N (2009) Interactive query refinement. In: International conference on EDBT, pp 862–873
21. Harman D (1988) Towards interactive query expansion. In: Proceedings of the 11th annual international ACM SIGIR conference on research and development in IR. ACM, pp 321–331
22. Wilson TD (1999) Models in information behavior research. J Doc 55(3):249–270

23. Ruotsalo T et al (2013) Directing exploratory search with interactive intent modeling. In: Proceedings of the 22nd ACM international conference on information & knowledge management, pp 1759–1764
24. Jones R et al (2006) Generating query substitutions. In: Proceedings of the 15th international conference on WWW. ACM, pp 387–396
25. Komlodi A (2007) Search history support for finding and using information: user interface design recommendations from a user study. IP&M 43(1):10–29
26. Golovchinsky G (1997) Queries? links? is there a difference? In: Proceedings of CHI. ACM Press, pp 404–414
27. Belkin NJ (2016) People, interacting with information. ACM SIGIR Forum 49(2):13–27
28. Kadlag A, Wanjari AV, Freire J, Haritsa JR (2004) Supporting exploratory queries in databases. In: International conference on DASFAA, pp 594–605
29. Billerbeck B et al (2003) Query expansion using associated queries. In: CIKM, pp 2–9
30. Bhatia S et al (2011) Query suggestions in the absence of query logs. In: SIGIR, pp 795–804
31. Jones R et al (2006) Generating query substitutions. In: Proceedings of 15th ACM international conference on WWW, pp 387–396
32. White RW, Kules B, Drucker SM (2006) Supporting exploratory search. Commun ACM 49(4):36–39
33. Hellerstein J et al (1997) Online aggregation. In: International conference on SIGMOD records, pp 171–182
34. Sadikov E et al (2010) Clustering query refinements by user intent. In: International conference on WWW, pp 841–850
35. Stohn C (2015) How do users search and discover? Findings from Ex libris user research
36. Cooper WS (1973) On selecting a measure of retrieval effectiveness. J Am Soc Inf Sci 24(2):191–204
37. Bi B et al (2015) Learning to recommend related entities to search users. In: Proceedings of WSDM, pp 139–148
38. Sanderson M et al (2009) NRT: news retrieval tool, vol 4(4). Electronic Publishing, pp 205–217
39. Tran QT et al (2009) Query by output. Technical report, School of Computing, National University of Singapore
40. Golovchinsky G et al (2012) The future is in the past: designing for exploratory search. In: Proceedings of the ACM 4th information interaction in context symposium. ACM, pp 52–61
41. Jeh G, Wisdom J (2003) Scaling personalized web search. In: Proceedings of WWW, pp 271–279
42. Rose DE, Levinson D (2004) Understanding user goals in web search. In: Proceedings of WWW, pp 13–19
43. Rodrygo LT et al (2011) Intent-aware search result diversification. In: Proceedings of the 34th international conference of ACM SIGIR, pp 595–604
44. Tsur G et al (2016) Identifying web queries with question intent. In: Proceedings of WWW, pp 783–793
45. Bharat K (2000) SearchPad: explicit capture of search context to support web search. In: Proceedings of WWW, pp 493–501
46. Zloof MM (1975) Query by example. In: International conference on AFIPS NCC, pp 431–438
47. Jansen Bernard J et al (2000) Real life, real users, and real needs: a study and analysis of user queries on the web. J Inf Process Manag 36(2):207–227
48. Elad YT et al (2005) Learning to estimate query difficulty: including applications to missing content detection and distributed information retrieval. In: Proceedings of SIGIR, pp 512–519

Regression-Based Approach to Analyze Tropical Cyclone Genesis

Rika Sharma, Kesari Verma and Bikesh Kumar Singh

Abstract This paper attempts to explore regression approach for the prediction of tropical cyclone genesis in North Indian Ocean. The cyclonic storms from the Bay of Bengal are the usual phenomenon in east coast of India. The present study mainly focuses on two such very severe cyclonic storms, PHAILIN and HUDHUD in 2013 and 2014, respectively. Various meteorological parameters that clearly explain the favorable situation for the occurrence of cyclonic storm are investigated and analyzed. The study emphasizes on selection of the potential predictor variables using stepwise regression method to analyze the efficiency of model. Regression analysis yields a predicted value resulting from a linear combination of the predictors. For the regression model, the predictor set consists of sea level pressure (SLP) and wind speed parameters. Different types of the regression methods have been used in the study and compared to find the efficient regression model for cyclone genesis prediction. The threshold-based anomaly detection algorithm is proposed for the potential predictors obtained using stepwise regression which is capable of successfully identifying the situation that led to the very severe cyclonic storm.

Keywords Regression model · Cyclonic storm · Wind speed · Wind direction Anomaly detection

R. Sharma · K. Verma (✉)
Department of Computer Applications, National Institute of Technology, Raipur, India
e-mail: kverma.mca@nitrr.ac.in

R. Sharma
e-mail: rikasharma678@gmail.com

B. K. Singh
Department of Biomedical Engineering, National Institute of Technology, Raipur, India
e-mail: bsingh.bme@nitrr.ac.in

© Springer Nature Singapore Pte Ltd. 2019
M. L. Kolhe et al. (eds.), *Advances in Data and Information Sciences*, Lecture Notes in Networks and Systems 39, https://doi.org/10.1007/978-981-13-0277-0_7

1 Introduction

A cyclonic storm is usually a rapidly rotating storm that is characterized by strong winds, a low pressure center, and the spiral arrangement of thunderstorms which is responsible for producing heavy rain. Cyclonic storm is considered to be one of the world's most destructive and costly natural hazards and thus the accurate estimates of future changes in their intensity, frequency, and location become more important [1]. In a study, it is found that the cyclonic storms from the Bay of Bengal are the usual phenomenon in east coast of India [2, 3] and its occurrence is an important concern to India. In literature, regression methods have been widely used for prediction and forecasting purposes. The purpose of present study is aimed at developing a regression model from such hazards through a critical analysis of the observed data. Several methodologies have been adopted for the seasonal tropical cyclone activity forecast such as Poisson regression models used for counting hurricanes [4], for predicting number of tropical cyclones that made landfall [5], and for predicting TC activity for East China sea [6]. Some researchers have worked on Bayesian regression models, such as Elsner et al. [7, 8] successfully developed a prediction model for annual U.S. hurricane count. The log-linear regression model is used by Choi et al. [9] to detect significant regime shifts in a time series that is further applied to the statistical change-point analysis (CPA) technique in order to determine the interdecadal change in the Korea tropical cyclones activity. Richman et al. [10] evaluated the predictive skill of a machine learning method that includes support vector regression (SVR) model for the seasonal prediction of tropical cyclone counts for the North Atlantic region and concluded that the SVR model predictions outperform with significant increased correlation and reduced MAE (RMSE) between observed and predicted TC count.

For regression model, the key phase is the selection of predictor variables. Emanuel et al. [11] concluded that only part of the observed increase in the TC's power was caused by increased sea surface temperature and the rest can be explained and explored by the changes in wind speed. Predicting wind speed and direction is one of the most important tasks for predicting cyclone genesis as wind speed determines the cyclone intensity and wind direction closely relates to wind behavior. Machine learning techniques have often been used to predict the nonlinear wind evolution. Further, it was reported by Choi et al. [12] that the wind–pressure relationship between the parameters mean sea level pressure (MSLP) and maximum wind speed is used for finding climatology of TC intensity. In this context, this paper presents sea level pressure and wind speed based regression model for prediction, which have the advantage of being simple and fast. The study shows the impact region in the Indian subcontinent. The effect of some climatic parameters is illustrated for cyclonic storm that occurred in Odisha (Phailin) and Andhra Pradesh (Hudhud) region during years 2013 and 2014, respectively. Figure 1 shows the mapping of most affected places of the east coast of India due to very severe cyclonic storm (Gopalpur, Odisha due to cyclone Phailin in 2013, and Visakhapatnam, Andhra Pradesh due to cyclone Hudhud 2014).

Fig. 1 Mapping of study regions showing cyclone-affected states from the Bay of Bengal toward east coast of India. Image drawn in Grads software

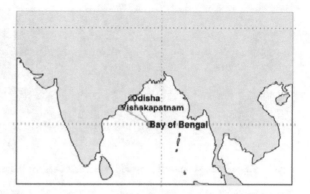

1.1 Contribution and Outline of the Paper

We summarize our contributions as follows:

1. The study and analysis of various meteorological parameters for identifying and selecting potential predictor are responsible for cyclone genesis prediction using stepwise regression method. Further, different regression techniques are applied to potential predictor and are compared to find the efficient regression model.
2. The threshold-based anomaly detection algorithm is proposed and applied to the selected potential predictors SLP and wind speed that plays an important role in cyclone genesis.

The fact is reproduced from the results obtained using the proposed algorithm for threshold determination that the simultaneous occurrence of relatively low pressure and strong wind speed accelerated the synoptic situation that leads to very severe cyclonic storm Phailin and Hudhud in 2013 and 2014, respectively (see Fig. 5).

This paper is organized as follows: Sect. 2 explains the datasets used. Section 3 describes the methodology adopted and proposed threshold-based anomaly detection algorithm. Section 4 explores the results and discussions and finally, Sect. 5 concludes the research work.

2 Dataset

Andhra Pradesh and Odisha region (shown in Fig. 1) witnessed unprecedented damage due to very severe cyclonic storm Phailin and Hudhud in October month of the year 2013 and 2014, respectively, that resulted in damage to life, property, and heavy infrastructure loss. So, the study gives emphasis on this particular region. From global data, Andhra Pradesh and Odisha regions with latitude range 15°–20°N and longitude range 80°–87.5°N in India has been selected, to explore the effect of

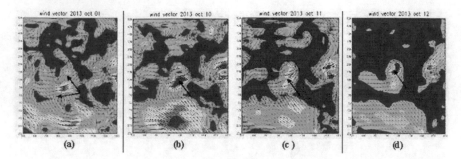

Fig. 2 Wind direction and wind speed toward east coast of India during cyclone Phailin 2013

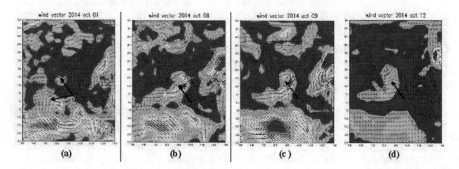

Fig. 3 Wind direction and wind speed toward east coast of India during cyclone Hudhud 2014, India. Image drawn in Grads software

various climate parameters on cyclonic storm that occurred on October 11–12, 2013 and 2014.

Dataset used in this study is downloaded from NCEP/NCAR reanalysis [13] in NetCDF format. The climate variables that have been explored for cyclone genesis analysis include precipitable water, surface pressure, sea level pressure, wind zonal, and wind meridional components. Cyclones develop due to significant wind speed and the wind direction that is deflected because of the Coriolis force. Figures 2 and 3 show different wind patterns during cyclonic storm Phailin (2013) and Hudhud (2014) from intense to landfall state. The cyclone refers to their cyclonic nature, in which the wind blows counterclockwise in the northern hemisphere and clockwise in the southern hemisphere as shown by arrow in Figs. 2a and 3a, respectively. Winds blow from areas of high pressure to areas of low pressure; the Coriolis effect influences the wind by deflecting its path to the right in the Northern Hemisphere and thus fails to arrive at the low pressure center. The opposite direction of circulation is due to the Coriolis effect. The effect of Coriolis force in wind direction can be visualized from Figs. 2b, c and 3b, c. Also, it can be clearly depicted that the intensity of wind speed is increasing and its direction is deflected toward the east coast of India during cyclone occurrence. Figures 2d and 3d show the wind speed and direction after cyclone landfall.

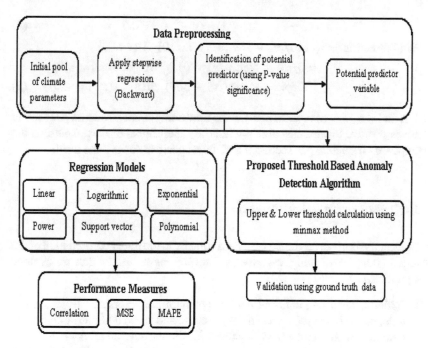

Fig. 4 Methodology

3 Methodology

The methodology adopted in the study is explained in Fig. 4. The study is twofold: Initially, the selection of potential predictor from various climatic parameters using stepwise regression. Then, comparing different regression techniques for the potential predictor variables and thus finding an efficient regression model for cyclone genesis prediction. Second, applying proposed threshold-based anomaly detection algorithm for selected potential predictors and validating the results with the ground truth data.

3.1 Data Preprocessing

Data normalization is performed to remove seasonality from the data using Eq. (1).

$$\text{Z-score Normalization}: Z = (X - \mu)/\sigma \tag{1}$$

where mean: $\mu = (\sum X_i)/N$ and standard deviation $\sigma = \sqrt{\{(\sum X_i - \mu)^2/N\}}$, $X_i = $ data (ith) and $N = $ Number of elements

Wind speed and direction are computed using Eqs. (2) and (3)

$$\text{Wind speed} = [(u \cdot {}^\wedge 2) + (v \cdot {}^\wedge 2) \cdot {}^\wedge 0.5 \tag{2}$$

$$\text{Wind direction} = \{((180/pi) * atan(u/v) + 180) \quad if(v > 0)$$
$$((180/pi) * atan(u/v) + 0) \quad if(u < 0 \text{ and } v < 0)$$
$$((180/pi) * atan(u/v) + 360) \quad if(u > 0 \text{ and } v < 0)\} \tag{3}$$

where U-wind (i.e., Zonal or east/west) is the flow component in meters per second that denotes positive toward east and V-wind (i.e., Meridional or north/south) is the flow component in meters per second that denotes positive toward the north.

3.2 Regression-Based Approach

Regression analysis gives predicted value resulting from the linear combination of the predictors. In the present study, we have used several regression models for cyclone genesis prediction.

Step 1: Initially, describe the initial pool of potential predictors as latitude, longitude, precipitable water, sea level pressure, and zonal and meridional wind component. To extract the best subset for forecasting model from the set of predictors, the backward stepwise is performed that start with all the variables in the model and proceed backward by removing the least significant variable for each step. At each step, P-value is calculated and the variables with higher P-values are removed. Further, select a subset sea level pressure and wind speed of those predictors for inclusion in the regression model using a backward stepwise regression procedure.

Step 2: In the next step, use a fitting procedure on the data to derive the MLR that consists of linear equations for prediction. For regression analysis, wind speed is explored taking into consideration sea level pressure as a predictor variable. The regression methods used in the study involve the following:

(a) Linear regression—It defines a relation between the dependent variable and one or more independent variables using a best fit straight line. It is represented using Eq. (4).

$$Y = a + b * X, \tag{4}$$

where a is the intercept and b is the slope of the line.
Regression of the following form was obtained:

$$y\,13 = -2.152 \times 10^{-3}x + 221.836, \text{rss} = 546.97$$
$$y\,14 = -2.90 \times 10^{-3}x + 297.69, \text{rss} = 399.83$$

where y13 and y14 = predicted wind speed obtained for year 2013 and 2014, respectively, x = slp and rss = Residual sum of squares.

(b) Logarithmic regression—It is used when a nonlinear relationship exists between the independent and dependent variables and is defined as in Eq. (5)

$$Y = a\ln(x) + b$$

$$y13 = -217.19 \ln(x) + 2507.051, \text{rss} = 546.41$$

$$y14 = -292.90 \ln(x) + 3379.49, \text{rss} = 399.83 \qquad (5)$$

(c) Exponential regression—A variable is said to grow exponentially if it is multiplied by a fixed number greater than 1 in each equal time period, whereas the exponential decay occurs when the factor is less than one. Equation (6) defines the exponential regression

$$Y = ae^{bx}$$

$$y13 = 5.85 \times 10^{19} e^{-4.356 \times 10-4x}, \text{rss} = 537.41$$

$$y14 = 9.15 \times 10^{24} e^{-5.542 \times 10-4x}, \text{rss} = 380.66 \qquad (6)$$

(d) Polynomial regression—The power of independent variable is more than 1 as defined in Eq. (7). It is a curve that fits into the data points.

$$Y = a_n x^n + a_{n-1} x^{n-1} + \ldots + a_2 x^2 + a_1 x + a_0$$

$$y13 = 2.76 \times 10^{-6} x^2 - 5.57 \times 10^{-1} x + 28197.32, \text{rss} = 473.79$$

$$y14 = 3.086 \times 10^{-6} x^2 - 6.25 \times 10^{-1} x + 31651.33, \text{rss} = 334.71$$
$$(7)$$

(e) Power regression—Eq. (8) defines that the response variable is proportional to the explanatory variable raised to a power.

$$Y = ax^b$$

$$y13 = 1.12 \times 10^{203} x^{-40.451}, \text{rss} = 548.27$$

$$y14 = 5.293 \times 10^{299} x^{-59.771}, \text{rss} = 380.44 \qquad (8)$$

(f) Support vector regression—The input X is mapped onto an m-dimensional feature space using some fixed nonlinear mapping. Further, a linear model is constructed in the feature space $f(x, w)$ and is defined using mathematical notation as shown in Eq. (9).

$$f(x, w) = \sum_{j=1}^{m} w_j g_j(x) \tag{9}$$

where $g_j(x)$, $j = 1, \ldots, m$ denotes a set of nonlinear transformations

3.3   Proposed Threshold-Based Anomaly Detection Algorithm

Algorithm 1 is used to find the anomalies of the predictor variables selected for regression model.

Algorithm 1 Approach for finding threshold value to detect anomaly

---
Input: Let *DS (L, Lat, Lon)* = Dataset of location L with latitude Lat and longitude Lon. Let *zscore(i)* in *DS(t)* be the daily zscore normalized values for *parameter i and t = 2002, 2006, 2013, 2014* of time series.
Let, *N* = Length of total time of study period (365 days).
Output: Anomaly in time series data
  Procedure: Repeat
    1. FOR *i = 1 to N* do
       //Compute the upper threshold for wind speed
    2. Upper-Threshold = mean(max_wind_speed (2002, 2006, 2013, 2014))
       //Compute the lower threshold for sea level pressure
    3. Lower-Threshold = mean(min_slp (2002, 2006, 2013, 2014))
    4. ENDFOR
       //Compute the Anomalies from the time series:
    5. FOR *i = 1 to N* do
    6. IF ((zscore(i) > Upper-Threshold) and (zscore(i) < Lower-Threshold))
    7. Detect zscore(i) to be anomaly
    8. ENDIF
    9. ENDFOR
    10. until convergence

Explanation. To determine threshold, we have considered the time series of the year 2013 (Phailin), 2014 (Hudhud), and two normal years 2002 and 2006 when no cyclone occurred. The reason for taking two normal years for finding the threshold values is to generalize the method and to estimate the deviation more accurately during anomalous years. We have used mean of maximum and minimum values of the wind speed and sea level pressure for computing the threshold values. The data values that are above and below threshold can be detected as an anomaly. The threshold values obtained can be observed from Fig. 5. The upper black line shows upper threshold value, and lower black line shows lower threshold value.

Figure 5 shows two curves representing the parameters wind speed (in red) and sea level pressure (in blue) of year 2013 and 2014. It is clear from the figure (pointed using arrow) that there is a significant decrease in sea level pressure that crossed

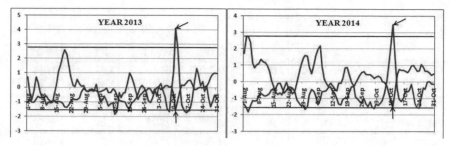

Fig. 5 Time series of wind speed (in red) and SLP (in blue)

Table 1 Performance measures

Pearson Correlation (R)	Mean Square Error (MSE)	Mean Absolute Percentage Error (MAPE)
$PC = \dfrac{\sum_{i=1}^{N}(X_i-\overline{X})(Y_i-\overline{Y})}{\sqrt{\sum_{i=1}^{N}(X_i-\overline{X})^2}\sqrt{\sum_{i=1}^{N}(Y_i-\overline{Y})^2}}$	$MSE = \dfrac{\left(Y-\hat{Y}\right)^2}{N}$	$MAPE = \sum \dfrac{\left|\left(Y-\hat{Y}\right)/Y\right|}{N}$

lower threshold line during October 11–12, 2013 and 2014. On the contrary, it is clear from the figure that the wind speed is increased significantly during October 11–12 crossing upper threshold value making situation prone for cyclonic storm and the obtained result is validated using ground truth data. Further, when strong winds with high wind speeds combine with very low pressure, it causes sea levels to rise and making favorable conditions for cyclones.

3.4 Performance Measures

Different performance measures that are evaluated and compared to find the efficiency of the regression models are defined in Table 1 [14].

4 Results and Discussions

The parameters responsible for extreme weather events and their effect on climate change are analyzed with reference to very severe cyclonic storm that occurred in Odisha (Phailin) and Andhra Pradesh (Hudhud) during October 2013 and 2014, respectively. Data of 2013 and 2014 are used to compare trend analysis. To compute threshold, we have taken data of the year 2002 and 2006 as normal year (no cyclones) to accurately identify anomalous pattern in 2013 and 2014 data. The results obtained for different regression methods are summarized in Table 2.

Table 2 Comparative analysis of regression methods

Regression methods	Year 2013			Year 2014		
	R (%)	MSE	MAPE	R (%)	MSE	MAPE
Linear	60.72	2.00	28.26	68.60	1.71	27.16
Log	60.77	2.00	28.24	68.66	1.71	27.13
Exponential	62.83	2.05	30.38	71.27	1.67	27.53
Polynomial	63.84	1.90	27.87	73.23	1.49	23.72
Power	62.73	1.96	26.00	71.43	1.52	24.12
SVR	**66.49**	**1.77**	**24.50**	**74.37**	**1.38**	**20.87**

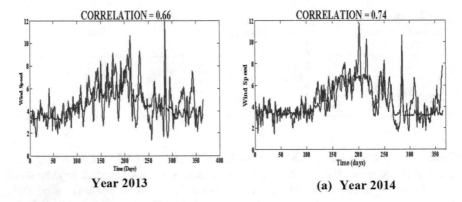

Year 2013   (a) Year 2014

Fig. 6 Correlation between actual and predicted wind speed

The correlation between actual and predicted wind speed using support vector regression can be visualized from Fig. 6.

5 Conclusion

Storm surge hazard is witnessed many times along the east coast of India. Stronger winds and low pressures lead to stronger storm surges. After the analysis using data-driven technique, it is clear that in the month of October–November most of the storms created in the Bay of Bengal move a long distance in the NW direction and finally take a NE turn due to Coriolis force and hit land in east coast of India. The storms Phailin and Hudhud are prototypes originated from identical area and moved almost in the parallel track, but there is wide difference in intensity, wind speed, and pressure. The aim of this study is the selection of potential predictor and to improve the predictive power of regression models. The experiments were carried out for different regression techniques and are compared to find the efficient regression-based model for cyclone genesis prediction and it is found that performance of support vector regression model is most efficient. The anomalous behavior of climate variables can be clearly seen from the graph results obtained using proposed threshold-based anomaly detection algorithm.

References

1. Abbs D (2012) The impact of climate change on the climatology of tropical cyclones in the Australian region. Accessed Sept 5, 2014
2. Shaji C, Kar SK, Visha T (2014) Storm surge studies in the North Indian ocean: a review. Indian J Geo-marine Sci 43(2):125–147
3. Elsner JB, Kossin JP, Jagger TH (2008) The increasing intensity of the strongest tropical cyclones. Nature 455(7209):92–95
4. Elsner JB, Schmertmann CP (1993) Improving extended-range seasonal predictions of intense Atlantic hurricane activity. Weather Forecast 8(3):345–351
5. Lehmiller GS, Kimberlain TB, Elsner JB (1997) Seasonal prediction models for North Atlantic basin hurricane location. Mon Weather Rev 125(8):1780–1791
6. Choi KS, Kim BJ, Kim DW, Byun HR (2010) Interdecadal variation of tropical cyclone making landfall over the Korean Peninsula. Int J Climatol 30(10):1472–1483
7. Elsner JB, Jagger TH (2004) A hierarchical Bayesian approach to seasonal hurricane modeling. J Clim 17(14):2813–2827
8. Jagger TH, Elsner JB (2006) Climatology models for extreme hurricane winds near the United States. J Clim 19(13):3220–3236
9. Choi JW, Cha Y, Kim JY (2017) Interdecadal variation of Korea affecting TC activity in early 1980s. Geosci Lett 4(1):1
10. Richman MB, Leslie LM, Ramsay HA, Klotzbach PJ (2017) Reducing tropical cyclone prediction errors using machine learning approaches. Procedia Comput Sci 114:314–323
11. Emanuel K (2005) Increasing destructiveness of tropical cyclones over the past 30 years. Nature 436(7051):686–688
12. Choi JW, Cha Y, Kim HD, Lu R (2016) Relationship between the maximum wind speed and the minimum sea level pressure for tropical cyclones in the western North Pacific. J Climatol Weather Forecast 4(180):2
13. Kalnay E, Kanamitsu M, Kistler R, Collins W, Deaven D, Gandin L, Zhu Y (1996) The NCEP/NCAR 40-year reanalysis project. Bull Am Meteorol Soc 77(3): 437–471
14. Tan PN, Steinbach M, Kumar V (2006) Introduction to data mining, Addison Wesley Publishers

Design and Evaluation of a Hybrid Hierarchical Feature Tree Based Authorship Inference Technique

R. Raja Subramanian and Karthick Seshadri

Abstract Authorship inference refers to the task of identifying the writer of a handwritten document. We propose a feature tree based supervised learning technique for solving the authorship inference problem. A combination of feature extraction techniques is used to carry out the experiments. The proposed technique provides a good recognition rate, when evaluated using standard handwritten datasets. An effective pruning technique is employed to restrict the number of comparisons before inferring the author of a handwritten sample. Empirical evaluations of the proposed technique reveal that the technique performs better than techniques like local gradient descriptors, run length features, and Arnold Transforms with respect to recognition rate.

Keywords Feature tree · Supervised learning · Authorship inference
Handwritten document classification · Feature extraction · Interest point detection

1 Introduction

Writer inference has applications in forensic research, historical documents analysis, plagiarism verification, and signature verification to quote a few. Writer inference also plays a major role in critical applications like settling court disputes and reconciling the era in which ancient manuscripts were written. Writer inference can be either dependent or independent of the document text, depending on whether the training and testing set are identical or not, respectively. Typically, there are two modes of writer inference: (i) Online and (ii) Offline. Online writer inference extracts features while the document is being written and uses these features for the inference task. Such features used for online inference include writing speed, stroke length, pen pressure, and allographic features. On the other hand, offline model uses structural

R. R. Subramanian · K. Seshadri (✉)
Department of Computer Science and Engineering,
Thiagarajar College of Engineering, Madurai, India
e-mail: skcse@tce.edu

R. R. Subramanian
e-mail: rajasubramanian.r@gmail.com

© Springer Nature Singapore Pte Ltd. 2019
M. L. Kolhe et al. (eds.), *Advances in Data and Information Sciences*, Lecture Notes in Networks and Systems 39, https://doi.org/10.1007/978-981-13-0277-0_8

features, which captures the writing style of the person encompassing direction of writing, degree of curvature, and texture features for writer inference. This paper focuses on devising an efficient technique for solving the writer inference task in the offline text-independent mode. Popular writer inference techniques [1–4] employ run length encoding to derive the binary representations of a handwritten document, which are then subsequently binned to construct a histogram per document. Two documents are deemed to be written by the same writer, if their histograms are approximately alike. Though these techniques are empirically shown to perform well at the task of writer inference, histogram construction is known to be a computationally intensive task. Further, the computational intensity grows nonlinearly as the number of writers to be discriminated increases. Several transform-based techniques [3, 5, 6] have been proposed in the literature to overcome the disadvantages of the techniques that use binning for writer inference. However, such techniques and others based on junction detection [7] and mixture models [8] have limited applicability with respect to the type of handwritten document. For instance, techniques based on radon transforms can deal with only signatures as it extracts features corresponding to exact matching. Similarly, techniques based on Arnold transforms can identify documents hosting a curly handwriting. Though transforms like discrete cosine transform have a generic applicability, it necessitates histogram construction to identify writers, which negatively impacts the computational complexity.

Further, to the best of our knowledge, research on writer inference in Tamil is scanty. Separate techniques need to be proposed and investigated for Tamil, as the language has characters with variegated shapes, curvatures, and curls. Also, as Tamil comprises 247 letters, the task of feature extraction and subsequent writer inference is challenging both from a computational and allographic standpoints. Hence, the problem addressed in this paper is to design and evaluate a semi-supervised offline algorithm based on hierarchical feature descriptors to solve the problem of writer inference with a specific focus on Tamil.

The key contributions of our work include the following:

(i) Creation of database for Tamil handwritten documents, for writer inference and verification.

(ii) We propose a novel hierarchical descriptor, with various feature extraction techniques, for efficient and effective writer inference.

(iii) Unlike existing encoding like run lengths and transformations like Arnold and Radon which deals with only specific textural features like curls or concavities, the proposed technique can handle a combination of several textural features.

(iv) We propose a hierarchical feature extraction technique that is insensitive to the scale of the handwritten characters.

(v) The proposed technique extracts feature descriptors that render the technique applicable to multiple languages.

Though we have employed more than one feature extraction technique, the time complexity of the proposed hierarchical descriptor is kept minimal through an intelligent orchestration of descriptors in various levels of the hierarchy. The paper is organized as follows: Sect. 2 depicts the formulation of the authorship inference problem.

Section 3 explains the state-of-the-art writer inference techniques. Section 4 explains the proposed methodology of writer inference, which is followed by a section that elaborates the feature extraction techniques used. Section 6 describes the experimental setup and results inferred from the experiments, followed by the conclusion and some directions for future research.

2 Problem Formulation

We are given with the set of handwritten documents $D = \{d_1, d_2, \ldots, d_M\}$ written by a set of writers $W = \{w_1, w_2, \ldots, w_N\}$. Let $\Psi(d_i)$ be a bijective mapping from D to W such that it maps the document d_i to its author w_j ($i\epsilon[1, M], j\epsilon[1, N]$). We assume that the following condition (1) holds:

$$\bigcup_{i=1}^{M} \Psi(d_i) = W \tag{1}$$

The above condition ensures that every document is written by one writer and each writer has written at least one document. The task is to build a model Θ that accepts D, W, and the mapping function Ψ as training inputs and upon getting a test input $d_t \notin D$, Θ returns a writer in W with the closest matching handwriting. Θ approximates and generalizes the mapping function Ψ to test samples which are not seen before.

3 Background Details and Related Work

Generally, the techniques used for writer inference involve two stages: (i) Extraction of descriptors for the handwriting; (ii) Given an unseen handwriting sample, identification of the writer of the sample by matching the sample's description with the descriptions extracted. Djeddi et al. [1] proposed a writer inference technique that uses run lengths to construct histogram descriptors. Histograms are typically computationally intensive to construct and have also been shown to exhibit an arguably inferior recognition rate when presented with cursive handwriting. The disadvantages of the underlying histogram construction can be overcome by a technique that is robust against such computational complexities.

Local gradient feature descriptors such as Scale-Invariant Feature Transform (SIFT) is employed by Surinta et al. [9] for feature extraction. The advantages of such gradient-based extraction techniques are the following: (i) accurate in computing descriptors and (ii) the extracted descriptors exhibit robustness to scale changes in the handwritten samples. Empirical evaluations of Surinta's algorithm revealed recognition rate of 96.12 and 98.32% on LATIN-C25 dataset while using k-Nearest

Neighbor (kNN) and Support Vector Machine (SVM) as the learning techniques, respectively. Surinta et al. also conducted experiments to compare the performance of SIFT against Histogram of Oriented Gradients (HOG). These experiments asserted that Surinta's algorithm with HOG as descriptors realized an inferior recognition rate of 95.17% and 98.25% while using kNN and SVM, respectively. Christlein et al. proposed a writer inference technique based on ensemble SVMs, which uses RootSIFT-based feature descriptors. RootSIFT [8] is a Hellinger normalization of SIFT, in which the feature vector is first l_1 normalized followed by a l_2 normalization.

The scale-invariant features are incapable of effectively capturing descriptors of handwritings with curls. Bulacu and Schomaker [10] proposed a technique based on run length encoding to capture structural features of the handwriting. Bulacu's technique was computationally expensive and empirically shown to exhibit a recognition rate of 97% on fire maker dataset subsequent researchers [1, 2, 11, 12] made use of computationally cheaper textural descriptors such as Local Binary Patterns (LBP), Local Ternary Patterns (LTP), and Local Phase Quantization (LPQ) to effectively capture shapes or interest point-related features. Hannad's algorithm works efficiently for discriminating among a limited number of writers. LBP binary encodes an image by comparing the intensities of central pixels with neighboring pixels after segmenting the image into windows of size 3×3 pixels. LTP [13] is a modification of LBP, which considers the variation between the central pixel and the neighboring pixel intensities, if and only if the variation is greater than a preset threshold. This is done to ensure that the writer inference algorithm is insensitive to intensity variations occurring due to noise. LPQ [14] captures phase information using short-term Fourier transform, and binary runs are created based on the signs of the real and imaginary parts of each Fourier component. Hannad's experiments [2] with these textural descriptors, LBP, LTP, and LPQ, show the recognition rates of 73.48%, 87.12%, and 94.89%, respectively, when evaluated with the IFN/ENIT database and the recognition rates of 65.54%, 84%, and 89.54%, respectively, when evaluated with the IAM database [15]. As can be seen from the above results, LPQ exhibits a better recognition rate as compared to LBP and LTP.

Transform-based features are used to quickly extract essential descriptors from the image as compared to the computationally intensive histogram construction techniques. Ooia et al. used Radon transform [5] for handwritten signature verification. Radon transform is an integral transform that projects the image in different angles, whose inverse reproduces the original image. Radon transform produces a large number of transform values, whose major axes of variation can then be revealed using Principal Component Analysis (PCA). Arnold transform [3] is another transformation technique, designed specifically to handle cursive samples. It extracts directional features from the samples and uses them as descriptors for writer inference. Dasgupta's algorithm using Arnold transforms was empirically shown to exhibit a recognition rate of $87.19 \pm 0.54\%$ with the CENPARMI English dataset and a recognition rate of $95.24 \pm 0.43\%$ with the lowercase version of the dataset. The recognition rate was lesser compared to the previously proposed methods [6, 9] but has the capability to learn curls.

Discrete Cosine Transforms (DCT), one of the image compression techniques, was used for feature extraction by Khan et al. [6]. Codebooks are constructed using the DCT features extracted from the handwritten samples. A randomly selected feature vector from each sample is used for the codebook generation. Multiple codebooks and subsequently descriptors are then generated with different randomly selected vectors from each sample. Khan's experiments empirically show recognition rates of 97.2%, 99.6%, 71.6%, and 76.0% for samples from IAM, CVL [16], AHTID/MW, and IFN/ENIT databases, respectively. From the abovementioned experimental results, Khan's technique performs poorly on the samples from the Arabic datasets: AHTID/MW and IFN/ENIT, because of the presence of curls and concavities in Arabic handwriting. As it is evident from the abovementioned literature, a single feature extraction technique cannot work generically for all types of handwritings. Hence, a combination of feature extraction techniques, each extracting prominent and unique features, has been employed in our proposed technique.

4　Proposed Writer Inference Methodology

The proposed methodology for writer inference named as Hybrid Hierarchical Feature Tree based Authorship Inference (HHFTAI) consists of the following three steps: (i) model generation, (ii) feature extraction from test sample, and (iii) writer inference. The methodology is shown in Algorithm 1.

(i) **Model Generation**: The model generation phase depicted in Algorithm 1.1 takes as input the training set $(D,\ W)$, the feature extraction techniques f_1, f_2, and f_3, the f_1 threshold I, and the f_2 threshold λ, and emits the model Θ as the output. Θ is multilevel model in which the root level is capable of discriminating samples into classes, based on coarser and easily computable features such as spacing between characters and size of characters. Each class corresponds to a child node of the root node and comprises training samples that are considered similar according to the parent node's feature. As we descend down from the root toward the leaf level, we find features that are progressively intensive to compute and can discriminate among classes using fine-grained details of the handwritten sample.

(ii) **Feature Extraction from Test Sample**: This part of the algorithm takes as input a test sample d_t and extracts the feature descriptors from d_t by traversing Θ from the root node, as in Algorithm 1.2. Based on the feature descriptor of d_t corresponding to the root node, we bucket d_t into one of the child classes of the root node. Similarly, the feature descriptors of d_t extracted at the successive levels help in determining the class in which d_t falls until we bucket d_t in one of the leaf level classes (C).

(iii) **Writer Inference**: Let C contain n samples. We project d_t onto the vector space containing n feature descriptors of the samples in C. The vector space is maintained as an R^+ tree to facilitate the quick retrieval of the sample (d_t)

that is nearest to d_t. The closeness between d_t and the samples is estimated using the Euclidean distance measure. We then return $\Psi(d_i)$ as the predicted writer of the document d_t, if the number of interest points which approximately match between d_i and d_t is greater than the threshold δ. Otherwise, we return no match. Writer inference phase is shown in Algorithm 1.3.

Algorithm 1: InferAuthor (Input: $(D, W), \Psi, d_t$, f$_1$, f$_2$, f$_3$, I, λ Output: w$_t$)

{

           $\Theta \leftarrow$ generateAuthorshipModel $((D, W), d_t$, f$_1$, f$_2$, f$_3$, I, λ)
           $(D_{out}, D_t) \leftarrow$ extractFeatureDescriptor (Θ, d_t)
           $w_t \leftarrow$ predictAuthor (D_{out}, D_t, Ψ)
           return w_t

}

Algorithm 1.1: generateAuthorshipModel (Input: (D, W), f$_1$, f$_2$, f$_3$, I, λ Output: Θ)

{

        List <Features> f $\leftarrow$ [f$_1$, f$_2$, f$_3$]
        Create Class C_{root}
        Class $C_k \leftarrow \Phi$
        C_{root} .child$\leftarrow$C$_k$
        for level l$\leftarrow$ 1 to f.length {
                if (l = 1){
                        for each d_i in D {
                                Compute f$_1(d_i)$ using Hough Circle Transform
                                childNum $\leftarrow$ f$_1(d_i)/I$;
                                if $(C_{childNum} = \Phi)$
                                      Create Class $C_{childNum}$
                                $C_{childNum} \leftarrow C_{childNum} \cup \{(d_i, w_i)\}$
                                C_{root} .child$\leftarrow$ C_{root} .child $\cup$ $\{C_{childNum}\}$
                      }
                }

```
                    if (l = 2){
                            Create classes C₁ and C₁'
                            C_{l-1}.child ← C_{l-1}.child ∪ {C₁,C₁'}
                            for each C in C_{l-2}.child {
                            for each dᵢ in C {
                                    Compute f₂(dᵢ) ← Vertical length of the black-
                                    pixel in the binary image
                                    if (f₂(dᵢ) < λ)
                                            C₁ ← C₁ ∪ {(dᵢ, wᵢ)}
                                    else
                                            C₁' ← C₁' ∪ {(dᵢ, wᵢ)}
                            }
                            }
                    }
                    if(l=3){
                            for each dᵢ ∈ C₁ ∪ C₁'{
                            Compute f₃(dᵢ) using SURF
                            if (dᵢ ∈ C_{l-1}) {
                                    C_{l-1}.child ← C_{l-1}.child ∪ {(f₃(dᵢ),wᵢ)}
                            } else {
                                    C_{l-1}'.child ← C_{l-1}'.child ∪{(f₃(dᵢ),wᵢ)}
                            }
                            }
                    }
            }
            return C_root
}
```

Algorithm 1.2: extractFeatureDescriptor (Input: Θ, d_t Output: (D_{out}, d_t))

```
{

        D₁ ← f₁(dₜ)
        D₂ ← f₂(dₜ)
        D₃ ← f₃(dₜ)
        k ← D₁/I
        if (D₂ < λ)
                    return (Θ_k.Θ₂', D₃)
        else
                    return (Θ_k.Θ₂', D₃)
}
```

Algorithm 1.3: predictAuthor (Input: D_{out}, D_t, Ψ Output: w_t)

{

        closestDist $\leftarrow \infty$

        closestDoc $\leftarrow \Phi$

        for each D_i in D_{out} {

                if (closestDist > euclidDist (D_i, D_t)) {

                        closestDist $\leftarrow$ euclidDist (D_i, D_t)

                        closestDoc $\leftarrow \{d_i, w_i\}$

                }

        if (numMatchingPoints (D_i, D_t) > δ)

                return $\Psi(d_i)$

        return Φ

}

---

5   Feature Extraction and Classification

Graphology [17] refers to the analysis of handwriting with an intent to reveal the writer's unique style. This analysis can be leveraged to extract descriptors for the handwritten text, which can then be used to infer authorship of an unseen handwritten sample. The following features are typically used for discriminating handwritings from different writers: (i) size of the characters, (ii) spacing between the characters, (iii) thickness of the characters, (iv) upward or downward drifting of sentences from a straight line, (v) curly writings, (vi) isolated characters, and (vii) cursive writings. Some of these features can act at a coarser level to discriminate between major classes of handwritings, whereas some of the features can be used as fine-grained features to distinguish minute shades of differences in handwriting. Using a combination of features will make up for each other's deficits and will enable the extraction of a robust hybrid descriptor from the handwritten sample. In the proposed algorithm, the model Θ is generated using three feature extraction techniques: Hough Circle Transform (f_1), Vertical length of the black pixel in a binary image (f_2), and Speeded-Up Robust Features (SURF) (f_3), each serving in a separate level in the feature tree. f_1, f_2, and f_3 are ordered such that it discriminates them from a coarser to a finer level, and at the same time, the former being computationally cheaper than the later ones. The following subsections discuss the existing feature extraction techniques f_1, f_2, and f_3.

5.1 Hough Circle Transform

Hough circle transform helps in extracting circular shapes from an input image. The circles identified by Hough are referred to as Hough circles. Hough circle transform outputs the triplet (x_c, y_c, r), with (x_c, y_c), the center and r, the radius that spans the maximum possible white space within the image without overlapping with any of the black pixels.

5.2 Calculation of Maximum Height of Characters

Height of the characters in the sample is one of the easiest to extract an effective feature, while performing the author inference task. This feature corresponds to the maximum vertical pen movements of a writer. The height of the sample is taken to be the maximum height across characters in the image. Height is usually computed by assessing the vertical length of black pixels in the binary image. The datasets typically contain less white space because of cropping extra white space during the preprocessing stage. We then classify the image into two subsets by comparing the height of the black pixels with the threshold λ. Images whose height is larger than λ are considered to be big, whereas images whose height is lesser than λ are considered normal. An appropriate value for λ has to be set after examining the datasets. For IAM, CVL, and Tamil datasets, we have set λ to 125, based on trial-and-error experiments.

5.3 Speeded-Up Robust Features

Speeded-up Robust Features (SURF) in essence is a variant of SIFT [18], which extracts features by constructing a histogram of locally oriented gradients of the interest points. Interest points may be defined as regular-sized subsections in the image which when moved will exhibit a high variation (σ^2) in the respective features. Harris Corner detection [19] can be employed for identifying the interest points. A window is affixed to the image at the position (x, y). The window is then displaced to a small distance of Δx units along the x-axis and Δy units along the y-axis. A window displacement is deemed as interesting, provided there is a significant variation in the pixel intensities of the window before and after its displacement:

$$\sigma^2(x, y, \Delta x, \Delta y) = N(x, y) \cdot \Delta I \tag{2}$$

The variation in the pixel intensities across window displacements $(\Delta x, \Delta y)$ is computed using (2). In (2), $N(x, y)$ is the Gaussian weighting factor which renders the interest point-scale invariant. A Hessian matrix [18] is then constructed to detect the interest point by maximizing sigma with respect to the displacements $(\Delta x, \Delta y)$.

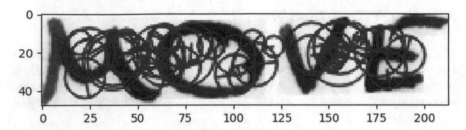

Fig. 1 Interest points detected using SURF

The interest points thus obtained are then normalized using the Gaussian derivative to render the interest point-scale invariant. If both the eigenvalues of the Hessian matrix are high, then we classify the corresponding point as interest point. For instance, Fig. 1 illustrates the inference of interest points for one of the samples collected from the IAM database. Typically, Haar wavelet transform [18] is employed for describing the interest points. Haar's descriptors are then used for inferring the authorship of the handwritten sample.

5.4 Nearest Neighbor Algorithm

Once the model is built, an unseen handwritten sample can be classified using the model into one of the classes each corresponding to the writer. The handwritten test sample d_t is made to traverse the hierarchical feature descriptor tree from the root to the particular leaf node representing one of the classes. To iterate our earlier discussion, the root level performs classification using the number of circle extracted out of Hough Transform as the feature. At the subsequent level, the height of the test sample will be used as the feature to classify d_t into one of the two categories (say C_t), as explained in the model construction phase. Finally, a feature descriptor (f_t) for d_t is formed using SURF. Let R^n denote the vector space spanned by the SURF feature vectors of each document in C_t. d_t is projected onto the vector space. Let d_t be the training sample which is closest to d_t as measured using the Euclidean distance metric in the space R^n as in (3):

$$dist(d_i, d_t) = \sqrt{\sum_{j=1}^{n}(d_i(j) - d_t(j))^2} \tag{3}$$

where $d_i(j)$ and $d_t(j)$ return the scalar value of the vector d_i, d_t along the jth dimension in R^n. We return $\Psi(d_t)$ as the writer of d_t. In case more than one sample is nearest to d_t, one can employ the majority voting scheme that returns the writer who has written the maximum number of nearest documents. In case of a tie, a writer is randomly selected with the uniform probability across writers in breaking the tie. Though

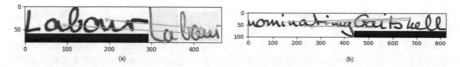

Fig. 2 SURF recognition **a** same word written by different writers, **b** different words written by the same writer

SURF is typically used for detection of regular objects from an image consisting of collection/superposition of objects, minor differences among the objects identified can be discriminated using SURF. This phenomenon can be seen in Fig. 2a in which though the same word "labor" has been written by two different writers, SURF is able to spot the differences between the images. On the other hand in Fig. 2b, two different words written by the same writer can be processed using SURF, to identify the commonalities in the handwritten style between the two words.

6 Experiments and Results

Empirical evaluation of the proposed algorithm was conducted using the IAM, CVL, and a Tamil dataset, which we created. The metadata about the datasets is as described below.

IAM Database: IAM handwritten database is among the popular databases used by researches working on the problem of author inference. About 657 writers have contributed to the IAM database with PNG images of resolution 300 DPI. We divide the set of writers into two subsets: (i) W, the set of writers who have contributed a single sample and (ii) W', the set of writers who have contributed more than one sample. From the metadata of IAM, we know that $|W| = 356$ and $|W'| = 301$. Let D be the set of samples in IAM. We construct three types of training set: T_{IAM}^1, T_{IAM}^2, and T_{IAM}^3. T_{IAM}^1 is constructed by sampling one handwritten sample from each of the writers in W'. T_{IAM}^2 is taken to be $\{d \in D | \Psi(d) = W\}$. The third training set T_{IAM}^3 is constituted from 50% of documents from T_{IAM}^1 and 50% of documents from T_{IAM}^2. The test set (T') is then orchestrated by $T' = \{d \in D | d \notin T \wedge \Psi(d) \in W'\}$.

CVL Database: CVL database comprises cursively written handwritten samples. About 310 writers have contributed to the CVL database with four documents each in English and a single document each in German. We have used two English samples from every writer to form the training set. As we only have a single sample per writer from German, we have resorted splitting each sample into two subsamples: one subsample is used for constituting the training set (T_{CVL}^4), whereas the other subsample is used as test. Four training sets are constituted for CVL: T_{CVL}^1, T_{CVL}^2, T_{CVL}^3, and T_{CVL}^4. Out of these four training sets, the first three are for English handwritten documents and T_{CVL}^4 is formed from German handwritten documents. If D represents the number of handwritten documents in the CVL database, then the formation of T_{CVL}^1, T_{CVL}^2, and T_{CVL}^3 are similar to the formation of T_{IAM}^1, T_{IAM}^2, and T_{IAM}^3, respectively.

Tamil Database: To the best of our knowledge, no standard author inference databases exist for Tamil. We have collected two handwritten samples each from 56 writers. The training set was constituted by picking one sample at random from each writer and the test was framed using the rest of the samples. The writers comprise people having different writing styles. Tamil comprises 247 letters each having considerable number of curls and curvatures in its shape. This makes the extraction of descriptor little harder when compared to other languages. T_{Tamil}^1, T_{Tamil}^2, and T_{Tamil}^3 are the three training sets formed using Tamil database similar to that of IAM.

Hough transforms are used to construct the feature set f_1 that forms the first level in the feature tree. As empirically observed by us, circles extracted by Hough transforms correspond to spaces between characters. Thus, it helps in the chunking of samples into subsamples or clusters. The portion of the sample within a cluster is expected to have lower variance between descriptors extracted from different portions of the subsample, whereas descriptors extracted from clusters typically have much higher variance. The scatter plot shown in Fig. 3 depicts the number of Hough circles extracted across the samples from different writers. Clearly, this scatter plot exhibits a well-defined pattern, thus suggesting the presence of commonality across samples written by some writers. The scatter plot constructed in Fig. 3 is for samples of 20 writers from IAM database.

Each cluster obtained after performing the Hough transform is subsequently split into subclusters based on the height of the characters in the image. After reaching the leaf level of the feature tree, we find group of training samples with similar samples as inferred by SURF. A test image falling inside the cluster will inherit the writer of the nearest sample. Experiments to evaluate the proposed technique HHFTAI have been controlled using two parameters: (i) number of SURF features and (ii) matching threshold—number of interest points required to be approximately identical between two images d_t and d_c, so that d_t can inherit its writer as $\Psi(d_c)$. Figures 4, 5a, and 5b depicts the recognition rate exhibited by the models learnt using the training sets T_{IAM}^1, T_{IAM}^2, and T_{IAM}^3, respectively. From Fig. 4, it can be inferred that, as the number of features increases, the recognition rate decreases. This could be attributed due to the fact that, as we increase the number of features, the model enforces a strict matching between the training and test description. Figure 5a shows that as the number of features increases, the recognition rate also increases. This is an expected behavior especially because the training and test samples did not share any common writer. Figure 5b illustrates the behavior of the model trained using T_{IAM}^3. This behavior could be due to the hybrid behavior of the models learnt from T_{IAM}^1 and T_{IAM}^2. The characteristics shown in Fig. 5b reach the maximum recognition rate of 97.15% when the number of features considered is 3000 and matching threshold is 21. Figure 6 illustrates the performance of model learnt using T_{CVL}^3. The characteristic shown in Fig. 6 exhibits a similar trend as those of the corresponding models learnt using IAM training sets. The maximum recognition rate achievable by our model for the CVL database is 95.78% when the number of features and the matching threshold are set as 3000 and 18, respectively. Figure 7 demonstrates a trend similar to that of

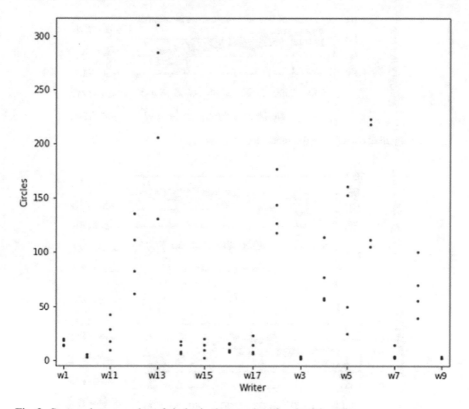

Fig. 3 Scatter plot on number of circles in the samples of each of the writers

Figs. 5b and 6, and shows a maximum recognition rate of 96.5% at 3000 features and at a threshold of 18 for German database. Figure 8 depicts the performance of proposed model HHFTAI on Tamil handwritten datasets. As can be seen from these figures, the model exhibits a similar pattern to that of the English and German datasets considered. The recognition rate achieved by our model HHFTAI (97.35%) for the Tamil dataset is considerably higher as compared to the English and German datasets. This is due to the fact that Tamil contains more curls and curvatures, causing spaces between and within characters, as compared to English and German. The feature transformation at the root level f_1 is very effective in discriminating samples with such varying spaces.

The recognition rates in Table 1 for the other models have been taken from the respective papers [2, 8, 10, 11].

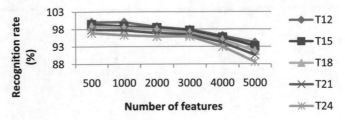

Fig. 4 Recognition rate of HHFTAI model learnt using T^1_{IAM}

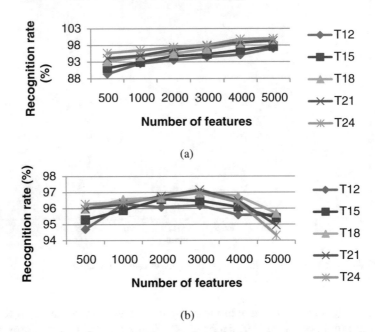

(a)

(b)

Fig. 5 Recognition rate of HHFTAI model learnt using **a** T^2_{IAM}, **b** T^3_{IAM}

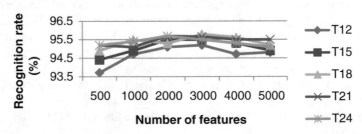

Fig. 6 Recognition rate of HHFTAI model learnt using T^3_{CVL}

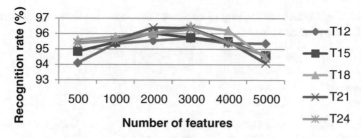

Fig. 7 Recognition rate of HHFTAI model learnt using T_{CVL}^4

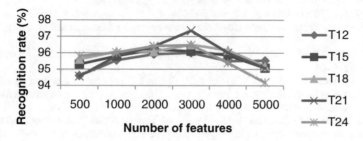

Fig. 8 Recognition rate of HHFTAI model learnt using T_{Tamil}^3

Table 1 Comparison of existing author inference studies with HHFTAI

Inference techniques	Recognition rate	
	IAM database	CVL database
By Hannad et al. [2]	89.54%	–
By Bertolini et al. [11]	96.70%	–
By Bulacu et al. [10]	89.00%	–
By Christlein et al. [8]	–	91.00%
Proposed technique HHFTAI	97.15%	95.78%

7 Conclusion

This paper proposes a hybrid hierarchical feature tree based model for solving the author inference problem. The performance of the model with respect to the recognition rate has been evaluated using two standard databases namely IAM and CVL. We have also evaluated the proposed model's performance using a self-curated Tamil handwritten dataset. Empirical evaluations reveal that HHFTAI model achieves a better recognition rate for the Tamil handwritten dataset as compared to IAM and CVL datasets. The proposed model also exhibits a better recognition rate as compared to the state-of-the art models like run length features, Arnold transforms, and local gradient descriptors. Further research direction might include incorporating some more computationally cheaper, yet effective feature descriptor while building the model. We are also attempting to construct theoretically time–space complexity bounds for the proposed model.

References

1. Djeddi C, Siddiqi I, Souici-Meslati L, Ennaji A (2013) Text-independent writer recognition using multi-script handwritten texts. Pattern Recogn 34(1):1196–1202
2. Hannad Y, Siddiqi I, El Youssfi M, Kettania E (2016) Writer identification using texture descriptors of hand written fragments. Pattern Recogn 34(1):1196–1202
3. Dasgupta J, Bhattacharya K, Chanda B (2016) A holistic approach for off-line handwritten cursive word recognition using directional feature based on Arnold transform. Pattern Recogn 79(1):73–79
4. He S, Schomaker L (2017) Writer identification using curvature-free features. Pattern Recogn 63(1):451–464
5. Ooia SY, Teohb ABJ, Panga YH, Hiewa BY (2016) Image-based handwritten signature verification using hybrid methods of discrete Radon transform, principal component analysis and probabilistic neural network. Appl Soft Comput 40(1):274–282
6. Khan FA, Tahir MA, Khelifia F, Bouridane A, Almotaeryi R (2017) Robust off-line text independent writer identification using bagged discrete cosine transform features. Expert Syst Appl 71(1):404–415
7. He S, Wiering M, Schomaker L (2015) Junction detection in handwritten documents and its application to writer identification. Pattern Recogn 48(12):4036–4048
8. Christlein V, Bernecker D, Hönig F, Maier A, Angelopoulou E (2017) Writer identification using GMM sSuper vectors and exemplar-SVMs. Pattern Recogn 63(1):258–267
9. Surinta O, Karaaba MF, Schomaker LRB, Wiering MA (2015) Recognition of handwritten characters using local gradient feature descriptors. Eng Appl Artif Intell 45(1):405–414
10. Bulacu M, Schomaker L (2007) Text-independent writer identification and verification using textural and allographic features. IEEE Trans Pattern Anal Mach Intell 701–717
11. Bertolini D, Oliveira LS, Justino E, Sabourin R (2013) Texture based descriptors for writer identification and verification. Exp Syst Appl 40(6): 2069–2080
12. Said H, Tan TN, Baker KD (2000) Personal identification based on handwriting. Pattern Recogn 33:149–160
13. Tan X, Triggs B (2007) Enhanced local texture feature sets for face recognition under difficult lighting conditions. In: Analysis and modeling of faces and gestures pp 168–182, Springer
14. Ojansivu V, Heikkilä J (2008) Blur insensitive texture classification using local phase quantization. In: Image and signal processing pp 236–243, Springer
15. Marti U, Bunke H (2002) The IAM-database: an english sentence database for off-line handwriting recognition. Int J Doc Anal Recogn 5:39–46
16. http://caa.tuwien.ac.at/cvl/research/cvl-database/index.html. Accessed 31 Aug 2017
17. https://en.wikipedia.org/wiki/Graphology. Accessed 31 Aug 2017
18. Bay H, Ess A, Tuytelaars T, Van Gool L (2008) Speeded-up robust features (SURF). Comput Vis Image Underst 110(3):346–359
19. Yi-bo L, Jun-jun L (2011) Harris corner detection algorithm on improved contourlet transform. Procedia Eng 15:2239–2243

Breast Cancer Detection Using Low-Computation-Based Collaborating Forward-Dependent Neural Networks

Karan Sanwal and Himanshu Ahuja

Abstract Considering the eminence of breast cancer detection, a variety of models have been proposed for its diagnosis. Although many of these models provide high-performance accuracy, they are computationally expensive. We propose a novel ensembling method, CFDNN which aims at reducing the computational expense, whilst maintaining the high-performance accuracy comparable to the state-of-the-art models in the breast cancer detection problem. To the best of our knowledge, the proposed CFDNN model is the computationally fastest model in breast cancer detection with an accuracy of 99.01%, making CFDNN an optimal choice for providing a second opinion in medical diagnosis.

Keywords Computer-aided diagnosis · Breast cancer detection · Neural network ensemble

1 Introduction

Cells are the smallest structural and functional units in any living being. Moreover, these cells have the capability to transfer their structural and functional characteristics into daughter cells. Cancer is caused by the abnormal cells, which divide uncontrollably and destroy the tissue of origin. This uncontrollable division of the cancer cells leads to the formation of a tumour. A tumour can be further classified into either benign or malignant. A benign tumour does not intrude on the surrounding tissue; hence, it is in a way controllable. Most benign tumours are rendered curable through modern medicinal techniques.

Around the world, breast cancer is the most prevalent cancer diagnosed in women. In the United States alone, an estimated 250,000 new patients and approximately 40,450 deaths were reported in 2016 [1]. These startling numbers remark the importance and the urgent need for early detection of breast cancer. In such a situation, machine-aided diagnosis can be utilised for a second opinion in medical.

K. Sanwal (✉) · H. Ahuja
Computer Science Department, Delhi Technological University, New Delhi, India
e-mail: karansanwal@gmail.com

© Springer Nature Singapore Pte Ltd. 2019
M. L. Kolhe et al. (eds.), *Advances in Data and Information Sciences*, Lecture Notes in Networks and Systems 39, https://doi.org/10.1007/978-981-13-0277-0_9

It is also of utmost importance that the methods for training these models to be computationally fast, so their access becomes more diverse and extensible. Most algorithms today are computationally expensive. To the best our knowledge, the novel ensembling method, CFDNN, is the computationally least expensive approach in breast cancer diagnosis. CFDNN provides an accuracy of 99.01% in just two layers of neural network ensemble on Wisconsin Breast Cancer Dataset.

2 Related Work

Many algorithms have been proposed for breast cancer diagnosis using the Wisconsin Breast Cancer Dataset (WBCD) in literature [2–6], which is a two class-classification problem in machine learning.

The AMMLP algorithm proposed by Marcano-Cedeno et al. [7] achieved an accuracy of 99.36% using artificial meta-plasticity which prioritises update of weights for less frequent activations over the frequent ones. Abdel-Zaher and Eldeib [8] used a deep belief network for pretraining followed by a supervised phase utilising a neural network with backpropagation. Their model achieved a classification accuracy of 99.68%.

Devi and Devi [6] used a three-step procedure for early diagnosis of breast cancer, by treating the problem statement as that of outlier detection. In the first step, they clustered the data using the farthest-first algorithm, after which they applied an outlier detection algorithm. Finally, the J48 classification algorithm was used to classify the instances. Their model achieved an overall accuracy of 99.9% via the tenfold cross-validation technique.

Akay [3] used support vector machines combined with an F-score based feature selection method to achieve a classification accuracy of 99.51% using 80–20 data split. Nahato et al. [9] presented the algorithm RS-BPNN which first constructed rough sets to fill in the missing data and then used backpropagation algorithm to train the neural networks, to achieve a classification accuracy of 98.6%. Sayed et al. [10] proposed the use of a meta-heuristic optimization technique to chose features for classification, called Whale Optimization Algorithm (WOA) and implemented it on the WBCD.

Genetic algorithms have been extensively used to modify the structure and parameters of the neural networks and selectively chose the best model structure in ensemble modelling or to select the most significant features. Alickovic and Subasi [5] achieved 99.48% accuracy through the implementation of rotation forests for classification, and feature selection through genetic algorithms to eliminate the insignificant features. Bhardwaj and Tiwari [11] defined their crossover and mutation operations and utilised genetic algorithms to produce neural network offspring according to their fitness. It obtains an overall accuracy of 99.26%.

3 Wisconsin Breast Cancer Database

Developed by Dr. Willian H. Wolberg at the University of Wisconsin Hospital, the Wisconsin Breast Cancer Dataset contains a total of 699 training instances but due to missing features, 16 examples were omitted. The resultant dataset contains 444 training examples of benign tumour and 239 examples of a malignant tumour. Each training example consists of 10 features. The sample code feature is the id of each instance and is not considered during training or prediction. All the other attributes are integers from 1–10 [12].

4 Proposed Approach

The proposed algorithm, CFDNN, is aimed at reducing the computation associated with the machine-based detection of breast cancer. The algorithm leverages the following criteria for a faster performance:

1. Reducing computations at the base units.
2. Implementing a parallelizable global structure that can scale and take advantages of a multiprocessor-based architecture.

With the above being mentioned, the sensitivity of the problem at hand cannot be overlooked. CFDNN aims at achieving high accuracy at the lowest possible computational expense through a novel ensembling algorithm. Ensembling refers to a popular technique in machine learning, where, rather than using a single complex classifier, multiple weak classifiers are used before arriving at a single solution to the problem [5, 13–15]. However, most ensembling techniques in literature [2, 8] are computationally expensive due to the following reasons:

1. Large amount of computations wasted in increasing the base classifier accuracy.
2. Large amount of computations wasted in the creation of the ensemble in a way that introduces variation among its members.

CFDNN introduces an ensembling technique that intuitively creates varied classifiers which aim to reduce the error intersections amidst them. A property is far more salient than targeting a single classifier accuracy as a metric for the entire ensemble. The algorithm takes inspiration from the way students may effectively collaborate towards a group test. Each neural network in the ensemble is seen as a student, and the training set is seen as the syllabus of a test that the neural networks have to perform well on. Rather than the popular technique of dividing the syllabus by volume (Dividing the training set amidst the various neural networks), CFDNN divides it by difficulty. Each successive neural network studies only that part of the syllabus where its predecessor was unable to learn effectively.

Following this section, we introduce three computationally inexpensive flavours of the algorithm, based on the architecture of the system the model is running and the user requirements.

1. Serial CFDNN, the baseline CFDNN architecture
2. Parallel CFDNN Version: 1
3. Parallel CFDNN Version: 2

4.1 Serial CFDNN

Figure 1 highlights the architecture of the CFDNN model. The algorithm's base unit utilises a neural net with a single hidden layer comprising five nodes each. We define the *depth* of a neural network as a step-incremented variable (Assigned to *one*, for the first neural network) which describes the layer depth of the neural network in the ensemble model. The neural network created at *zero depth* is termed as the *Central Classifier (CF)* which is made to train on the entire dataset. The number of iterations (α) for learning is optimally set as per the dataset through k-cross-validation techniques. Due to the mentioned restrictions on its structure and training, when evaluated on the dataset, it is thence expected to misclassify a set of instances it could not learn. Once the *zero depth* neural network is trained on the entire dataset, a second neural network (with *depth* = 1) is initialised and is made to train only on the misclassifications of the first neural network. This pattern is repeated along the layer depth of the ensemble where the ith neural network trains solely on the misclassifications of the $(i - 1)$th neural network. It is hence easy to intuit that the CF neural network has the most general idea of the dataset and the subsequent neural networks keep on specialising on the instances that are seen as an anomaly by the patterns learnt by the lower depth neural networks. It can be observed that the central

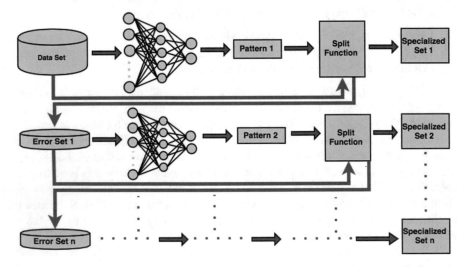

Fig. 1 The *n-depth* Serial CFDNN architecture

classifier splits the dataset into two sets: One which can be classified by the general pattern learnt by a weak-restricted CF and then another can be seen as an anomaly with respect to it. This relatively anomalous dataset is fed to the subsequent neural network, which, being a weak-restricted classifier itself, divides it into two sets in a similar fashion and the pattern is repeated along the depth. The general dataset division equation is given by

$$D_i = G_i \cup A_i \quad \text{and} \quad D_{i+1} = A_i \quad \forall i \in \{1, 2, 3, \ldots, n\}, \tag{1}$$

where n is the total depth of the neural network. G_i (General dataset for the ith neural network) contains the training examples in the dataset (D_i) in which the ith neural network correctly classifies and A_i (Anomalous dataset for the ith neural network) contains the training examples in which the ith neural network misclassifies.

This novel technique in ensembling is seen better for the following reasons:

1. To force the weak neural network to learn the anomaly dataset (A_i), the optimiser might shift its parameters in a way that it loses some accuracy over the general dataset.
2. Since the next neural network is only trained on the anomaly dataset (A_i), it is relatively specialised in that dataset compared to its predecessor neural network which is trained on the superset $G_i \cup A_i$. This property helps to create a low-computation-based mapper function which helps in increasing the effectiveness of parallelization as we will see in Sect. 4.2.
3. This also aids the idea of focussing on reducing the intersection between error sets rather than increasing sole accuracies of classifiers. It is easy to see that the higher depth neural networks will be weaker than the predecessor neural networks, but it will be able to solve instances that the predecessor neural network was unable to solve.

During the prediction phase, the central classifier CF's confidence (Δ) and prediction (y) are taken. A hyperparameter threshold (β) is optimally chosen to compare the pairwise difference between the confidence of the CF and every other subsequent neural network of the ensemble ($\Delta' - \Delta$). If the confidence crosses β, the prediction of that neural network is returned.

4.2 Parallel CFDNN

The above model acts as a baseline model for the classification of the breast cancer tumour but CFDNN also realises the scenario where more than one processing unit (node) is available in the system architecture on which it is being executed. We propose two parallel variants (Fig. 2) of the serial CFDNN which leverage themselves based on user requirement. The presence of multiple nodes can either be used to make the system faster or to further optimise the model's performance. In the subsequent subsections, we propose:

1. Parallel CFDNN Version: 1, a learning system that utilises the computing power of multiple nodes to make the system faster with minimum drop in accuracy.
2. Parallel CFDNN Version: 2, a learning system that utilises the computing power of multiple nodes with increase in accuracy with minimum increase in execution/train time.

Algorithm 1: Serial CFDNN

---

    input : dataset, $depth$, n
    output: A trained neural network ensemble

1 **Initialization:** $gen = dataset$, α $ensemble = \phi$, $depth = 0$;
2 **Function** $Serial(gen, \alpha, depth, n)$
3   **if** $gen == NULL$ **then**
4 | return;
5   **end**
6   **if** $depth > n$ **then**
7 | return;
8   **end**
9   Create Neural Network = NN;
10   Train NN with gen using α iterations;
11   Adam Optimizer, Softmax Activation;
12   Sigmoid Hidden Layer Activation;
13   A_{depth} = NULL;
14   **for** $each (x, y) \in gen$ **do**
15   **if** $NN.evaluate != y$ **then**
16 | $A_{depth} = A_{depth} \cup \{(x, y)\}$;
17   **end**
18   **end**
19   ensemble[$depth$] = NN;
20     $Serial(A_{depth}, \alpha, depth+1, n)$;
21 **end**

Algorithm 2: Confidence Mapper

---

    input : ensemble, $depth$, n, β
    output: Predicted Classification : y

1 **Function** $Mapper(ensemble, n, \beta, x)$
2   Δ, y = ensemble[1].evaluate(x);
3   **for** $i \leftarrow 2$ **to** n **do**
4   Δ', y' = ensemble[i].evaluate(x);
5   **if** $\Delta + \beta < \Delta'$ **then**
6 | return y';
7   **end**
8   **end**
9   return y;
10 **end**

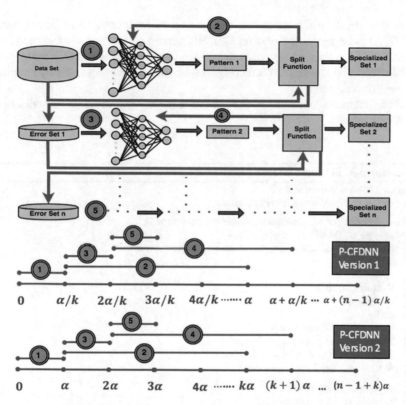

Fig. 2 The *n-depth* parallel CFDNN model. The above timelines represent how the iterations are run simultaneously in the two flavours of the parallel model. Here, 1, 2, 3, 4 and 5 are representative of steps indicating training of the corresponding neural networks with the specified dataset

4.2.1 Parallel CFDNN Version: 1

The central principles of serial CFDNN are adapted with slight modifications to fit the new parallel architecture. Rather than running α iterations before generating the training set of the subsequent neural network, a parent neural network splits the dataset into two sets $D_{\alpha/k}$ and $D'_{\alpha/k}$, where $D_{\alpha/k}$ comprises the training examples that the neural network correctly classifies after just α/k iterations and $D'_{\alpha/k}$ represents the misclassified examples after α/k iterations. Here, k is the scaling factor which controls the number of nodes running in parallel and is set according to computational requirements. The parent neural network then pipelines its remaining iterations ($\alpha - \alpha/k$) with the training of the subsequent neural networks that can now train in parallel with the dataset received earlier ($D'_{\alpha/k}$). This early splitting of the data introduces the following changes in the way CFDNN behaves:

1. At any depth of the ensemble, the size of the misclassified instances dataset $(D'_{\alpha/k})$ is larger for the parallel CFDNN variant when compared to the size of misclassified training instances at the same depth in the serial CFDNN.
2. Likewise, the size of the correctly classified dataset is comparatively smaller at any given depth.
3. Different neural networks of the ensemble are trained on different nodes of the device used.

Algorithm 3: Parallel CFDNN Version: 1

---

   input : dataset, $depth, n, k$
   output: A trained neural network ensemble

1 **Initialization:** $gen = dataset$, α ensemble $= \phi$, $depth = 0$;
2 **Function** $PCFDNN1(gen, \alpha, depth, n, k)$
3   **if** $gen == NULL$ **then**
4 | return;
5   **end**
6   **if** $depth > n$ **then**
7 | return;
8   **end**
9   Create Neural Network $= NN$;
10   Train NN with gen using α/k iterations;
11   Adam Optimizer, Softmax Activation;
12   Sigmoid Hidden Layer Activation;
13    $D'_{\alpha} = NULL$, $D_{\alpha} = NULL$;
14   **for** $each (x, y) \in gen$ **do**
15   **if** $NN.evaluate != y$ **then**
16 | $D'_{\alpha} = D'_{\alpha} \cup \{(x, y)\}$;
17   **else**
18 | $D_{\alpha} = D_{\alpha} \cup \{(x, y)\}$;
19   **end**
20   **end**
21   Machines[$depth + 1$].execute($PCFDNN1(D'_{\alpha}, \alpha, depth+1, n, k)$);
22   Goto (10): Train with $\alpha - \alpha/k$ iterations with $gen = D_{\alpha}$;
23   ensemble[$depth$] = NN;
24 **end**

These two changes are intuitive results obtained due to the fact that it is a partially learnt neural network that is splitting the data. Since $D'_{\alpha/k}$ can now be a dataset of considerable size, training on it repeatedly may now lead to a loss of accuracy of the neural network over the partial dataset $D_{\alpha/k}$.

This goes against the basic intuition of CFDNN where each neural network should focus to be highly specialised on its assigned dataset $D_{\alpha/k}$ and not to improve its classification accuracy on the entire dataset. Therefore, another introduced difference in the Parallel CFDNN Version: 1 is that the parent network trains only on $D_{\alpha/k}$ after the data split point (i.e. α/k iterations), whereas, in the serial CDFNN, the parent network would train completely on $D_{\alpha/k} \cup D'_{\alpha/k}$, for the complete α iterations. This strengthens its confidence over $D_{\alpha/k}$ and allows us to use the same low-computation-based mapper function that we used in serial CFDNN.

4.2.2 Parallel CFDNN Version: 2

The second parallel variant of CFDNN aims to leverage the increased computation power to increase the performance of the learning system without an exponential increase in execution/training time. The variant 2 increases the number of iteration per neural network to $k\alpha$, instead of α/k like in the variant 1. Here again, k is the scaling factor, but rather than scaling down the number of iterations, it scales up the number of iterations. The split point like serial CFDNN is created after α iterations. The subsequent $(k-1)\alpha$ iterations are pipelined along with the subsequent neural networks like in parallel CFDNN variant 1. The effects of this scaling are as follows. For the breast cancer detection classification problem, $k = 2$ was found to be the optimal scale-up factor.

Algorithm 4: Parallel CFDNN Version: 2

---

    input : dataset, $depth, n, k$
    output: A trained neural network ensemble

1   **Initialization:** $gen = dataset, \alpha$ ensemble $= \phi, depth = 0$;
2   **Function** $PCFDNN2(gen, \alpha, depth, n, k)$
3   **if** $gen == NULL$ **then**
4 | return;
5   **end**
6   **if** $depth > n$ **then**
7 | return;
8   **end**
9   Create Neural Network $= NN$;
10   Train NN with gen using α iterations;
11   Adam Optimizer, Softmax Activation;
12   Sigmoid Hidden Layer Activation;

13     $D'_\alpha = NULL, D_\alpha = NULL$;
14   **for** *each* $(x, y) \in gen$ **do**
15   **if** $NN.evaluate \mathrel{!=} y$ **then**
16 | $D'_\alpha = D'_\alpha \cup \{(x, y)\}$;
17   **else**
18 | $D_\alpha = D_\alpha \cup \{(x, y)\}$;
19   **end**
20   **end**
21   Machines$[depth + 1]$.execute($PCFDNN2(D'_\alpha, \alpha, depth+1$, n, k));
22   Goto (10): Train with $(k-1) \times \alpha$ iterations with $gen = D_\alpha$;
23   ensemble$[depth] = NN$;
24 **end**

Table 1 The comparison of the various flavours of CFDNN with state-of-the-art learning algorithms on WBCD Database (Data Distribution: 70–30)

Algorithm	Average Accuracy	Algorithm	Average Accuracy
TF-GNNE [16]	99.90%	Devi et. al [6]	99.60%
AMMLP [7]	99.26%	**S-CFDNN**	**99.01%**
DBN-NN [8]	99.59%	**P-CFDNN Version: 1**	**98.50%**
RS - BPNN [9]	98.60%	**P-CFDNN Version: 2**	**99.01%**

5 Results

The three variants of the CFDNN algorithm were evaluated on the Wisconsin Breast Cancer Dataset. The models were implemented in Python using TensorFlow. We used 70–30% training to test examples ratio to experiment our models. We used $k = 2$ for both the parallel variants of the CFDNN Algorithm. Each member of the neural network ensemble contained five units in the hidden layer.

An average accuracy of 99.01% was obtained over 10 runs of the serial CFDNN model. The parallel CFDNN version 1 model approaches the serial CFDNN with an average accuracy of 98.5%. The parallel CFDNN version 2 model gets the maximum clocked accuracy of 99.5%, but averages exactly as the serial CFDNN, most probably owing to overfitting due to a small dataset size. Table 1 compares CFDNN with various algorithms tested on WBCD database. Each of these algorithms is computationally highly expensive compared to CFDNN.

6 Conclusions

All the three variants of CFDNN seem to be competing with the state-of-the-art algorithms in literature despite being computationally inexpensive. The mapper function proposed to map the multiple unit outputs to a single ensemble output seems to be fitting the intuitive sense on which the learning model is based; however, given that experimentation has shown considerably low error intersection set even when trained with less data, it is a salient task to implement an intelligent learning-based mapper function as well as a part of our future work. Also, since the amount of data decreases with the depth of the ensemble, computationally inexpensive algorithms that train on sparse data efficiently definitely show a promise in increasing the performance of the system.

References

1. Siegel RL, Miller KD, Jemal A (2016) Cancer statistics, 2016. CA: Cancer J Clin 66:7–30
2. Azami H, Escudero J (2015) A comparative study of breast cancer diagnosis based on neural network ensemble via improved training algorithms. In: 37th Annual International conference of the IEEE engineering in medicine and biology society (EMBC). pp 2836–2839
3. Akay MF (2009) Support vector machines combined with feature selection for breast cancer diagnosis. Expert Syst Appl 36:3240–3247
4. Derya E, Ubeyli (2007) Implementing automated diagnostic systems for breast cancer detection. Expert Syst Appl 33:1054–1062
5. Alickovic E, Subasi A (2017) Breast cancer diagnosis using GA feature selection and rotation forest. Neural Comput Appl 28(4):753–763
6. Devi RDH, Devi MI (2016) Outlier detection algorithm combined with decision tree classifier for early diagnosis of breast cancer. Int J Adv Eng, Chicago. Tech/Vol VII/Issue II/April–June 93:98
7. Marcano-Cedeo A, Quintanilla-Domnguez J, Andina D (2011) WBCD breast cancer database classification applying artificial metaplasticity neural network. Expert Syst Appl 38(8):9573–9579
8. Abdel-Zaher AM, Eldeib AM (2016) Breast cancer classification using deep belief networks. Expert Syst Appl 46:139–144
9. Nahato KB, Nehemiah HK (2015) Kannan A (2015) Knowledge Mining from clinical datasets using rough sets and backpropagation. Neural Netw Comp Math Methods Med 460189(460181–460189):460113
10. Sayed GI et al (2016) Breast cancer diagnosis approach based on meta-heuristic optimization algorithm inspired by the bubble-net hunting strategy of whales. In: International conference on genetic and evolutionary computing, Springer International Publishing
11. Bhardwaj A, Tiwari A (2015) Breast cancer diagnosis using genetically optimized neural network model. Expert Syst Appl 42(10):4611–4620
12. Bache K, Lichman M (2013) UCI machine learning repository, 2013 [Online]. http://archive.ics.uci.edu/ml
13. Opitz DW, Shavlik JW (1996) Generating Accurate and Diverse Members of a Neural-Network Ensemble. Adv Neural Inf Process Syst 8:535–541
14. Rivero D, Dorado J, Rabual J, Pazos A (2010) Generation and simplification of artificial neural networks by means of genetic programming. Neurocomputing
15. Koza JR, Rice JP (1991) Genetic generation of both the weights and architecture for a neural network. In: International joint conference on neural networks (IJCNN-91), vol 2. pp 397–404
16. Singh I, Sanwal K, Praveen S (2016) Breast cancer detection using two-fold genetic evolution of neural network ensembles. In: IEEE 2016 International conference on data science and engineering (ICDSE). IEEE

A Modified Genetic Algorithm Based on Improved Crossover Array Approach

Stuti Chaturvedi and Vishnu P. Sharma

Abstract From several years, genetic algorithms are being used for solving many types of constrained and unconstrained optimization problems. Type of crossover operator, rate of crossover, replacement scheme and mutation rate are some of the essential factors that contribute to the efficiency of GA. Evidently, not one algorithm is suited for every problem according to the free lunch theorem but one form of crossover can solve a large number of problems. So many variants of GA are proposed in the past years with different types of crossover operators. So in this paper, we are introducing a new modified GA with improved crossover by dynamically choosing the type of crossover operator which is going to be used for the problem. Randomized mutation is used and a new virtual population is created which contains the best population so far. The modified GA is tested on 40 benchmark functions and the results are compared with the basic GA and one other GA variant.

Keywords GA · Evolutionary algorithms · Crossover array

1 Introduction

Many things and also essential ones are provided to us by nature in many different forms. There are several complex mechanisms that are taken out in nature through very efficient and simple procedures. The idea of nature-inspired genetic algorithm comes from there. It is probabilistic stochastic search technique which imitates the process of natural selection and natural genetics to solve a complex problem.

S. Chaturvedi (✉) · V. P. Sharma
Government Engineering College, Ajmer, India
e-mail: stutichaturvedi12@gmail.com

V. P. Sharma
e-mail: vp.ecajmer@gmail.com

© Springer Nature Singapore Pte Ltd. 2019
M. L. Kolhe et al. (eds.), *Advances in Data and Information Sciences*, Lecture Notes in Networks and Systems 39, https://doi.org/10.1007/978-981-13-0277-0_10

Genetic algorithm is an evolutionary algorithm. Evolutionary algorithms are those which are inspired by the natural selection process where the species are selected to survive in a better environment or changing environment. Genetic algorithms (Abbreviated as GA) were first developed by JOHN HOLLAND in the early 1970s. The uses of mutation operator, selection strategy and the genotype mapping are the significant differences that we can conclude from the two algorithms. Genotype and phenotype are the two spaces in which we can divide the working of GA. Genotype or coding space is used to create phenotype, and phenotype or solution space represents an individual's external characteristics. There is a mapping which exists between the two through which phenotypes are developed from the information extracted from genotype. The remainder of the paper is organized as follows. Section 2 portrays the related work. Section 3 defines the basic and the proposed algorithm. Section 4 describes the experiments conducted and their results. Section 5 states the conclusion.

2 Related Work

Crossover is the primary operator in GA. It explores the solution space more thoroughly so it brings a great degree of exploration to the GA [1]. Mostly changes done to the crossover operator in GA are problem dependent. As travelling salesman problem (TSP) is the most common optimization problem, many crossovers are recommended for it like partially mapped crossover (PMX), cycle crossover [2], edge recombination (ERX) [2], distance preserving crossover (DPX) [2], Greedy crossover [3], Moon crossover and many more. Other crossovers are designed for general-purpose problems. In selective crossover, a dominance value is associated with each gene and a chromosome is a real-valued vector. The dominance value of two parents is compared. If one of the parents has gene with higher dominance value, then that gene goes through crossover otherwise not and that gene's dominance value is increased only [1]. Affenzeller proposed a new approach in which dynamic usage of multiple crossover operators in parallel is done [4]. The new population is taken out by choosing the best of all the populations got by different crossover operators after applying mutation on each one of them. The idea taken is from species conservation [5] where population is divided into subpopulation based on their similarities. So subpopulations instead of populations are developed by applying different crossover operators on them so that each species can survive.

Li describes the use of crossover matrix instead of crossover operator and its control parameter in DE algorithm [6]. Crossover matrix is a random binary integer-valued matrix composed of 0 and 1. Convergence matrix is also used to guide the production of new generation.

3 Proposed Algorithm

Basic algorithm

The basic algorithm has the following steps:

1. Begin
2. Initialize the population Pop (of chromosomes) with random candidate solution Pop[i] = Frand_range (low, high) where i = 1: N, N is population size;
3. Calculate the fitness Ft of each candidate solution:

    (i) Ft = obj_value (for maximum function)
    (ii) Ft = 1/(1 + obj_value) (for minimum function);

4. Repeat until (termination criteria is satisfied) Do;
5. Select the candidate solution with Pop[i].Ft > Pop[q].Ft, where q is the randomly generated number between 1:N;
6. Crossover of parents (selected candidate solution);
7. Mutate the resulting offspring;
8. Calculate the fitness Ft of each new candidate;
9. Now take the entire new candidate solutions for new generation or compare the old population fitness (OldPop.Ft) with new population fitness (NewPop.Ft) and take the best between the two.
10. End.

Proposed Crossover:

The proposed algorithm has same steps as the basic algorithm; it is just the crossover operator that is modified. A job of crossover is to produce new possible solutions which are inherited from the selected parents of previous population. But there is no guarantee that the new produced child will be better than its parents. That is why mutation is introduced after crossover to exploit the solution space. When multiple crossovers are applied on same population in parallel, then there is a possibility of overuse or underuse of resources. Then after that, we have to apply a sorting mechanism on each one of them and then have to merge the best solutions of all into one population which is termed as new population. And if we divide the population into subpopulation, then the most difficult part is to decide the parameter on which the classification must be done. Also, it is the most difficult task to generate subpopulation in case of multiple objective functions. In subpopulation case, communication between them is another big task and possibility of reduction of diversity is also there. So an idea is proposed to take the benefit of multiple crossovers without applying different crossovers on full population in parallel. In our ecosystem, we also see different species evolve through different processes or ways. Same goes with our algorithm; it develops a new population using different processes or crossovers.

The basic idea is taken from crossover matrix [6]. Instead of crossover matrix, a crossover array is generated and that array will decide the type of crossover to be implemented on the individual. In this, we have implemented the crossover array in GA instead of DE and we have used two crossover types instead of changing the one and full chromosome will undergo the change, whereas in DE it is applicable to particular gene or dimension in a vector. So the type of crossover implemented

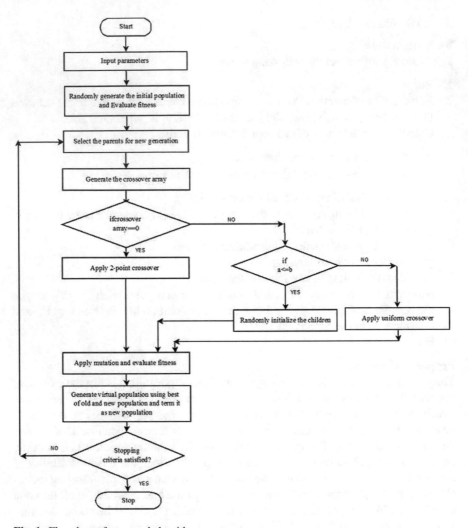

Fig. 1 Flowchart of proposed algorithm

is decided dynamically. Crossover array is randomly generated binary-valued array with size defined by crossover probability. As it is a randomly generated array, there will be a fair chance for all the types of crossovers. Here in this we implemented only two crossovers: one is two-point crossover and other is uniform crossover; and some randomization is also induced in it as lastly GA is a randomized algorithm with controlled parameters. Two random numbers are compared and their output will decide whether randomization is done or uniform crossover is done. Here, we have used only two crossovers as our choice but it can be extended to include multiple crossovers, we just have to change the crossover array according to it then. The working of crossover array is shown in Fig. 1.

4 Experiment Setup and Results

Here in this we have used tournament selection method with size $= 4$. Randomized mutation is used with mutation rate $= 0.01$ (same for all tests) and crossover rate is 0.9. Number of runs is 20 and population size is 50 and number of generation $=$ (total number of function evaluation (F_eval)/population size). F_eval $= 10,000$ [7]. So the generation is 200 for population size of 50. We have tested the algorithm on 40 benchmark problems of minimization from dimension 2–30. The results are compared with the basic GA algorithm and shown in Table 1. The basic GA algorithm uses one-point crossover and tournament selection with same size $= 4$ as our modified algorithm. All other parameters are same in both the algorithms. We have taken the best of old population to fill up the rest of the current population in the modified algorithm instead of randomly taking the individuals from old population. We have used the Mersenne Twister random number generator [8]. It is frequently used because it has a cycle length of $2^{19937} - 1$. Recursion is the ground for any random number generator which is typically of the following form: $X_n = F(X_{n-1}, X_{n-2}, ..., X_{n-k})$.

Table 1 shows the comparison of basic GA and modified GA on 40 benchmark problems. In all tables, obj_fun stands for objective function, D for dimension and Gmin for global minima. Figures 2 and 3 show the results of table in graph. Table 2 shows the working of proposed algorithm with different population sizes of 30, 50 and 100 with generations 334, 200 and 100, respectively, with crossover rate $= 0.9$ and mutation rate $= 0.01$. From the result, we can conclude that the algorithm works fine with population size of 50 most of the time. Some other experiments conducted with crossover rates 0.7, 0.8 and 0.9 on problems with population size of 50 and 200 generations show that crossover rate 0.9 gives best output among the three all the time. In Table 3, comparisons with a variant of GA called homologous gene scheme replacement genetic algorithm are done. In [9], the HGR operator in HGRGA aims towards propagating the good and short schema present in the best gene of elite with above-average fitness from local gene to the full chromosome to improve its overall functional capacity. In Table 3, comparison between them is shown which clearly states that our algorithm is better than HGRGA. The results are carried out for dimension $= 10$, crossover rate $= 0.8$, mutation rate $= 0.05$, population size $= 200$, runs $= 20$ and generations $= 2000$. The parameters are same for all three variants of GA. The results for function 1, 2, 5 and 6 show that the best value of proposed GA is better than that of HGRGA and simple GA. For other functions like 9, 7, 10 and 11, our results are better but with very little difference. A very large difference is there in the best value of our proposed GA and HGRGA in function 10. For some functions, our proposed algorithm value is same as that of HGRGA. From Fig. 4, it can be concluded that our algorithm performs better than HGRGA and simple GA for maximum functions on which it is tested.

Results:

Table 1 Comparison of basic GA and modified GA on benchmark problems

S. no.	Function name	D	Obj_fun global minima	Obj_fun basic GA value	Obj_fun modified GA value
1.	Parabola function	30	0.00	1.648950	0.345243
2.	De Jong's function 4	30	0.00	4.841780	0.421360
3.	Griewank's function	30	0.00	9.111661	2.101324
4.	Ackley function	30	0.00	6.121386	5.045128
5.	Michalewicz function	10	−9.66015	−8.354758	−8.364350
6.	Salomon problem	30	0.00	5.224607	2.963811
7.	Axis parallel hyper-ellipsoid function	30	0.00	20.852633	6.105120
8.	Pathological function	30	0.00	6.803973	5.588649
9.	Sum of different power functions	30	0.00	0.002012	0.025808
10.	Levy montalvo 1 function	30	0.00	0.072074	0.017179
11.	Levy montalvo 2 function	30	0.00	0.295214	0.066449
12.	Beale function	2	0.00	0.077642	0.003353
13.	Colville function	4	0.00	23.419301	1.210247
14.	Branin's function	2	0.3979	0.467246	0.403407
15.	Kowalik function	4	0.000308	0.005142	0.001681
16.	Compression spring function	3	2.625422	3.112128	2.023952
17.	Goldstein-Price function	2	3	6.868339	3.079909
18.	Exponential function	30	−1	−0.96035	−0.657114
19.	Quartic function	30	0.00	10.817500	10.867970
20.	Minibench mark 4-problem	2	0.00	4.285349	0.636868
21.	Shifted parabola/sphere (CEC 2005 benchmark) function	10	−450	−297.393772	−438.863478
22.	Shifted CEC 2005 Rastrigin's function	10	−330	−315.298393	−327.023693
23.	Shifted CEC 2005 Griewank's function	10	−180	−176.398734	−178.908891
24.	Shifted CEC 2005 Ackley function	10	−140	−132.662772	−137.483341
25.	Inverted cosine wave function	10	−9	−6.525224	−6.077571
26.	6-hump camel back problem	2	−1.0316	−0.869201	−1.030467

(continued)

Table 1 (continued)

S. no.	Function name	D	Obj_fun global minima	Obj_fun basic GA value	Obj_fun modified GA value
27.	Easom's function	2	−1	−0.184868	−0.937793
28.	Dekker's and Aart's function	2	−24777	−20869.401799	−24617.556438
29.	McCormick problem	2	−1.9133	−1.903995	−1.912621
30.	Meyer and Roth function	3	0.4e−4	0.002361	0.002648
31.	Shubert problem	2	−186.7309	−175.032817	−186.451391
32.	Sinusoidal problem	10	−3.5	−2.287321	−3.342565
33.	Lennard_jones problem	5	−9.103852	−8.648523	−9.355349
34.	Hasaki problem	2	−2.3458	−2.342008	−2.345490
35.	Moved axis parallel hyper-ellipsoid function	30	0.00	114.933904	18.640403
36.	Parameter estimation for FM sound waves	6	0.00	23.840125	13.935329
37.	Constraint problem 2	7	680.630057	2737.441467	696.098918
38.	Schwefel's function	30	0.00	8.876296	5.356117
39.	Welded beam design optimization problem	4	1.724852	3.396050	2.613710
40.	Rastrigin's function	30	0.00	42.805440	4.738283

Fig. 2 Comparison of basic and modified GA objective function values for function from 1 to 20. Red (basic GA), green (modified GA) and blue (global minima)

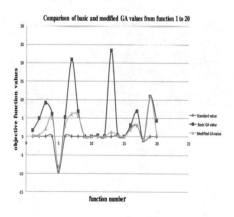

Table 2 Comparison of different population sizes on benchmark problems

S. no.	Function name	Obj_fun value with pop size = 30	Obj_fun value with pop size = 50	Obj_fun value with pop size = 100
1.	Parabola function	10.296685	0.345243	0.597970
2.	De Jong's function 4	1.280542	0.421360	0.932704
3.	Griewangk's function	2.532371	2.101324	3.385906
4.	Ackley function	14.899287	5.045128	14.487143
5.	Michalewicz function	−7.659069	−8.364350	−7.733831
6.	Salomon problem	12.358868	2.963811	10.108134
7.	Axis parallel hyper-ellipsoid function	5.213035	6.105120	8.541973
8.	Pathological function	10.073333	5.588649	9.757534
9.	Sum of different power functions	0.014574	0.025808	0.017836
10.	Shifted parabola/sphere (CEC 2005 benchmark) function	−434.404948	−438.863478	−426.906841
11.	Shifted CEC 2005 Rastrigin's function	−285.992935	−327.023693	−290.891071
12.	Beale function	0.002762	0.003353	0.003142
13.	Colville function	12.640655	1.210247	28.154001
14.	Shifted CEC 2005 Griewank's function	−157.653738	−178.908891	−169.737778
15.	Shifted CEC 2005 Ackley function	−126.690409	−137.483341	−129.022063
16.	Sinusoidal problem	−1.960438	−3.342565	−1.958045
17.	Lennard_jones problem	−5.802796	−9.355349	−6.294068
18.	Exponential function	−0.588147	−0.657114	−0.711652
19.	Quartic function	16.214041	10.867970	13.655242
20.	Minibench mark 4-problem	0.838546	0.636868	1.193034

Table 3 Comparison of proposed GA with HGRGA and simple GA

S. no.	Function name	Gmin	HGR GA obj_fun value		Simple GA obj_fun value		Proposed GA obj_fun value	
			Best	Mean	Best	Mean	Best	Mean
1.	Zakharov	0	0.0220	11.5021	0.4209	0.8854	0.0155	0.0282
2.	Schwefel's 1.2	0	0.3423	399.01	2.0139	3.1443	0.0079	0.0098
3.	Rastrigin	0	0	0	0.1549	0.8505	0	0.0006
4.	Schwefel's 2.6	0	1.27e-04	0.0601	0.6734	1.5469	2.18e-04	0.0036
5.	Michalewicz	-9.66	-9.6595	-9.6078	-8.8948	-8.6059	-9.6620	-9.6201
6.	Griewangk	0	0	0	0.2455	0.3459	0	0
7.	Ackley	0	8.8e-16	8.8e-16	3.26367	5.5867	5.1e-04	5.1e-04
8.	Rosenbrock	0	0	0.0433	10.309	12.1623	1.2e-04	0.0122
9.	Hybrid 1-Bent Cigar+Rastrigin+Schwefel 2.6	0	778.1345	1070	1671.406830	3267.5195	131.7177	159.2108
10.	Hybrid 2-Schwefel's 2.2+Rastrign+Griewangk	0	0	0	51.4749	52.4565	0.1288	1.2317
11.	Hybrid 3-Rosenbrock+Griewangk+Discuss	0	1.1309	1.4272	48.0688	50.8423	1.0112	1.0239

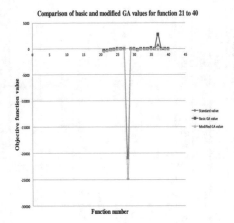

Fig. 3 Comparison of basic and modified GA objective function values for function from 21 to 40. Red (basic GA), green (modified GA) and blue (global minima)

Comparison of simple GA,proposed GA and other GA variant

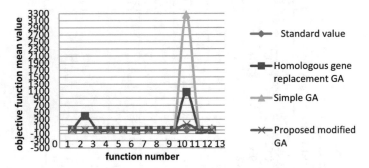

Fig. 4 Comparison of simple GA, proposed GA and other GA variant (HGRGA) for 11 benchmark functions

5 Conclusion

The algorithm works well for all the 40 problems of dimensionality from 2 to 30 on which it is tested. Comparison with basic GA shows that our algorithm is better than the basic GA for all the tested problems and gives values closed to global minima. The comparison with other GA variant shows that our algorithm is better for hybrid functions. Second, for all the 11 functions, our mean optimized objective function value is better than the HGRGA; it shows that the whole population is being optimized towards the global minima. The best objective value of our algorithm is better than HGRGA or closed to global minima. On the basis of the observations, the algorithm is good for unimodal and multimodal single and multi-objective functions.

Proposed work: Simulation on some real-world problem and comparison with other variants of GA is yet to be done. The algorithm can be extended to include the convergence matrix to direct our search to find the solution faster.

References

1. Vekaria K, Clack C (1998) Selective crossover in genetic algorithms: an empirical study. In: Parallel problem solving from nature—PPSN V. Springer, Berlin, Heidelberg
2. Nguyen HD et al (2007) Implementation of an effective hybrid GA for large-scale traveling salesman problems. IEEE Trans Syst Man Cybern Part B (Cybernetics) 37(1):92–99
3. Prasad JS, Jain V (2016) An optimized algorithm for solving travelling salesman problem using greedy cross over operator. In: 2016 3rd international conference on computing for sustainable global development (INDIA Com), IEEE
4. Affenzeller M (2002) New variants of genetic algorithms applied to problems of combinatorial optimization.na
5. Li J-P et al (2002) A species conserving genetic algorithm for multimodal function optimization. Evol Comput 10(3):207–234
6. Li YL, Feng JF, Hu JH (2016) Covariance and crossover matrix guided differential evolution for global numerical optimization. SpringerPlus 5(1):1176
7. Genetic and evolutionary algorithm toolbox for use with MATLAB documentation. http://www.geatbx.com/docu/fcnindex-01.html#P86_3059
8. Instructions for the random number generator libraries. http://www.agner.org/random/ran-instructions.pdf
9. Iqbal S, Hoque MT (2017) HGRGA: a scalable genetic algorithm using homologous gene schema replacement. Swarm Evol Comput 34:33–49

Analysis and Information Retrieval from Electroencephalogram for Brain–Computer Interface Using WEKA

Jayesh Deep Dubey, Deepak Arora and Pooja Khanna

Abstract In the recent years, there has been huge development in the area of brain–computer interface. Brain–computer interface technology provides a new method of communication for people with neuromuscular damages and problems due to which they are unable to use usual communication methods. A brain–computer interface system makes it possible for an individual to direct the outside world through his thoughts without relying on muscle activity. Current brain–computer interface systems use either the EEG activity in which the brain signals are recorded through electrodes by placing them on the scalp of the individual or invasive procedures in which the electrodes are placed inside the brain. In this paper, we have used EEG data recorded from 8 selected channels out of 64 available channels. The eight channels were selected by using principal components analysis and association rule mining using apriori algorithm. K-means clustering was then implemented on the data obtained from eight selected channels and the number of cluster instances was set to three because the number of events observed in the data while observing it in the EEGLAB was three. The study analyzed electroencephalogram signal using WEKA and presented a set of information that can be retrieved. This information is important for the brain–computer interface systems.

Keywords EEG · BCI · PhysioBank · EEGLAB · EDFBrowser

J. D. Dubey · D. Arora · P. Khanna (✉)
Amity Institute of Engineering and Technology, Amity University, Noida,
Uttar Pradesh, India
e-mail: inkhanna@gmail.com

J. D. Dubey
e-mail: jayesh.d.dubey@gmail.com

D. Arora
e-mail: deepakarorainbox@gmail.com

1   Introduction

Brain–computer interface is defined as a technical system that provides direct inter-action between the brain of a human being and external devices. BCI translates the user's objective which is reflected by brain signals into a desired output: controlling an external device or computer-based communication. The name "Brain–Computer Interface" was first used in the early 1970s by the works of Dr. J. Vidal, which was inspired by the optimism to create an alternate output method of communication for disabled individuals and enhancing human ability to control the external systems [1]. In brain–computer interface, there is no requirement for any kind of muscle activity for issuing the commands and completing the interaction. BCI was initially developed by research community for creating assistive devices for biomedical appli-cations [2]. These devices have been able to restore the movement activity for people with physical disabilities and replacing lost motor functionality.

BCI systems bypass the muscular output pathways and use digital signal pro-cessing and machine learning to translate signals from the brain into actions [3]. A brain–computer interface system work by recognizing patterns in the data that has been extracted from the brain and associating those patterns with the commands [4]. The most important and interesting application of BCI is medical rehabilitation. Since a BCI system does not include any motor activities, people with severe motor damages [such as individuals suffering from paralysis] can also operate a BCI-based communication system. The objective of BCI this situation is to record the brain sig-nals and convert them into a form such that computers and other devices can easily understand it.

2   Literature Review

Brain–computer interface is a system that is based on electroencephalography and is used to establish communication channel between brain of a human being and external environment. BCI systems use electrical activity in the brain to allow users to control external devices through their thoughts without relying on muscle activity [5]. These systems are very helpful in enhancing the quality of life of the disabled individuals who are suffering from motor disorders and neuromuscular damages due to which they are unable to use usual communication methods. BCI systems use EEG activity in which the brain signals are recorded through electrodes by either placing them on the scalp of the individuals according to 10-10 or 10-20 electrode placement system or by placing the electrodes inside the brain [6]. BCI systems work by recognizing patterns in the data that has been extracted from the brain and associating these patterns with the commands for the external devices.

In the recent years, there has been a great deal of research and development in the area of brain–computer interface. Brain–computer interface has been suc-cessfully applied in many fields such as military, entertainment, smart house, etc.

However, the most important and interesting application of BCI is medical rehabilitation. Researchers at the University of Minnesota have achieved a breakthrough that enables individuals to control a robotic arm through their minds. This invention can help paralyzed patients and those suffering from neurodegenerative disorder [7]. In one of the researches, the research team placed a brain implant in a patient suffering from ALS disease which prohibited the patient to move and speak. The brain implant enabled the patient to operate a speech computer through her mind and she was able to spell her first word [8]. Researchers at the University of Utah and Oregon State University College of Engineering have carried out a research that has created hope for patients suffering from spinal cord injuries. The research suggests that a wearable, smartphone-sized control box might enable at least some level of movements in these patients by delivering impulses to implant electrodes in their peripheral nervous system [9]. Researchers have developed advanced electrodes known as "glassy carbon" electrodes that are capable of transmitting clearer, robust signals than the currently used electrodes. These electrodes could allow restoration of movements in patients with damaged spinal cords [10].

3 Dataset Description

The data used in this research article has been obtained from PhysioNet database. The data is electroencephalographic motor movement/imagery data, and the dataset contains about 1500 electroencephalographic recordings of 1 and 2 min duration. These recordings have been procured from 109 subjects. The experiment conducted to obtain the data has been described below [11, 12].

Different motor/imagery tasks were performed by the subjects, while 64-channel electroencephalography was recorded using the BCI2000 system. 14 tasks were carried out by each subject: two baseline tasks of 1 min each (one in which subject kept his eyes opened and the other in which subject's eyes were closed), and three 2-min tasks of each of the following:

1. A cue is displayed on either the left or right side of the monitor. The subject opens and closes the corresponding hand until the cue vanishes. Then, the subject relaxes.
2. A cue is displayed on either the left or right side of the monitor. The subject imagines opening and closing the corresponding hand until the cue vanishes. Then, the subject relaxes.
3. A cue is displayed on either the bottom or top of the monitor. The subject opens and closes either both feet (if the target is on the bottom) or both fists (if the target is on the top) until the cue vanishes. Then, the subject relaxes.
4. A cue is displayed on either the bottom or top of the monitor. The subject imagines opening and closing either both feet (if the target is on the bottom) or both fists (if the target is on the top) until the cue vanishes. Then, the subject relaxes.

14 tasks were conducted for each subject but in our research work, we have considered 4 tasks for each subject which is described as follows.

1. Task 1 (opening and closing left or right fist),
2. Task 2 (imagine opening and closing left or right fist),
3. Task 3 (opening and closing both fists or both feet), and
4. Task 4 (imagine opening and closing both fists or both feet).

In our study, we have used data from first 14 subjects out of the 109 subjects available in the dataset.

4   Data Extraction and Preprocessing for Experimental Setup

We have used **PhysioBank ATM** online toolbox to export the data in EDF (European data format) format. PhysioBank ATM is an online toolbox for exploring PhysioBank through web browser. The toolbox contains software that can display annotated waveforms, RR interval time series, and histograms, and convert WFDB signal files to text, CSV, EDF or .mat files, and more.

After obtaining the data in EDF format, we imported it in **EEGLAB** toolbox. EEGLAB is a Matlab toolbox that is used for processing continuous and event-related electroencephalographic, magnetoencephalographic, and electrophysiological data. EEGLAB operates under Windows, Linux, UNIX, and Mac OS. It provides graphical user interface that enables users to process EEG and other brain data. After importing the data in EEGLAB, the data was bandpass filtered between 3 and 30 Hz to remove unwanted frequency components from the data. The data was filtered using EEGLAB, and then **EDFBrowser** was used to convert the data from EDF format to text format.

5   Analysis of EEG Data

5.1   Channel Selection

The EEG data obtained from PhysioBank ATM was recorded from 64 channels according to 10-10 electrode system. In order to obtain the channels that are capturing majority of the motor imaginary data, we have used principal component analysis for each of the runs of all the 14 subjects. This resulted in a list of channels for each of the experimental tasks.

After obtaining the list of channels, we have applied association rule mining to obtain correlated channels. Apriori algorithm was used for this task. Weka was used as a tool to employ principal components analysis and apriori algorithm. The channels that we have obtained on the basis of this analysis are **FC5, FC3, FC1, FCz, FC2,**

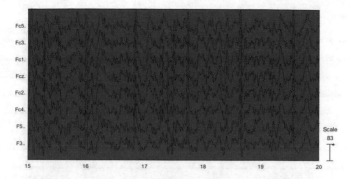

Fig. 1 EEGLAB plot for baseline run with only one type of event marked in red color

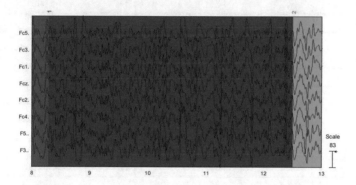

Fig. 2 EEGLAB plot for motor movement task with three types of events marked in red, pink, and green colors

FC4, F3, and F5. We will use the data recorded from the above mentioned channels for further analysis.

5.2 Data Analysis in the EEGLAB

When data for the baseline run [in which the subject is in relaxed state and not performing any movement] was analyzed in the EEGLAB, then only one type of event was observed which corresponds to the relaxed state (Fig. 1). Then, we performed analysis of data for Task 1, Task 2, Task 3, and Task 4 as mentioned above in the dataset description. For this data, three types of events were observed, one of which is the same event as observed in the baseline run which corresponds to relaxed state and two more events were observed which corresponds to the movement or imagination of movements of fist or feet (Fig. 2).

Table 1 Clusters obtained for each of the 4 tasks for all the 14 subjects

Subject	Task	Cluster 1 (%)	Cluster 2 (%)	Cluster 3 (%)
Subject 1	Task 1	22	48	30
Subject 1	Task 2	30	22	47
Subject 1	Task 3	20	30	50
Subject 1	Task 4	21	49	30
Subject 2	Task 1	26	48	26
Subject 2	Task 2	25	47	28
Subject 2	Task 3	52	23	25
Subject 2	Task 4	26	26	48
Subject 3	Task 1	35	52	13
Subject 3	Task 2	21	29	50
Subject 3	Task 3	8	57	35
Subject 3	Task 4	37	57	6
Subject 4	Task 1	37	2	61
Subject 4	Task 2	36	3	61
Subject 4	Task 3	3	75	22
Subject 4	Task 4	56	42	2
Subject 5	Task 1	22	50	28
Subject 5	Task 2	54	29	17
Subject 5	Task 3	48	27	26
Subject 5	Task 4	26	24	50
Subject 6	Task 1	32	14	54
Subject 6	Task 2	28	53	19
Subject 6	Task 3	32	55	13
Subject 6	Task 4	51	28	21
Subject 7	Task 1	54	9	37
Subject 7	Task 2	27	52	21
Subject 7	Task 3	10	32	58
Subject 7	Task 4	18	53	29
Subject 8	Task 1	9	37	54
Subject 8	Task 2	30	54	16
Subject 8	Task 3	5	39	56
Subject 8	Task 4	19	28	53
Subject 9	Task 1	11	35	54
Subject 9	Task 2	31	53	17
Subject 9	Task 3	37	9	54
Subject 9	Task 4	17	30	53

(continued)

Table 1 (continued)

Subject	Task	Cluster 1 (%)	Cluster 2 (%)	Cluster 3 (%)
Subject 10	Task 1	48	29	23
Subject 10	Task 2	52	29	19
Subject 10	Task 3	27	51	22
Subject 10	Task 4	51	22	27
Subject 11	Task 1	28	51	22
Subject 11	Task 2	49	23	28
Subject 11	Task 3	23	29	48
Subject 11	Task 4	28	24	48
Subject 12	Task 1	40	55	4
Subject 12	Task 2	38	4	58
Subject 12	Task 3	6	61	34
Subject 12	Task 4	31	61	8
Subject 13	Task 1	20	28	52
Subject 13	Task 2	28	22	50
Subject 13	Task 3	28	53	19
Subject 13	Task 4	50	29	21
Subject 14	Task 1	47	23	30
Subject 14	Task 2	30	48	22
Subject 14	Task 3	26	50	24
Subject 14	Task 4	28	23	49

5.3   Clustering of Data

WEKA was used as a tool to run **K-means clustering** algorithm on the data recorded from the selected channels as mentioned above. The k-means algorithm was run for each of the four tasks for all the subjects. Since we have considered 14 subjects and 4 experimental tasks for each subject, the algorithm was run for 56 times. The data was divided into three clusters because during analysis of the data in EEGLAB, three types of events were observed. The clusters that we have obtained can be seen in the table below. It can be observed in Table 1 that for most of the subjects and tasks two clusters are uniformly distributed, while the third cluster is almost twice the size of other two clusters. For example, for task 1 of subject 2, the clusters obtained are 26, 48, and 26%; for task 1 of subject 10, the clusters are 48, 29, and 23%; and for task 4 of subject 14, the clusters are 28, 23, and 49%. Similar pattern can be found for many tasks for the subjects in the table.

Figure 3 shows the graphical illustration of the three types of clusters for task 1, task 2, task 3, and task 4 for all the 14 subjects as described below.

1. 3a is the cluster graph for task 1 for all the 14 subjects.

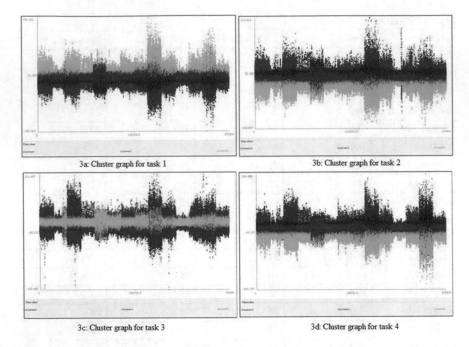

| 3a: Cluster graph for task 1 | 3b: Cluster graph for task 2 |

| 3c: Cluster graph for task 3 | 3d: Cluster graph for task 4 |

Fig. 3 Cluster graphs for each of the 4 tasks for all the 14 subjects, **a** is the cluster graph for task 1 for all the 14 subjects, **b** is the cluster graph for task 2 for all the 14 subjects, **c** is the cluster graph for task 3 for all the 14 subjects, **d** is the cluster graph for task 4 for all the 14 subjects

2. 3b is the cluster graph for task 2 for all the 14 subjects.
3. 3c is the cluster graph for task 3 for all the 14 subjects.
4. 3d is the cluster graph for task 4 for all the 14 subjects.

6 Conclusions

Analysis of EEG data is the most important part of BCI and many different methods have been used by researchers around the world for analysis of EEG data. In our experiment, we have used Weka as a tool for analysis of executed and imagined EEG data. The data from eight selected channels [out of 64 channels] has been considered in the experiment. The channel selection was performed using principal component analysis and apriori algorithm. K-means clustering algorithm has been used for clustering the data into three clusters. During the analysis of data in EEGLAB, three types of events were observed, and therefore we have selected the number of clusters as three while performing k-means clustering. EEG signal can be clustered on the basis of type of task brain is performing. Though EEG signals are captured using/from many electrodes point, few plays significant role in task identification.

However, further analysis of data is required for exact classification of data into left- and right-hand motor movements which will be performed in future experiments.

References

1. Vidal JJ (1977) Real-time detection of brain events in EEG. Proc IEEE 65(5):633–641
2. Rao RPN, Scherer R (July 2010) Brain-computer interfacing [In the Spotlight]. IEEE Signal Process Mag 27(4):152–150. https://doi.org/10.1109/msp.2010.936774, http://ieeexplore.ieee.org/stamp/stamp.jsp?tp=&arnumber=5484181&isnumber=5484155
3. Wolpaw JR, Birbaumer N, Heetderks WJ, McFarland DJ, Peckhalm PH, Schalk G, Donchin E, Quatrano LA, Robinson CJ, Vaughan TM Brain computer interface technology: a review of the first international meeting
4. Rohani DA, Henning WS, Thomsen CE, Kjaer TW, Puthusserypady S, Helge BD (2013) Sorensen BCI using imaginary movements: the simulator. Comput Methods Programs Biomed 111:300–307
5. Abdulkader SN, Atia A, Mostafa M-SM Brain computer interfacing: applications and challenges. HCL-LAB, Department of computer Science, Faculty of computers and information, Helwan University, Cairo, Egypt
6. Niedermeyer E (1999) The normal EEG of the waking adult, Niedermeyer E, Lopes da Silva FH (eds) Baltimore, MD, Williams and Wilkins
7. Meng J, Zhang S, Bekyo A, Olsoe J, Baxter B, He B (2016) Noninvasive electroencephalogram based control of a robotic arm for reach and grasp tasks. Sci Reports 6:38565. https://doi.org/10.1038/srep38565
8. Vansteensel MJ, Pels EGM, Bleichner MG, Branco MP, Denison T, Freudenburg ZV, Gosselaar P, Leinders S, Ottens TH, Van Den Boom MA, Van Rijen PC, Aarnoutse EJ, Ramsey NF (2016) Fully implanted brain–computer interface in a locked-in patient with ALS. New Engl J Med. https://doi.org/10.1056/nejmoa1608085
9. Frankel MA, Mathews VJ, Clark GA, Normann RA, Meek SG (2016) Control of dynamic limb motion using fatigue-resistant asynchronous intrafascicular multi-electrode stimulation. Front Neurosci. https://doi.org/10.3389/fnins.2016.00414
10. Vomero M, Castagnola, FC, Maggiolini E, Goshi N, Zucchini E, Carli S, Fadiga L, Kassegne S, Ricci D (2017) Highly stable glassy carbon interfaces for long-term neural stimulation and low-noise recording of brain activity. Sci Reports 7:40332. https://doi.org/10.1038/srep40332
11. Goldberger AL, Amaral LAN, Glass L, Hausdorff JM, Ivanov PCh, Mark RG, Mietus JE, Moody GB, Peng C-K, Stanley HE PhysioBank, PhysioToolkit, and PhysioNet: components of a new research resource for complex physiologic signals
12. Schalk G, McFarland DJ, Hinterberger T, Birbaumer N, Wolpaw JR (2004) BCI2000: a general-purpose brain-computer interface (BCI) system. IEEE Trans Biomed Eng 51(6):1034–1043

Using Artificial Neural Network for VM Consolidation Approach to Enhance Energy Efficiency in Green Cloud

Anjum Mohd Aslam and Mala Kalra

Abstract Cloud computing is a popular on-demand computing model that provides utility-based IT services to the users worldwide. However, the data centers which host cloud applications consume an enormous amount of energy contributing to high costs and carbon footprints to the environment. Thus, green cloud computing has emerged as an effective solution to improve the performance of cloud by making the IT services energy and cost efficient. Dynamic VM consolidation is one of the main techniques in green computing model to reduce energy consumption in data centers by utilizing live migration and dynamic consolidation. It minimizes the consumption of energy by monitoring the utilization of resources and by shifting the idle servers to low power mode. This paper presents a VM selection approach based on artificial neural network (ANN). It uses backpropagation learning algorithm to train the feedforward neural network to select a VM from an overloaded host. Thus, it optimizes the problem of VM selection by learning training dataset and enhances the performance of selection strategy. To simulate our proposed algorithm, we have used MATLAB and the simulation result depicts that our proposed method minimizes the energy consumption by 30%, SLA violation by 3.51%, the number of migrations by 10%, and execution time by 29.7%.

Keywords Artificial neural network (ANN) · Dynamic virtual machine consolidation · Energy efficiency · Green computing

1 Introduction

Cloud computing is the most popular computing model as well as a recent trend in information technology that has moved computing and data resources away from small companies or systems to large-scale virtualized data centers. Cloud computing

A. M. Aslam (✉) · M. Kalra
Department of Computer Science & Engineering, NITTTR, Chandigarh, India
e-mail: anjumquest@gmail.com

M. Kalra
e-mail: malakalra2004@gmail.com

© Springer Nature Singapore Pte Ltd. 2019
M. L. Kolhe et al. (eds.), *Advances in Data and Information Sciences*, Lecture Notes in Networks and Systems 39, https://doi.org/10.1007/978-981-13-0277-0_12

utilizes pay-as-you-go model to deliver the computing resources to its users on-demand [1]. Thus, instead of spending a huge amount on purchasing IT infrastructure and managing the upgradation and maintenance of software and hardware, organizations can utilize the services of cloud according to their need.

The cloud data centers have grown rapidly in the past several years to serve the increasing need for computing. The leading IT organizations, such as Google, Facebook, Microsoft, and IBM, are rapidly increasing their data centers around the whole world to provision the cloud computing services and to manage the increasing cloud infrastructure demand. Moreover, this emerging demand has significantly raised the energy consumption of the cloud data centers, which is becoming a vital issue [2]. In addition to that, high consumption of power by the servers and cooling systems leads to CO_2 emissions which are contributing to the greenhouse effect and making the environment unfriendly. It contributed to around 2% of global CO_2 emission in 2007 [3]. The reason for the high consumption of energy is the inefficiency in the usage of the computing resources by the servers, which increases the expenses on the over-provisioning of resources, and further maintaining and managing these over-provisioned resources increase the total cost of ownership [4].

According to McKinsey report, the power consumed by the cloud data centers has doubled in the year from 2000 to 2006 and today, the energy consumption by an average data center is as much as consumed by 25,000 households, thus contributing to high energy consumption [5]. Based on the survey report by the American Society of Heating, Refrigerating and Air-Conditioning Engineers (ASHRAE) [6] in 2014, it has been evaluated that the IT equipment cost contributes only to 25% and the infrastructure and energy cost contributes to 75% of the total expenditure of the data center [7]. Therefore, managing the resources in an energy- and cost-efficient manner is the biggest challenging issues in the data centers which will grow unless energy-aware resource management techniques are developed and utilized. Thus, there arises a need to optimize the energy efficiency by adopting various energy-efficient solutions while maintaining the high service performance levels.

To overcome this problem, dynamic server consolidation is one such effective technique utilized to reduce the power consumption of data centers while ensuring the quality of service at the desired level. It is based on the observation that the server in the data center does not use the available resources at their peak rate most of the time. Thus, according to their current requirement of resources, the virtual machines can be consolidated to the minimum number of physical machines using live migration which will then lead to a minimum number of the active host at a given time as well as it will improve the utilization of resources. It will further minimize the energy consumption by shutting down or shifting the idle servers to power-saving modes, i.e., hibernation, sleep. However, large consolidation of VMs can raise various problems related to performance degradation when there is a sudden fluctuation in VMs footprint which can cause congestion and when these requirements are not fulfilled; it can lead to time-outs, increased response time, and failures. Thus, to ensure the quality of service while reducing the consumption of energy, the decision-making process for VM selection, overload, and underload detection must be designed to obtain an optimal result to deal with the variability in the workload of different applications. In our research work, we have applied artificial neural network for making an intelligent decision to select an appropriate VM from an overloaded host for migration. ANN works on the principle of learning by example and adjusting to new knowledge. The simulation results depict that our proposed technique works very efficiently and achieve 30% higher energy efficiency than fuzzy VM selection algorithm. Thus, it successfully reduces the consumption of energy and achieves increased performance with less SLA violation.

The remainder of this paper is organized as follows. Section 2 discusses the background and related work. Section 3 describes the artificial neural network; Sect. 4 will describe the proposed method and algorithms. In Sect. 5, experimental setup, results, and comparisons have been presented followed by a conclusion and future work.

2 Background Details and Related Work

In recent years, a significant number of researchers have been done using various meta-heuristic and heuristic techniques which aimed to minimize the power consumption in the data centers. These techniques are applied to VM consolidation problem which deals with the four subproblems as studied in the previous works—overload detection, underload detection, VM selection, and placement. Moreover, researchers have utilized various methods to predict the workload requirements of a server in future to ensure that the request is serviced properly with a minimum amount of power consumption. Ferreto et al. [8] proposed a dynamic consolidation with migration control method to minimize the number of migrations. They presented a linear programming (LP) formulation and heuristics approach to control the migrations by adding a constraint of not migrating a VM with the steady workload, and VM with varying capacity can be migrated to lower the number of physical machines and the migrations. This will directly reduce the electric power consumption.

Beloglazov et al. [9] discussed the concept of managing cloud environment in an energy-efficient manner. They defined green cloud architectural framework and presented dynamic resource provisioning and allocation algorithms based on modified best-fit decreasing, minimization of migration, and power management with the aim to achieve the quality of service, high performance, and cost savings. In [10], the same authors have presented various VM selection policies for VM migration allowing less VM to be migrated, thus reducing the migration overhead. They described various energy-efficient policies for the virtualized data center to reduce energy consumption, and proposed optimal online and offline deterministic algorithms for static and dynamic virtual machine consolidation problem. They discussed adaptive heuristics for VM consolidation using median absolute deviation, a robust local regression-based approach for host overload detection, and maximum correlation policy for VM selection. The experimental results depicted performance gain and energy reduction than other traditional dynamic VM consolidation algorithms. Fahimeh et al. [11] presented a meta-heuristic algorithm utilizing ant colony system for dynamic consolidation of the virtual machine to find an optimal solution for VM migration. This approach minimizes the energy consumption, QoS degradation, and number of migrations.

Monil et al. [12] have proposed three VM selection algorithms with migration control strategy. The migration control strategy applied with these algorithms improves performance by saving energy and reducing the number of VM migrations. It further reduces the network traffic and gives better performance than other proposed heuristic approaches. Monil and Rahman [13] have incorporated fuzzy logic in a VM selection approach. The author has proposed fuzzy VM selection with migration control algorithm to make an intelligent decision based on dynamic workload, uncertainties. The proposed algorithms incorporating fuzzy logic offer various benefits which help to handle uncertainty in the real world with intelligent decision-making to achieve an optimal result. With respect to existing methods, the proposed approach saves energy and maintains QoS with less SLA violation.

Farhnakian et al. [14] proposed a linear regression technique (LiRCUP) for predicting the CPU utilization. The linear regression function forecasts the future CPU utilization for overload detection, as well as it predicts an underloaded host. The proposed technique is implemented in CloudSim and results depict the significant reduction in energy cost and achieve a high level of QoS. In [15], reinforcement learning based dynamic consolidation (RL-DC) method is proposed by the same author to reduce the active set of physical servers according to the workload requirement. This method uses an agent to obtain an optimal solution. According to current resource requirements, it makes an intelligent decision to determine which host is to be shifted to sleep mode or an active mode. Comparing with other existing dynamic consolidation approaches, this proposed solution minimizes energy and SLA violation rate efficiently.

Sheng et al. [16] designed a Bayes model for cloud load prediction. This method predicts a mean load over a long-term and consecutive future time intervals. It achieves greater accuracy in load prediction comparing with auto-regression, noise filter methods. Prevost et al. [17] proposed an architectural framework which combines load demand prediction and stochastic state transition models. They utilize neural networks and auto-regression linear prediction algorithm to forecast future network load. The simulation result shows that the linear prediction provides the most accurate results.

3 Artificial Neural Network (ANN)

Neural network is an information processing system which imitates the human brain neural. It is designed in such a way that behaves likes a human brain to execute any task by learning examples without the task-specific programming [18]. This adaptative nature of the neural network is considered as one of the salient features to solve any complex problems or the problem which is difficult to solve by conventional computer or human beings. It can be used for solving a wide variety of tasks such as speech recognition, pattern recognition, social network filtering, medical diagnosis, prediction of the workload in data centers, etc.

ANN is comprised of a network of artificial neurons (nodes) which are highly connected to each other and processes the information to give the desired output. It consists of three neurons as shown in Fig. 1 with an input layer, hidden layer, and output layer units. It links the input to the output layer with weight connections. To train the ANN, the actual workload is provided to the input unit and then the outputs are calculated at each layer and thus compared with the desired output. Then, the errors are calculated and the weights are adjusted. This process continues and is repeated with large datasets until the network is fully trained with the accepted error rate. Once the ANN is fully trained, it is then tested with different datasets. At the end, when ANN is learned, the training is stopped.

Neural networks work on the learning principle, and thus initially these networks are trained so that they can perform the computation on the input to give the desired output. By adjusting the weights, i.e., value of connection between the processing elements, these networks are trained to process any task. After training, ANN can be used to predict any new situation and to model various nonlinear applications [19]. Then, at the end of the learning phase, the desired output will be generated at the output layer. Thus, it achieves better results with an appropriate selection of the input variables and training dataset. Figure 2 describes the working model, where the ANN is adjusted based on the comparison of target and output values until the results are obtained.

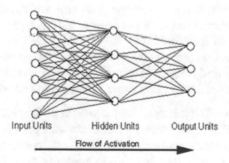

Fig. 1 Schematic view of a neural network [18]

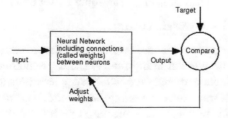

Fig. 2 Working model of neural network [20]

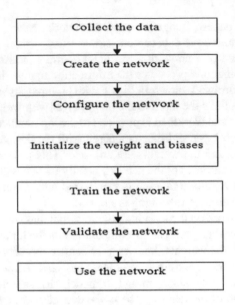

Fig. 3 Design steps of artificial neural network

ANN is trained by using different learning algorithms such as feedforward and backpropagation [21]:

1. **Feedforward ANN**: It is a non-recurrent network which moves the information in only one direction, from the input layer to hidden layer and then through the output layer. From the input layer, the data is passed to the processing units where it performs the computation on the basis of the weighted sum of the input values. The calculated values will be fed to the next layers where it processes and determines the output. It contains linear and nonlinear transformation functions.

2. **Backpropagation ANN**: It modifies the weight of the neural network to find a local minimum of the error function. It optimizes the performance of the network by going backward and calculating the weights of the connection, again and again, to achieve a minimum error value, thus to minimize the loss function. It utilizes a fitting tool to select data, create, and train the network and then assess the performance of the network using mean square error and regression analysis.

Design steps of artificial neural network:

For designing neural problems, it has seven primary design steps as shown in Fig. 3.

Neural networks are good at fitting function, i.e., predicting the value from the training dataset available and apply the intelligence as a human brain to solve any complex problem. It learns from the training dataset and from this learned knowledge it tries to solve various problems with increased performance and accuracy, i.e., it fits to match with the target achievable value.

4   Proposed Model

VM consolidation problem is mainly categorized into four subproblems, i.e., host underload detection, overload detection, VM selection, and VM placement. In our research work, we have designed a VM selection algorithm based on artificial neural networks. Algorithm 1 given below describes a basic concept of VM consolidation approach. Initially, the hosts are created and VMs are allocated to the available hosts and tasks are allocated to the VM. At every specified time interval, check for underload detection and after detecting the less utilized host, put that host into sleep mode and migrate the VM from that host to other active machines. Identify overloaded host, select the VM to be migrated, and place the VMs to the available hosts. At the end of the simulation experiment, calculate the energy consumption, QoS, and a number of migrations taken place.

Algorithm 1: VM consolidation Algorithm

---

1. Input the number of hosts
2. Create VMs and assign it to hosts.
3. Create tasks and assign it to hosts.
4. Identify underloaded host, overloaded host at every scheduled time interval
5. After detecting overloaded host, select a VM which is required to be migrated
6. Determine the host where the VM is to be placed
7. Calculate the Energy consumption and QoS value.
8. END

4.1   VM Selection Using Artificial Neural Network

VM selection using ANN is proposed which helps in intelligent decision-making to select a VM from an overloaded host. As ANN is built on the principle working of the human brain, it solves the complex problems with new sets of classifications and the desired output value. The main feature of using artificial neural network is its simplicity and efficiency to solve a problem based on learning and prediction ideas. It also works efficiently for providing a solution to incomplete information by applying its learning knowledge. Training is done using feedforward and backpropagation learning algorithm to get the desired VM to be selected from an overloaded host. Initially, using feedforward algorithm, the information travels from one end to other and then backpropagation learning algorithm trains the ANN by propagating backward to the hidden layer and then to the input layer. It calculates the weight again and again until the target output is not achieved with minimum error rate. Then, again it feedforwards to the output layer to give the desired result. It helps in the optimal utilization of resources, and hence increases the performance of the system.

To develop the VM selection method based on ANN, we have selected three input variables which are fed into ANN.

4.1.1 Standard Deviation

It describes the steady-state resource consumption of a VM [13]. We can neglect the high resource consuming VM from migration or restrict the steady resource consuming VM. If the workload demand on a VM varies highly, it can become overloaded. In dynamic VM consolidation, the VMs can be resized in every iteration and SLA violation may occur when a VM requests for high CPU and host may not have that amount of CPU available at that point of time. Thus, if a VM is consuming the resources at a steady rate over some iteration, then it will be considered as the least possible VM to make the host overloaded. If the standard deviation is high, it means the CPU request changes suddenly and if it is low it is a steady resource consuming VM. The standard deviation of CPU usage of particular VM "A" can be calculated by using the following equation:

$$Stdev = \sqrt{\frac{1}{n} \sum_{i=1}^{n} \left(CPU_i - CPU_{average}\right)^2} \tag{1}$$

4.1.2 RAM

RAM is the random access memory, which is the memory utilization of VM. The migration time depends on the memory utilized by the VM. $RAM(a)$ is the memory utilization of VM "a" and the $RAM(b)$ is the memory utilization of VM "b" NET_h is the bandwidth available for migrating the VMs and V_h is the set of VMs of host h.

$$a \in V_h | \forall b \in V_h, \frac{RAM(a)}{NET_h} \leq \frac{RAM(b)}{NET_h} \tag{2}$$

Thus, it selects the VM which will take minimum time among all the VMs placed on the host.

4.1.3 Correlation

The higher the correlation among the resource utilization of the applications running on a server, the higher is the probability of the server being overloaded [13]. It provides the information about VMs that can cause the host to become overloaded. It will be good to migrate a VM who is not having a higher correlation with other VMs.

Algorithm 2: VM selection Algorithm using Artificial Neural Networks

---

Input: Overloaded host h
Output: Virtual machine to b migrated, VM_m

1. Input overloaded host h;
2. VM_h = GetMigratableVM(h);
3. Utilm(VM) = UtilizationMatrix(VM);
4. Metric(n)= Corr-Coeff(Utilm(VM));
5. For each VM V_i of Metric(n)
6. CPU_{hist} = GetCPUHistory(V_i);
7. STDEV(V_i) = StandardDeviation(CPU_{hist});
8. RAM(V_i) = Ram(V_i);
9. CORR(V_i) = Corr-Coeff(V_i);
10. $Output_{neural}$ = trainingdata(STDEV(V_i), RAM(V_i), CORR(V_i));
11. If the mean square error of $Output_{neural}$ is minimum then VM_m = V_i;
12. End;
13. Return VM_m;

The Algorithm 2 describes the working of VM selection using artificial neural network. The overloaded host is given as an input to the algorithm which is identified in the previous phase of overload detection [13]. At step-1, after having the host h, getmigratablevm(h) at step-2 fetches the VMs which are currently placed on the host "h". At step-3, Utilm(VM) calculates the utilization matrix of the VMs. At step-4, correlation coefficient matrix is calculated for each VM. At step-6, for each VM V_i,, CPU usage history is fetched. At step-7, standard deviation is calculated to determine the steadiness in the CPU usage of VM. Current utilization of RAM will be calculated at step-8. At step-9, correlation of VMs will be fetched. At step-10, this calculated training data will be passed to the neural network and it will follow feedforward and backpropagation learning method to generate an output with VM which has least mean square error. The VM with least mean square error will be selected for migration. Finally, step-13 returns the VM to be migrated.

5 Experimental Setup and Results

To evaluate the proposed algorithm, we have implemented it in MATLAB to show the relevance and correctness of results. MATLAB offers various functions and is an optimized tool for solving various engineering and scientific problems. The simulation parameters are shown in Table 1. We have considered 200 hosts and 20 VMs; the experiment is conducted for 20 iterations to find the accurate data.

Training is done through feedforward and backpropagation learning algorithm which is implemented in MATLAB as shown in Fig. 4, where the snapshot of code and the result of the neural network training tool are shown. Neural network maps between the set of inputs and target values. The neural network fitting tool helps to select data, create, and train the network, and then it assesses the performance using mean square error and regression analysis. This two-layer feedforward network has sigmoid hidden neurons and linear output neurons. The neurons are highly connected to each

Table 1 Simulation parameters

Tool	MATLAB R2012a
No. of host	200
No. of VMs	20
Processor	Xenon
Iterations	20
Host capacity	16 GB
Vm size	2500 MB

other carrying associated weights which contain knowledge. Initially, the data center takes input which we have provided, i.e., RAM, correlation, and standard deviation as described in Algorithm 2, and then this data is passed over the network. The network learning from the given dataset and data flows between the neurons from input to hidden to the output layers. Then, the ANN will be trained by backpropagation method to minimize the error rate and to achieve the target value. Thus, it is considered as a supervised learning method. In our experiment, we have taken the epoch value = 50. Epoch describes the number of times the algorithm sees the entire training dataset. So, each time the algorithm sees the sample in the dataset to compute the output, an epoch gets completed. It has taken six iterations as shown in Fig. 4, which means the number of times the data passes through the algorithm by a forward pass and by backward pass and achieves the desired output.

Figure 5 shows the performance graph which describes that as the network learns, mean square error rate decreases rapidly. Training shows decreasing error rate on decreasing data. Here, the best validation performance is at epoch 1, i.e., the iteration at which it has reached a minimum.

The graphs shown in Fig. 6 give the description of datasets which are used for the training purpose. The four graphs represent training, validation, testing, and the output of training. There are two lines which are present in the graph. First is the straight line and second is the dotted line which represents the accuracy of training. The straight line denotes the best-fit linear regression line between outputs and targets. The dashed line represents a target as the perfect result − outputs = targets. The R-value is an evidence of the relationship between the outputs and targets. If the value of R is equal to 1, it denotes that there is an exact linear relationship between outputs and targets. But there is no linear relationship between outputs and targets if the value of R is close to zero.

Comparison of fuzzy VM selection algorithm [13] with the proposed VM selection algorithm using ANN is given below. Fuzzy is stateless in nature and manually we need to define the rules set for it to function. In fuzzy, outside of its defined linguistic variables and inference rules, it does not retain state. It cannot work if unexpected workload, volume, or density of data is increased. It can only implement the defined rules and based on that it gives the output but in ANN, the system learns as it goes along by adjusting its own values. ANN incorporates the human thinking process to solve any complex problem.

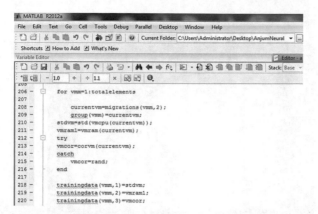

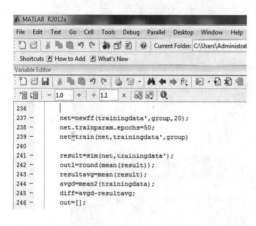

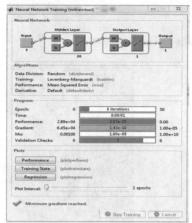

Fig. 4 Neural network training

5.1 Performance Metrics

For the comparison, we have taken four metrics. (1) Energy consumption, (2) SLA violations, (3) Number of VM migrations, and (4) Execution time.

(1) **Energy consumption (MilliJoules):** The essential objective to design a VM consolidation algorithm is to reduce the energy consumption. This metric represents the total energy consumption of the physical resources in the data centers. By comparing the existing and proposed algorithm as shown in Fig. 7a, the energy consumption is reduced significantly. By using the proposed technique, on an average 30%, energy consumption is reduced.

Energy is calculated as

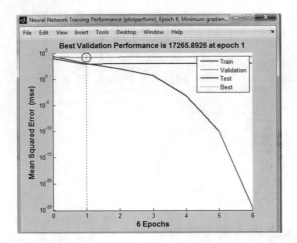

Fig. 5 Performance graph

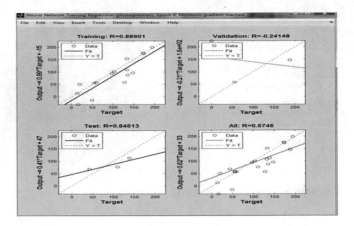

Fig. 6 Regression graph

$$E = \sum_{i=1}^{n} \left(\frac{\sum_{v=1}^{k} VM_{conf} + \sum_{j=1}^{p} Host_{ce}}{Iterations} \right)$$

where

n	total number of allocations
k	total allocated VMs.
p	total number of hosts.
VM_{conf}	configuration energy.
$Host_{ce}$	Energy consumed to contain VM

(2) **SLA Violation (SLAV)**: Service-level agreement is a mutual agreement between a cloud user and service provider. SLA violation causes degradation of QoS and thus it should be minimized. The main contribution to the SLA violation is the overloaded host and a large number of migrations when the VM becomes incapable of serving the needs of the user. SLA violations are calculated as the total MIPS requested by all VMs U_r and the allocated MIPS U_a to the total MIPS requested by the VMs over the lifetime. From Fig. 7b, it has been seen that SLA violation is reduced by 3.51% using this proposed technique. SLAV is calculated as

$$SLAV = \sum_{i=1}^{n} \left(\frac{U_r - U_a}{\sum_{i=1}^{n} U_r} \right), \quad \text{Where n} = \text{number of VMs.}$$

(3) **Number of VM migrations**: VM migration is a costly operation which includes link bandwidth, CPU processing, and downtime. Thus, less number of migrations is desirable with less traffic on the network and less degradation of quality of service. The proposed technique reduces number of migrations by 10% as shown in Fig. 7c.

(4) **Average execution time**: The execution time is the running time of the framework to obtain the output. By using the proposed technique, on an average 29.7% of execution time is reduced.

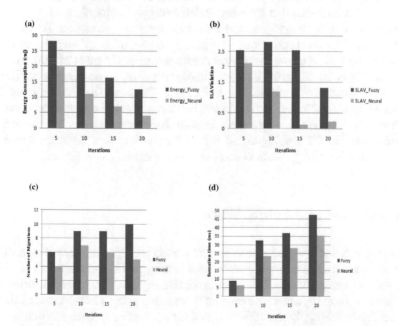

Fig. 7 Comparison based on energy consumption, SLA violation, number of VM migrations, and execution time

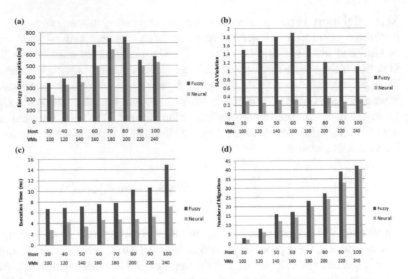

Fig. 8 Comparison based on energy consumption, SLA violation, number of VM migrations, and execution time on different numbers of hosts and VMs

We have also compared the performance of our proposed algorithm with the existing algorithm [13] by varying the number of hosts and the number of VMs. Thus, it is analyzed that by using ANN, energy consumption, SLA violation, number of VM migrations, and execution time are decreased. As shown in Fig. 8a, ANN can better reduce the energy consumption by varying the size of hosts and VM. It shows that the minimum energy achieved is 240 mJ when the host size is 30 with 100 VMs and maximum energy consumption is 700 mJ when the host size is 80 with 200 VMs. Figure 8b shows that the minimum SLA violation is 0.12% and the maximum is 0.37%. Figure 8c shows the execution time taken by ANN where the minimum time was taken as 2.8 ms and the maximum is 5.2 ms. Figure 8d shows the number of migrations where the reduction is clearly visible as compared to the existing technique.

6 Conclusion and Future Work

In this research work, we have proposed VM selection algorithm based on artificial neural network. It takes an intelligent decision to select a VM from an overloaded host based on the learning it has achieved from the training datasets. After performing the simulation and making a comparison with the existing fuzzy VM selection method, it has been found that applying artificial neural network achieves better results in terms of energy saving, less SLA violation, and increased system performance. Thus, the objective to achieve energy–SLA balance has been achieved efficiently. In future,

ANN can be applied to improve the underload detection and VM placement algorithm to achieve more energy savings.

References

1. Buyya R, Yeo CS, Venugopal S (2011) Cloud computing and emerging IT platforms: vision, hype, and reality for delivering computing as the 5th utility. Future Gener Comput Syst 25(5):599–616
2. Beloglazov A, Buyya R, Lee YC, Zomaya A (2011) A taxonomy and survey of energy-efficient data centers and cloud computing systems. Adv Comput 82:47–111
3. Pettey C (2016) Industry accounts for 2 percent of global CO_2 Emissions. http://www.gartner.com/it/page.jsp?id=503867,2007. Accessed 25 Dec 2016
4. Barroso LA, Holzle U (2007) The case for energy-proportional computing. Computer 40(12):33–37
5. Kaplan J, Forrest W, Kindler N (2009) Revolutionizing data center energy efficiency. McKinsey
6. ASHRAE Technical Committee 99 (2005) Datacom equipment power trends and cooling applications
7. Belady C (2007) In the data center, power and cooling costs more than the it equipment it supports. http://www.electronics-cooling.com/articles/2007/feb/a3/
8. Ferreto TC, Netto MAS, Calheiros RN, De Rose CAF (2011) Server consolidation with migration control for virtualized data centers. Future Gener Comput Syst 27(8):1027–1034
9. Beloglazov A, Abawajy J, Buyya R (2012) Energy-aware resource allocation heuristics for efficient management of data centers for Cloud computing. Future Gener Comput Syst 28(5):755–768
10. Beloglazov A, Buyya R (2012) Optimal online deterministic algorithms and adaptive heuristics for energy and performance efficient dynamic consolidation of virtual machines in Cloud data centers. Concurr Comput Pract Exp 24(13):1397–1420
11. Farahnakian F, Ashraf A, Liljeberg P, Pahikkala T, Plosila J, Porres I, Tenhunen H (2014) Energy-aware dynamic VM consolidation in cloud data centers using ant colony system. In: 2014 IEEE 7th international conference on cloud computing (CLOUD), pp 104–111
12. Monil MAH, Qasim R, Rahman RM (2014) Incorporating migration control in VM selection strategies to enhance performance. Int J Web Appl (IJWA) 6(4):135–151
13. Monil MAH, Rahman RM (2016) VM consolidation approach based on heuristics, fuzzy logic, and migration control. J Cloud Comput 5(1):1–18
14. Farahnakian F, Liljeberg P, Plosila J (2013) LiRCUP: linear regression based CPU usage prediction algorithm for live migration of virtual machines in data centers. In: 39th EUROMICRO conference on software engineering and advanced applications (SEAA), pp 357–364
15. Farahnakian F, Liljeberg P, Plosila J (2014) Energy-efficient virtual machines consolidation in cloud data centers using reinforcement learning. In: 22nd Euromicro international conference on parallel, distributed and network based processing (PDP), pp 500–507
16. Di S, Kondo D, Cirne W (2012) Host load prediction in a google compute cloud with a Bayesian model. In: Proceedings of the international conference for high performance computing, networking, storage and analysis (SC), Salt Lake City, UT, 10–16 Nov 2012
17. Prevost J, Nagothu K, Kelley B, Jamshidi M (2011) Prediction of cloud data center networks loads using stochastic and neural models. In: Proceedings of IEEE system of systems engineering (SoSE) Conference, pp 276–281

18. Kumar EP, Sharma EP (2014) Artificial Neural networks-a study. IJEERT 2(2):143–148
19. Sozen A (2009) Future projection of the energy dependency of Turkey using artificial neural network. Energy Policy 37:4827–4833
20. https://in.mathworks.com/help/nnet/gs/neural-networks-overview.html
21. Vora K, Yagnik S (2015) A survey on backpropagation algorithms for feedforward neural networks. IJEDR, 193–197

Bangla Handwritten Character Recognition With Multilayer Convolutional Neural Network

B. M. Abir, Somania Nur Mahal, Md Saiful Islam and Amitabha Chakrabarty

Abstract Handwritten character recognition from a natural image has a large set of difficulties. Bangla handwritten characters are composed of very complex shapes and strokes. Recent development of deep learning approach has strong capabilities to extract high level feature from a kernel of an image. This paper will demonstrate a novel approach that integrates a multilayer convolutional neural network followed by an inception module and fully connected neural network. The proposed architecture is used to build a system that can recognize Bangla character from different writers with varied handwriting styles. Unlike previous handcrafted feature extraction methods, this CNN-based approach learned more generalized and accurate features from a large-scale training dataset. 1,66,105 training images of Bangla handwritten character of different shapes and strokes have been used to train and evaluate the performance of the model, and thus allows a higher recall rate for the character in an image and outperforms some current methodologies.

Keywords Bangla · Handwritten character · Recognition · Neural network Inception module · CNN

1 Introduction

Recognizing handwritten text from document has received a lot of attention as it allows any system using it to be robust and the use of such models in variable real-world scenarios. In recent times, the field of computer vision has seen great strides toward expert systems that can be trained using large image dataset due to the rise of high-performance computing (HPC) systems. High performing parallel processing GPUs has enabled the computer vision researchers to turn the problem of image recognition into a problem of numerical optimizations. So, now there exists a wide

B. M. Abir · S. N. Mahal (✉) · Md. S. Islam · A. Chakrabarty
BRAC University, Mohakhali, Dhaka, Bangladesh
e-mail: somanianurmahal@gmail.com

B. M. Abir
e-mail: bmabir17@gmail.com

© Springer Nature Singapore Pte Ltd. 2019 155
M. L. Kolhe et al. (eds.), *Advances in Data and Information Sciences*, Lecture Notes in Networks and Systems 39, https://doi.org/10.1007/978-981-13-0277-0_13

range of algorithms and architecture to tackle such problems. One such architecture to create models and solve those problems is neural networks. It has been designed on the basis such that it imitates some function of the neural pathways of human brain. And to maximize the performance of such architecture, numerous methodologies have been proposed and one such implementation of a system of detector and recognizer that uses machine learning techniques is convolutional neural network or CNN [3, 9, 11]. In CNN, the state-of-the-art detection performance is 80% on F-measure with ICDAR 2011 [11]. And for end-to-end recognition, the result is 57% accurate on SVT dataset [10]. The abovementioned work clearly focuses on detection and recognition of English alphabets. For Bangla character recognition, the current state-of-the-art classifier underperforms in the real-world recognition systems as they mainly focus on simple Bangla characters and numerals. Our main goal is to improve the accuracy of the recognition system that can classify simple and compound characters as well as numerals and expand the recognition capability of classification systems. Such improved recognition system integrated with text localization will improve end-to-end Bangla text recognition systems. The real-world application of these systems is groundbreaking as it can be used as an alternative to braille and thus help the visually impaired people to read handwritten Bangla texts.

In this study, we will be showing the results of experimentation on multilayer convolutional neural network architecture which will be appropriately evaluated in the context of Bangla alphabet datasets like BanglaLekha-Isolated [1]. By analyzing this dataset, we will establish a base ground truth for a superior CNN architecture that can be used in handwritten text recognition in Bangla. And ultimately with the chosen CNN architecture, we will implement a Bangla character recognizer that will be able to recognize characters from documents and notes image. With the experimentation results, we determine which architecture is performing better for our current problem. In order to improve the accuracy of a CNN-based architecture, it is imperative to fine-tune the hyperparameters according to the specific needs of the problem domain. And using the tuning of hyperparameters is one of the most powerful advantages the neural network architecture, making any problem set to be solved. This general-purpose usage of neural networks makes it so powerful that a complex problem in any field can now be solved with ease. Currently, one of the most popular CNN-based architectures is called Inception or GooLeNet [8]. This general-purpose architecture contains blocks of CNN-based module called inception module. We have used a modified version of inception architecture, using a combination of inception module and CNN layers. Ultimately, we will demonstrate the improvement in accuracy for Bangla handwritten characters with convolutional neural network.

2 Background Details and Related Work

End-to-end text recognition in images has two key components: (I) text spotting and (II) text recognizing [9]. The main objective of ours is to recognize the actual characters. Many research and work have been done for solving the character recognition problem [9]. Much approach and methods are implemented and showed higher performance. However, in a complex image locating, recognizing character still remains a problem. A wide range of approaches are used for text segmentation and recognition. In an end-to-end system, text detection system identifies the text region which is the input of character or word recognizer [9]. Then, the recognizer recognizes each character in the word or the whole word. Deep convolutional neural network models have higher performance rate for recognizing and computing high-level deep features [2]. LeNet is a traditional CNN model, powerful for digit and handwritten character recognition [4]. In [6], a CNN-based approach BHCR-CNN is proposed for recognizing Bangla character, where CNN is employed to classify individual character. In [7], a deep belief network is introduced to recognize Bangla numerals and alphabet, where an unsupervised feature learning followed by a supervised fine-tuning of the network parameters is used.

3 Proposed Approach

Main objective of our work is to build a Bangla character recognizer which can recognize word or character from any simple or complex image components. In this section, our proposed multilayer inception-based convolutional model is explained in detail. Following subsections give a brief idea of dataset, image preprocessing, system architecture, and classifier.

3.1 Dataset

The text recognition system is trained on character level by using BanglaLekha-Isolated [1] dataset. BanglaLekha-Isolated has total 1,66,105 handwritten character images. It consists of a sample of 84 different Bangla handwritten numerals, basic characters, and compound characters which are collected from different geographical locations of Bangladesh and different age groups. In Fig. 1a, collected from [1], a sample of Bangla handwritten character of a particular age is shown, where first 11 characters are vowel followed by 39 consonant characters, 10 characters are Bangla numerals, and rest 24 are Bangla compound character. The dataset provides multiple labels per character/character group. It contains a wide variety of frequently used Bangla compound character which is very complex shaped. In Fig. 1b, last two characters are compound characters.

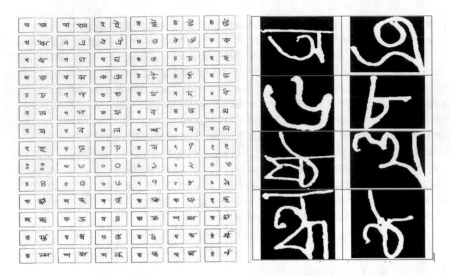

Fig. 1 Samples of Bangla handwritten characters on left (**a**) on right (**b**)

3.2 Image Preprocessing

Bangla character data from dataset has different image sizes, so it was a bit challenging to train them as we used fixed image size which is 28×28 pixel for each image. Therefore, image preprocessing is necessary to handle this large size of image data. For resizing the images, we used python imaging library PIL and anti-aliasing filter is used so that quality of the image does not degrade while resizing. After setting up the training environment, we converted the images into number and saved these images in separate class file. Then, images of each class are shuffled to form random validation and training set and merge them into a single data file. Duplicate data between training and test set can skew the results. So, we needed to remove all the overlap between training and test data. We then searched for the duplicate image files in test, train, and validation sets, and found that 1081 samples overlap between valid and train data, 1278 samples overlap between test and train data, and 185 samples overlap between test and valid data. Then, we created a sanitized test and valid datasets. This dataset cleaning process will ultimately lead to better accuracy for training.

3.3 Network Architecture

Our system architecture consists of several layers of convolutional neural network. As shown in Fig. 2, the preprocessed data with a size of 28×28 image is taken as an input into the first convolutional neural network layer with patch or kernel size of

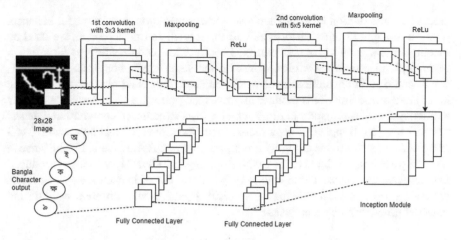

Fig. 2 Data flow between the proposed CNN architecture for Bangla character recognition

3×3 and "same" padding property. After max-pooling, the output is again fed into a second convolution layer with patch size of 5×5 and "same" padding. The output is again fed into max-pooling layer of same configuration as before. The max-pooling from the previous layer is used as the input of the inception module, where CNN is stacked not only sequentially but also parallelly. Then, the output of the inception module is fed into a fully connected network where every single node is connected to each other. This layer flattens the high-level feature outputs from the inception module and converts it into a nonlinear combination of these features. After that, another fully connected layer is added to flatten the data again.

3.4 Classification Using CNN and Inception Module

Classifying Bangla handwritten character has high-dimensional complexity which requires a multilayer, hierarchical neural network [6]. For visual recognition, three principle factors are very important, for instance, local receptive fields, weight sharing, and subsampling layer which makes a difference in CNN from other simple feed-forward neural network [9]. In our proposed CNN-based approach, we used two convolution layers followed by one inception module. Unlike other traditional MLP-based method, here CNN itself captures local features from the input image by forcing each filter of CNN layer that depends on spatially local patch of the previous layer. Therefore, a feature map is generated in the next layer. In CNN, a great advantage of weight sharing is it significantly reduces the number of free parameter [9] by sharing the same set of edge weights across all features in the hidden layers. Subsampling or max-pooling layer is another powerful building block of CNN [5]. The main objective of pooling layer is to reduce the dimensionality of the response map created by convolution layer [9], and also allows translation invariance, for instance, rotation, scale, and other distortions into the model. In our approach, we used an inception module on top of convolutional layers. This module works as filter inputs of multiple convolutions, which works on the same input. For instance, a 1×1, 3×3, and 5×5 convolution layer works on same input and also employ pooling layer at the same time. This module improves the performance of overall model by extracting multilevel feature in Fig. 2 which shows structure of convolutional neural network used for Bangla character recognition: Two convolutional layers with 3×3 and 5×5 receptive fields along with max-pooling and ReLu layers. For instance, a 28×28 pixel image is input to the CNN. Applying first CNN layer on input produced first-level feature maps. These feature maps are produced in such a way that it can extract a variety of local features where distinct patches with different weights and biases from other patches are used.

4 Experimental Setup and Results

4.1 Architecture for Training the Model

The model architecture sometimes also known as logits takes some placeholder inputs such as weights and biases for each convolution and fully connected layer. The weights and biases are the main tuning parameter that is used to make a model learn from its training dataset. When the training session is first initiated, the weights are randomly generated with a standard deviation of 0.3 and biases with zero. Then, the logits use these initial placeholder weights and biases to calculate a predicted value for the initial input of character image. With the predicted output of the initial values from the logits, a SoftMax function is used to normalize the prediction output between 0.0 and 1.0. Now, with this prediction, we can compare it to the true label of the input data and update the value of weights and biases accordingly. In order for our system to determine how much of the weights and biases value have to be changed in order to improve the accuracy of the prediction, we have calculated the loss function. To calculate the loss function, we have used the cross-entropy formula. Cross-entropy is a method of comparing the scores predicted from each training or testing step and comparing it with the ground truth label from the dataset. It is considered to be the distance between the predicted scores and labels.

$$L(s, i) = - \sum Li * log(S_i)$$

Or simply, loss (score, label) = −Labels. Log (Scores). To train the parameters of the logits, an optimizer is used to reduce the loss function of the model in a particular

Fig. 3 A data flowchart for our training process

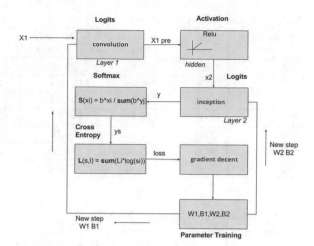

training instance using a technique called gradient descent optimization. First, the gradient of the score is calculated using the loss function and objective function.

$$\Theta = \Theta - \eta \cdot \nabla j(((L(s, l)), \Theta)$$

J() function is the objective function which is considered to be the goal of the gradient descent algorithm. Gradient descent algorithm will try to reduce the loss function by comparing it with the objective function. In order to reduce the loss function, it will change the weights and biases of the logits on each training step. But the gradient descent algorithm is only suitable to reduce the loss function only when the training dataset is small. This is because the training time for systems using gradient descent increases rapidly when the dataset size increases. So in order to use large datasets, we have followed a modified version of gradient descent called stochastic gradient descent algorithm. The main difference is that the dataset is randomly shuffled and a mini-batch of data is picked that is to be used to train the model. The whole training process is illustrated in Fig. 3.

Our dataset has been divided into three parts: training, validation, and testing. Training set is the sole set of data that is used to train our model. But as we are using stochastic gradient descent, the training set was divided into random mini-batch to decrease the training time for large datasets.

4.2　Experiments

We performed a thorough comparison on different hyperparameters that can be tuned to measure the performance of our proposed architecture for Bangla handwritten character recognizer. We started with 128 hidden layers and 0.005 learning rate and

Table 1 Variable hidden layer experiments

Hidden layer	Steps	Mini-batch accuracy (%)	Validation accuracy (%)	Test accuracy (%)	Mini-batch loss
128	30000	88.00	89.80	87.60	0.587701
384	30000	90.00	92.70	88.60	0.583768
512	30000	94.00	92.60	88.80	0.503671
720	30000	88.00	94.20	**89.30**	**0.397282**
750	30000	88.00	**94.70**	88.70	0.348352
800	30000	94.00	94.20	88.20	0.568419
900	30000	**98.00**	93.80	89.10	0.345489

Table 2 Variable learning rate experiments

Learning rate	Steps	Mini-batch accuracy (%)	Validation accuracy (%)	Test accuracy (%)	Mini-batch loss
0.055	30000	84.00	93.60	88.40	0.480084
0.006	30000	92.00	**94.70**	88.80	0.305358
0.005	**30000**	**98.00**	93.80	**89.10**	**0.345489**
0.003	30000	92.00	94.60	88.05	0.367663

got a decent mini-batch, validation, and test accuracy of 88.0%, 89.80%, and 87.60%, respectively, as shown in Table 1. One of the two main hyperparameters that need most attention is numbers of hidden layers in the subsampling layer or ReLu layer that is found immediately after the convolution layer. Therefore, we intuitively changed the number of hidden layers and trained for the optimum test accuracy. It seems that increasing the hidden layers up to 720 increases the test accuracy result.

After 720 hidden layers, the test accuracy starts to decline. But it seems that mini-batch accuracy and validation accuracy still rises even if the test accuracy falls. This proves that after 720 hidden layers, the network starts to memorize the training dataset and thus there is a stark difference between the test accuracy and mini-batch accuracy. With 900 hidden layers, the network has fully memorized the mini-batch training set. Therefore, with 720 hidden layer hyperparameter, we tested different variations of the learning rate hyperparameter, obtaining the results in Table 2. We started with a high learning rate of 0.055 and it provides a very low testing accuracy compared with the previous value of 0.005 learning rate. Therefore, we significantly decreased the learning rate and found a better test accuracy with learning rate of 0.006 and best of the class validation accuracy. But increasing the learning rate more than 0.005 yields no more better results.

Figure 4 shows a graphical comparison of the variable hidden layer experiments. This illustrates that the performance of the model provides the maximum test accuracy of 89.3% with 720 hidden layers and a learning rate of 0.006.

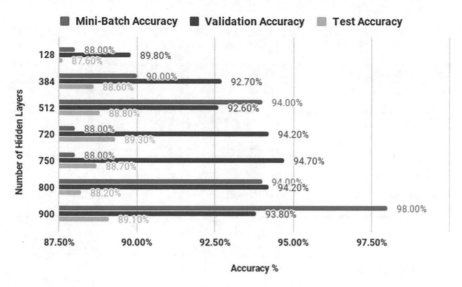

Fig. 4 Comparison of accuracy with variable hidden layer experiments

Table 3 shows comparison of test accuracy with proposed method, CNN-based BHCH-CNN [6], and unsupervised deep belief network. The most significant part of BHCR-CNN and proposed method is that no feature selection techniques are used here. In [6], 50 classes (Bangla basic characters) are only used in classification and the classification gives 85.96% recognition accuracy, while our proposed method with same 50 classes of character has classification accuracy of 91.1%. In [7], an unsupervised deep belief network is used where 60 classes (10 Bangla numerals and 50 basic characters) are used and have 90.27% recognition accuracy but our proposed classifier has performed 90.8% recognition accuracy. No compound character was used for training this classifier.

Then to have a wide range of classification, we have used the BanglaLekha dataset with 84 classes to train our classifier, which consists of Bangla basic character, numerals, and compound characters. Compound characters are complex in shape, so it is much more challenging for a classifier to recognize compound character. This classifier successfully classified commonly used compound Bangla characters with a test accuracy of 89.3%. Therefore, our proposed classifier, with simple convolutional and inception network, shows overall improved performance while recognizing more classes. This shows the power of convolutional in generalizing a large number of handwritten classes and further improvement of recognition is possible with deeper convolutional networks.

Table 3 Performance comparison with another CNN-based approach and unsupervised approach

Other works	Number of class	Classification	Compound character	Accuracy (%)
BHCR-CNN [10]	50	CNN	No	85.96
Our proposed method	50	CNN	No	**91.1**
DBNs [11]	60	Unsupervised deep belief network	No	90.27
Our proposed method	60	CNN	No	90.8
Our proposed method	84	CNN with inception	Yes	89.30

5 Conclusion

Convolutional neural network has powerful and robust ability to recognize visual pattern from pixel image. In this paper, we have considered a multilayer convolutional neural network without any feature selection for Bangla handwritten character classification. The proposed method gives competitive performance with the existing other methods in terms of test accuracy. And it also tested on a large dataset. Adding more convolutional layer and hyperparameter tuning may give improved performance in future. And as machine learning networks can be used as a general-purpose learning model, we also want to conclude that the same multilayer convolutional network can be used for Bangla scene-text detection and recognition with state-of-the-art performance.

References

1. Biswas M, Islam R, Shom GK, Shopon M, Mohammed N, Momen S, Abedin MA (2017) Banglalekha-isolated: a comprehensive bangla handwritten character dataset. arXiv:1703.10661
2. He T, Huang W, Qiao Y, Yao J (2016) Text-attentional convolutional neural network for scene text detection. IEEE Trans Image Process 25(6):2529–2541
3. Huang W, Qiao Y, Tang X (2014) Robust scene text detection with convolution neural network induced mser trees. In: European conference on computer vision. Springer, pp 497–511
4. LeCun Y, Boser BE, Denker JS, Henderson D, Howard RE, Hubbard WE, Jackel LD (1990) Handwritten digit recognition with a back-propagation network. In: Advances in neural information processing systems. pp 396–404
5. Masci J, Giusti A, Ciresan D, Fricout G, Schmidhuber J (2013) A fast learning algorithm for image segmentation with max-pooling convolutional networks. In: 2013 20th IEEE international conference on image processing (ICIP). IEEE, pp 2713–2717
6. Rahman MM, Akhand M, Islam S, Shill PC, Rahman MH (2015) Bangla handwritten character recognition using convolutional neural network. Int J Image Gr Signal Process 7(8):42

7. Sazal MMR, Biswas SK, Amin MF, Murase K (2013) Bangla handwritten character recognition using deep belief network. In: 2013 International conference on electrical information and communication technology (EICT). IEEE, pp 1–5
8. Szegedy C, Liu W, Jia Y, Sermanet P, Reed S, Anguelov D, Erhan D, Vanhoucke V, Rabinovich A (2015) Going deeper with convolutions. In: Proceedings of the IEEE conference on computer vision and pattern recognition. pp 1–9
9. Wang T, Wu DJ, Coates A, Ng AY (2012) End-to-end text recognition with convolutional neural networks. In: 2012 21st International conference on pattern recognition (ICPR). IEEE, pp 3304–3308
10. Yin XC, Yin X, Huang K, Hao HW (2014) Robust text detection in natural scene images. IEEE Trans Pattern Anal Mach Intel 36(5):970–983
11. Zhang Z, Shen W, Yao C, Bai X (2015) Symmetry-based text line detection in natural scenes. In: Proceedings of the IEEE conference on computer vision and pattern recognition. pp 2558–2567

Lung Segmentation of CT Images Using Fuzzy C-Means for the Detection of Cancer in Early Stages

Satya Prakash Sahu, Priyanka Agrawal, Narendra D. Londhe
and Shrish Verma

Abstract Lung segmentation is the most crucial step toward the design of computer-aided diagnosis (CAD) system for detecting the lung cancer at prior stages. It is required to do preprocessing and take necessary step for getting the accurate region of interest (ROI). In this paper, the proposed method for lung segmentation of CT images is based on fuzzy c-means with automated thresholding and morphological operations. We have taken 10 CT scans or subjects from LIDC-IDRI which includes 5 juxtapleural cases. The adopted method achieved overall 99.9388% accuracy for overlap ratio 0.969 Dice similarity coefficient values.

Keywords Computed Tomography · Lung segmentation · CAD system · FCM
Nodule detection · Thresholding

1 Introduction

As per various statistics surveyed by different agencies, the lung cancer has been observed as most common cancer worldwide. In 2012, 1.8 million new lung cancer cases were estimated [1]. Deaths due to lung cancer are more as compared to other cancer-related death worldwide. It contributes 13% for the cases registered as cases and in the field of deaths due to cancer its contribution was 19% in 2012, reported by GLOBOCAN 2012 (IARC) Section of Cancer Surveillance (Fig. 1).

Lung cancer leads to highest mortality rate and this encourages researchers for diagnosis cancer in early stages. The survival rate can be increased to 70–80%, if the lung cancer is detected in early stages [2]. In the first screening test, only very large size nodules can be detected using thorax X-ray [3] but for nodules size up to 2 cm in diameter it is incapable due to their unfavorable location and positions.

In this paper, automated method by using FCM algorithm is used for determining lung boundary and to extract the field of Region of Interest with the purpose of detection analysis of various types of lung nodules irrespective of their location,

S. P. Sahu (✉) · P. Agrawal · N. D. Londhe · S. Verma
National Institute of Technology, Raipur, India
e-mail: spsahu.it@nitrr.ac.in

© Springer Nature Singapore Pte Ltd. 2019
M. L. Kolhe et al. (eds.), *Advances in Data and Information Sciences*, Lecture Notes
in Networks and Systems 39, https://doi.org/10.1007/978-981-13-0277-0_14

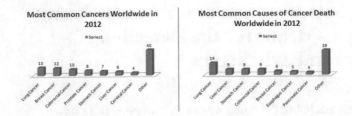

Fig. 1 Incidence rate and mortality rate worldwide (Courtesy of GLOBOCAN 2012 (IARC) Section of Cancer Surveillance)

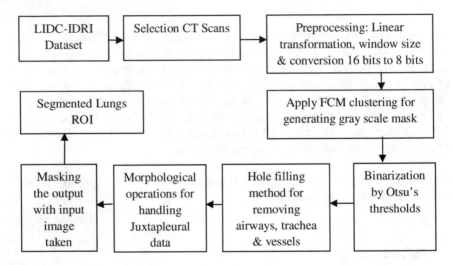

Fig. 2 Process flow of proposed method

size, and dimensions. After completing segmentation, all the broken boundaries are corrected with the help of morphological operations for including the juxtapleural cases; this step also helps in reducing over-segmentation rate. The process flow of proposed method has been shown in Fig. 2.

2 Related Work

For image segmentation of object, the most common method is gray level or intensity value and the technique through which this concept has been utilized effectively is *Threshold-based segmentation*. As the portion of lung lobes is having lower gray level (approximately −500 HU) as compared to other anatomical structure in region of thorax; due to this reason, authors have adopted optimum threshold as their basis for segmenting lungs in their research articles. For smoothing the irregular boundaries of lungs, Hu et al. [4] have given the automatic method for segmenting the lungs using iterative threshold along with some morphological operations. Gao et al. [5] have given similar threshold-based approach which includes preprocessing followed by region growing and morphological smoothing. *Active contour-based segmentation* as one of the important method for lung segmentation that has been used in number of research articles [6–10]. The main fundamental concept behind this method is minimizing energy function. This energy function gets minimized at each iteration through dynamic contour. The most challenging part of this approach is the problem of initialization and convergence; several authors addressed these issues in their research articles. Xu and Prince [6] introduced Gradient Vector Flow (GVF) as an external force to guide the snake so that better convergence can be obtained. Wang et al. [7] have further worked for improving the initialization and convergence by Normally Biased Gradient Vector Flow (NBGVF). *Shape-based segmentation* is another useful technique. This technique uses the prior shape information of lungs for extraction of ROI from thorax region. In this method, an energy framework is created to guide and help for deformable models in lung field segmentation. To deal with local minima issues in lung segmentation, Annangi et al. [11] adopted region-based active contour technique for X-ray images by using prior shape and low-level features. Pu et al. [12] proposed "break-and-repair" technique for segmentation of a medical image this approach based on geometric modeling and shape.

3 Materials and Methods

3.1 Data

The CT images of lungs have been taken from LIDC-IDRI dataset [10]. This database holds 1018 cases of lung CT scans under The Cancer Imaging Archive (TCIA) Public Access [13].

This database contains CT scan with an associated XML file having annotation done by four expert radiologists [14]. The adopted method holds 10 patients, which have been taken from the database with specifications as follows: DICOM image of size 512×512, tube voltage 120 kVp, tube current in the range of 265–570 mAs, and intensity value ranging from -600 to 1600 HU [11].

3.2 Data Preprocessing

Preprocessing is one of the most important steps in lung segmentation. It starts with the selection of suitable datasets from LIDC-IDRI database. In this experiment, 10 patient CT scan images have been taken including five cases of juxtapleural lung nodules. The pixel value of the images is converted into double format so that its accuracy can be improved.

Image enhancement is another crucial step in lung segmentation. In this optimum level of contrast can be set for selected data. For setting the window level, the pixel value required to be converted into Hounsfield Unit (HU). In the last step, 16-bit images are converted into 8-bit images.

3.3   Lung Segmentation

The segmentation of lung is implemented by the following steps:

A. Applying fuzzy c-means clustering-based segmentation and generating the grayscale-masked image of the CT images. The explanation and derivation of this algorithm are given in Sect. 3.3.1.
B. By using automatic thresholding algorithm (the method will be discussed in Sect. 3.3.2), the output of fuzzy c-means masks is converted into binary image followed by the background subtraction for ROI.
C. For the removal of the large airways, trachea, and other vessels, the hole-filling algorithm is applied. This algorithm involves dilation, intersection, and complementation.
D. The morphological closing operations are applied for handling the juxtapleural dataset cases.
E. Finally, masking the output image with input image for getting the segmented lung region.

3.3.1   Fuzzy C-Means Algorithm

Fuzzy C-Means (FCM) is a flexible method of clustering. In this method, data can be the member of two or more clusters at a time. This method is proposed by Dunn [15]. It is based on minimization of the following objective function:

$$J_t = \sum_{i=1}^{N} \sum_{j=1}^{C} v_{ij}^t \, ||x_i - c_i|| \tag{1}$$

Here, t is the real number that lies between 1 and infinity, v_{ij} represent degree of membership of x_i in cluster j, x_i represents ith of d-dimensional measured data, c_j is the d-dimensional center of the cluster, and $||*||$ express the similarity between any measured data and the center of the cluster.

Fuzzy partitioning is done with the help of iterative optimization of the objective function given in Eq. 1, with the update of membership v_{ij} and the centers of the cluster c_j. The membership and cluster center can be defined as

$$v_{ij} = \frac{1}{\sum_{k=1}^{c} \left(\frac{||x_i - c_j||}{||x_i - c_k||} \right)^{\frac{2}{i-1}}} \qquad (2)$$

$$c_j = \frac{\sum_{i=1}^{N} v_{ij}^t \cdot x_i}{\sum_{i=1}^{N} v_{ij}^t} \qquad (3)$$

The stopping condition of this iteration is $max_{ij} = \left\{ \left| v_{ij}^{(k+1)} - v_{ij}^k \right| \right\} < \varepsilon$, where ε is stopping criterion which ranges between 0 and 1, and k is number of iterations. This procedure converges to a saddle point or a local minimum of J_t.

The algorithm can be given by the following steps:

a. *Initialize matrix A with v_{ij}, and represent it by A(0).*
b. *At k-step: calculate the center vectors C(k) =[c_j] with A(k) as per Eq. 3.*
c. *Update A(k) and A(k) + 1) according to Eq. 2.*
d. *If the condition || A(k+ 1) − A(k) ||< ε then STOP; else go to step (b).*

3.3.2 Automatic Threshold Algorithm

The automatic threshold algorithm is based on Otsu's method [16]. The estimation of threshold method is obtained by using simple and effective approach of finding the gray level value (t^*) that helps in minimizing the weight within-class variance and maximizing the between-class variance with the only assumption that the histogram is bimodal. For any given CT image, let us assume that the pixels are represented in L gray levels [1, 2 ..., L], and the normalized histogram may be considered as probability distribution (p_k):

$$p_k = \frac{n_k}{N}, \quad p_k \geq 0, \quad \sum_{k=1}^{L} p_k = 1 \qquad (4)$$

where n_k denotes the number of pixel having the gray level k, and $N = n_1 + n_2 + \cdots + n_L$ is the total number of pixel.

Maximizing between-class variance may be rewritten by the following equation that defines the optimum threshold

$$\sigma_B^2(t) = \frac{[\mu_{TS}(t) - \mu(t)]^2}{s(t)[1 - s(t)]} \qquad (5)$$

Thus, general form of optimal threshold t^* is obtained by analyzing the histogram and testing each gray level for which the possibility of the threshold t that maximizes $\sigma_B^2(t)$ is given by

$$\sigma_B^2(t^*) = \max_{1 \leq t \leq L} \sigma_B^2 \qquad (6)$$

4 Experimental Results and Discussion

Proposed method has been tested by 10 subjects, which includes 5 subjects having juxtapleural nodules; these subjects are taken from the database of LIDC-IDRI. For the support of manual segmentation tasks, an open-source software system "The Medical Imaging Interaction Toolkit" (MITK 2016.11 Release Workbench and Toolkit 2016.11) has been used in this experiment [17]. This toolkit is a combination of the Visualization Toolkit (VTK) and the Insight Toolkit (ITK) and it has been applied in our work as an annotation tool.

4.1 Evaluation Measures

Performance of the lung segmentation has been measured based on the following metrics: over-segmentation rate, under-segmentation rate, overlap ratio, and Dice similarity coefficient.

The over-segmentation is defined as the region or volume (number of voxels) that is included as a part of the segmented ROI but are absent in ground-truth image or reference standard, whereas under-segmentation is defined as the region or volume (number of voxels) that are considered in reference standard, but are absent in segmented ROI generated by proposed method [18, 19].

The volume overlap ratio (VO) is defined as the relative overlap between two volumes, i.e., binary segmentation mask generated by our automated method (V_b) and reference standard (V_r) and computed by taking the intersection of these two volumes (V_b and V_r) divided by their union [18].

$$VO(V_b, V_r) = \frac{|V_b \cap V_r|}{|V_b \cup V_r|} \tag{7}$$

The other evaluation measures which are used for detecting the segmentation accuracies are Dice similarity coefficient [20]. This method is used to compare the similarity between the segmented ROIs (number of pixels or voxels) generated by adopted method to the reference standard region.

4.2 Experiment and Results

For evaluating the performance of our proposed method, it has been tested on CT scans of 10 subjects (approximately 1500 CT slices) including 5 subjects of juxtapleural nodules from the database of LIDC-IDRI of National Cancer Imaging

(a)   (b)   (c)   (d)   (e)   (f)

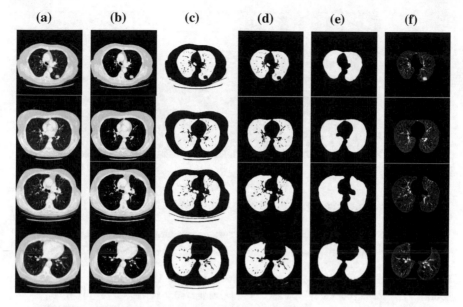

Fig. 3 The results of lung segmentation for proposed method. Column **a** four original lung CT images where row 1 is the image containing juxtapleural nodule; **b** grayscale-masked images after applying FCM; **c** binary images of (**b**) using otsu's threshold; **d** background subtracted images; **e** the final lung parenchyma mask after removing the trachea, airways, and vessels; and **f** segmented lungs

Archive [21]. Figure 3 shows the sample results and process of our method for lung segmentation. Using the above evaluation measures (Sect. 4.1), Fig. 4 shows the bar plots of over-segmentation, under-segmentation ratio, and overlap ratio difference. The proposed method achieved average over-segmentation of 0.037282, average under-segmentation of 0.023896, and average overlap ratio difference of 0.061178, Dice similarity coefficient of 0.969588. Also, for the patient having juxtapleural nodules, our method is showing the effectiveness by including all the juxtapleural nodules inside the ROIs and the segmentation accuracies with average overlap ratio of 99.93882% and Dice similarity coefficient value of 0.969588.

Various authors presented number of methods and approach for automated lung segmentation, but only a few are effective to handle juxtapleural nodules cases. Table 1 shows the comparison of performance for proposed method with the other lung segmentation approach given by the different authors.

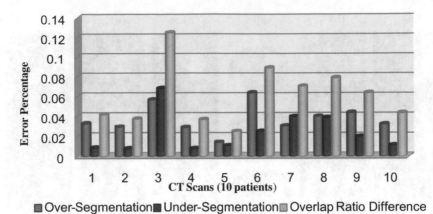

Fig. 4 Bar plots of proposed method of 10 patients: over-segmentation, under-segmentation, and overlap ratio difference

Table 1 Performance comparison of proposed method for lung segmentation

Work	Methodology	Datasets	Overlap ratio (%)	Over-segmentation	Under-segmentation
Wei et al. [22]	Iterative threshold, 3D region growing, and Bresenham algorithm	97 lung nodules CT scans 25 juxtapleural nodule scans	95.24	–	–
Zhou et al. [23]	FCM clustering, iterative weighted averaging, and threshold	20 CT scans with 65 juxtapleural nodules	95.81	–	–
Proposed method	FCM clustering, automatic threshold, and morphological operations	10 cases of LIDC-IDRI datasets with 5 cases of juxtapleural nodules having the total of 1500 CT images approximately	99.9388	0.0372	0.0238

5 Conclusion

The efficiency of our approach is demonstrated on 10 cases of lung thoracic CT scans from publically available datasets LIDC-IDRI which includes 5 cases of juxtapleural nodules having the total of approximately 1500 CT images. The proposed method can achieve the overall performance of 99.9388% overlap ratio with over-segmentation and under-segmentation rate of 0.0372 and 0.0238, respectively. The average processing time for the proposed method is 0.85 s per CT slice using MATLAB software tested on desktop with CPU Intel Core i5 3.1 GHz and 8 GB RAM. The proposed approach can fulfill the requirements of CAD system for lung cancer detection and provide the accurate ROIs for further processing toward the diagnosis of lung cancer.

References

1. Ferlay J, Soerjomataram I, Ervik M, Dikshit R, Eser S, Mathers C et al (2013) Lyon, France: international agency for research on cancer. GLOBOCAN 2012 v1.0, cancer incidence and mortality worldwide: IARC CancerBase No. 11. http://globocan.iarc.fr. Accessed 04 Oct 2016
2. Swensen SJ, Jett JR, Sloan JA, Midthun DE, Hartman TE, Sykes AM, Aughenbaugh GL, Zink FE, Hillman SL, Noetzel GR, Marks RS (2002) Screening for lung cancer with low-dose spiral computed tomography. Am J Respir Crit Care Med 165(4):508–513
3. Flehinger BJ, Melamed MR, Zaman MB, Heelan RT, Perchick WB, Martini N (1984) Early lung cancer detection: results of the initial (prevalence) radiologic and cytologic screening in the memorial sloan-kettering study 1–3. Am Rev Respir Dis 130(4):555–560
4. Hu S, Hoffman EA, Reinhardt JM (2001) Automatic lung segmentation for accurate quantitation of volumetric X-ray CT images. IEEE Trans Med Imaging 20(6):490–498
5. Gao Q, Wang S, Zhao D, Liu J (2007) Accurate lung segmentation for X-ray CT images. In: Third international conference on natural computation, ICNC 2007, Aug 24, vol 2. IEEE, pp 275–279
6. Xu C, Prince JL (1998) Snakes, shapes, and gradient vector flow. IEEE Trans Image Process 7(3):359–369
7. Wang Y, Liu L, Zhang H, Cao Z, Lu S (2010) Image segmentation using active contours with normally biased GVF external force. IEEE Signal Process Lett 17(10):875–878
8. Chan TF, Vese LA (2001) Active contours without edges. IEEE Trans Image Process 10(2):266–277
9. Cui W, Wang Y, Lei T, Fan Y, Feng Y (2013) Local region statistics-based active contour model for medical image segmentation. In: Image and graphics (ICIG), Jul 26. IEEE, pp 205–210
10. Athertya JS, Kumar GS (2014) Automatic initialization for segmentation of medical images based on active contour. In: 2014 IEEE conference on biomedical engineering and sciences (IECBES), Dec 8. IEEE, pp 446–451
11. Annangi P, Thiruvenkadam S, Raja A, Xu H, Sun X, Mao L (2010) A region based active contour method for x-ray lung segmentation using prior shape and low level features. In: 2010 IEEE international symposium on biomedical imaging: from nano to macro, Apr 14. IEEE, pp 892–895
12. Pu J, Paik DS, Meng X, Roos J, Rubin GD (2011) Shape "break-and-repair" strategy and its application to automated medical image segmentation. IEEE Trans Visual Comput Graphics 17(1):115–124

13. Clark K, Vendt B, Smith K, Freymann J, Kirby J, Koppel P, Moore S, Phillips S, Maffitt D, Pringle M, Tarbox L (2013) The cancer imaging archive (TCIA): maintaining and operating a public information repository. J Digit Imaging 26(6):1045–1057

14. Armato SG, McLennan G, Bidaut L, McNitt-Gray MF, Meyer CR, Reeves AP, Zhao B, Aberle DR, Henschke CI, Hoffman EA, Kazerooni EA (2011) The lung image database consortium (LIDC) and image database resource initiative (IDRI): a completed reference database of lung nodules on CT scans. Med Phys 38(2):915–931

15. Dunn JC (1973) A fuzzy relative of the ISODATA process and its use in detecting compact well-separated clusters. J Cybern 3:32–35

16. Otsu N (1975) A threshold selection method from gray-level histograms. Automatica 11(285–296):23–27

17. MITK: The Medical Imaging Interaction Toolkit, German Cancer Research Center, Division of Medical and Biological Informatics (2016). http://mitk.org/wiki/MITK

18. Pu J, Roos J, Chin AY, Napel S, Rubin GD, Paik DS (2008) Adaptive border marching algorithm: automatic lung segmentation on chest CT images. Comput Med Imaging Graph 32(6):452–462

19. Hoover A, Jean-Baptiste G, Jiang X, Flynn PJ, Bunke H, Goldgof DB, Bowyer K, Eggert DW, Fitzgibbon A, Fisher RB (1996) An experimental comparison of range image segmentation algorithms. IEEE Trans Pattern Anal Mach Intell 18(7):673–689

20. Sampat MP, Wang Z, Gupta S, Bovik AC, Markey MK (2009) Complex wavelet structural similarity: a new image similarity index. IEEE Trans Image Process 18(11):2385–2401

21. Armato III SG, McLennan G, Bidaut L, McNitt-Gray MF, Meyer CR, Reeves AP, Clarke LP (2015) Data from LIDC-IDRI. The cancer imaging archive. http://doi.org/101.7937/K9/TCIA.2015.LO9QL9SX

22. Wei Y, Shen G, Li JJ (2013) A fully automatic method for lung parenchyma segmentation and repairing. J Digit Imaging 26(3):483–495

23. Zhou S, Cheng Y, Tamura S (2014) Automated lung segmentation and smoothing techniques for inclusion of juxtapleural nodules and pulmonary vessels on chest CT images. Biomed Signal Process Control 30(13):62–70

An Efficient Approach for Classifying Social Network Events Using Convolution Neural Networks

Ahsan Hussain, Bettahally N. Keshavamurthy and Seema Wazarkar

Abstract The rise of social media and social platforms has led to enormous information dissemination. Images shared by social users at any moment convey what they see or where they have been to. Social images express far more information than texts which may involve individuals' characteristics like personality traits. Existing methods perform event classification based on fixed temporal and spatial image resolutions. In this paper, we thoroughly analyse social network images for event classification using Convolution Neural Networks (CNNs). CNN captures both important patterns and their contents, to extract the semantic information from the images. We collect shared images from Flicker specifying various sports events that users attend. Images are divided into three event classes, i.e. bikes, water and ground. After extensive experimentation using CNN, for training and classifying images, we obtain an accuracy of 98.7%.

Keywords Convolution neural networks · Event classification · Social image recognition · Online social networks

1 Introduction

Real-life interpersonal relationships pave the way for Online Social Networks (OSN), where social users act as nodes and the relationships form the network edges. OSN is a modern-day system whose history is moderately short, however unsettled. The enormous use of social data has caused a move on (a) sharing and imparting information between social users (b) competition and working in organizations and (c)

A. Hussain (✉) · B. N. Keshavamurthy · S. Wazarkar
Department of Computer Science and Engineering,
National Institute of Technology Goa, Farmagudi, Goa, India
e-mail: ahsan.hussain@nitgoa.ac.in

B. N. Keshavamurthy
e-mail: bnkeshav.fcse@nitgoa.ac.in

S. Wazarkar
e-mail: wazarkarseema@nitgoa.ac.in

© Springer Nature Singapore Pte Ltd. 2019   177
M. L. Kolhe et al. (eds.), *Advances in Data and Information Sciences*, Lecture Notes
in Networks and Systems 39, https://doi.org/10.1007/978-981-13-0277-0_15

political challenges and influences. Social Network Analysis (SNA) has changed almost completely the way various social surveys, interviews and questionnaires are used nowadays. SNA helps to investigate the business sectors and groups, keeping in mind the end goal to oversee learning, change, participation and hazards.

Social big data has influenced the way various sports and other related activities are influenced. The big data era has seen a great surge due to the impressive growth of OSNs. Mining and interpretation of huge amounts of personal, social relationship, specific event related and social–behavioural data, can be of enormous use in developing new applications. The fast spreading of news on OSN platforms like Twitter needs a thorough analysis that helps in managing various natural catastrophes and terrorist attacks. The challenges that OSNs have to deal with include privacy concerns and preservation of social data, and automation of unstructured social data communications.

Recently, CNNs with computer vision have become very popular because large amounts of data can be processed and great results can be achieved in classical recognition problems of object [1] and image styles [2]. Our approach adjusts CNNs pre-trained for classification of images with the expectation of co-selecting their compelling authentic power to indirectly capture the basic image features for predicting the sports events attended by a social user. This helps to obtain more related features for generalizing patterns in an image. CNN captures both important patterns and their contents, to extract the semantic information.

Sports analysis and movie reviews are hugely followed by the social users on OSN platforms. People share their pictures that they click at various events like social gatherings, movie places and indoor/outdoor sports with big crowds. The classification of these images carries much importance for SN analysts. In this paper, we use Convolution Neural Network (CNN) for classifying the images shared by social users about various sports events. The events are divided into three classes: racing, field sports and water sports. We obtained an accuracy of 97% for classifying the social events into the given image classes using CNN.

The organization of the remainder of the paper is as follows. In Sect. 2, we review the related work on CNN and image classification. Our proposed social image classification using CNN is presented in Sect. 3, followed by the implementation procedure and experimental results in Sect. 4. We conclude our paper in Sect. 5.

2 Background and Related Work

Convolution is the process of creating feature maps from the input data. CNN has become very popular for image data, in extracting the details and constituents of an image. Image captions can also be generated automatically using CNNs. CNN is unlike those neural networks that convert the input data into a one-dimensional vector of neurons. CNN matches the structural data it captures using the spatial structure that helps to classify images, videos, etc. [3, 4]. One of the most prominent features of using CNN is the sharing of weights between the neuron receptive field structures and neurons. Neural networks use supervised and unsupervised learning methods for the purpose of adjusting the weights for obtaining the best classification results. Various benefits of improved extension distance using CNNs include short training time, easier maintenance and strong adaptability among other features [5].

Effectiveness and robustness of learned CNN features have helped in achieving breakthroughs in the fields of image classification [6], contour detection [7] and object detection [8]. Authors in [9] performed event classification in event-driven multi-sensor networks using CNN and compared the results with SVM and KNN classifiers, getting improved accuracy. Xingyi et al. analyse the impact of tweet images on point-of-interest (POI) recommendation using different visual attributes of images for different POI categories [10]. They jointly modelled five Twitter features to obtain POIs from tweet images.

CNNs basic structure alternates between a convolutional and a pooling layer [3, 11]. This pairing continues until the data reduction is performed using this coupling and all the feature maps are combined into a Fully Connected Layer (FCL). The CNN structure can be divided into feature extraction and classification modules. The first part contains convolution layers alternating with the subsampling layers where the convolution layer has many randomly generated two-dimensional shared weights. This convolution layer operation, using the shared weight of the input data, can be expressed as follows:

$$h_j = tanh\left(\sum x_i * k_{ji} + bias\right) \qquad (1)$$

where x denotes the previous layer data and the weight matrix is represented by k. As per the shared weight matrix, reduction of input data follows the convolution operation. An activation function (hyperbolic tangent) is used for data calculation. So, each convolution layer output is normalized from -1 to 1 [12]. Average pooling and max pooling are two operations of the subsampling layer. The performance of the max pooling method is generally better than average pooling for some particular datasets. h is used for getting the max value that is a previous layer image. The convolution and subsampling layer operations are repeated, and finally, this data is given as input to the classification layer in the feature extraction part.

Authors in [13] claim that ResNet reduces the top-1 error by 3.5% when it is compared to plain networks, because of the training error reduction. The comparison verifies the effectiveness of residual learning on extremely deep systems. In addition, the 18-layer plain/residual nets are comparably accurate but the 18-layer ResNet converges faster that can ease the optimization at times. Samar M. Alqhtani et al. in their work have used KNN for event detection in twitter streams and obtained an accuracy of 86% when only images are used [14]. Authors in [15] have developed a global classification model based on the Naive Bayes classification scheme trusted third party. They performed experiments for different distributed database scenarios such as horizontal, vertical and arbitrary partitions and obtained an accuracy of 85%.

3 Proposed Methodology

In the proposed approach, we classify images of various sports events obtained from the social network Flicker using CNN. The schematic diagram of the approach is shown in Fig. 1. Our method consists of 17 convolution layers followed by a *Fully Connected Layer* having a few neurons. The design we used, helps to minimize the overfitting risk and at the same time, efficiently solves our problem of image classification between the three classes. The baseline architectures are similar to the normal plain residual networks, expect that a shortcut connection is added to each pair of intermediate filters as in Fig. 1. ResNet-18 has no extra parameter compared to the plain counterparts. ResNets can be considered as ensembling subsets of residual modules in a network.

The proposed framework consists of 17 convolution layers followed by a classification layer in a sequential order. Each convolution layer is separated by particular combinations of the normalization, ReLU, Sum and Pooling until the final layer (classification layer) is reached, where all four operations are applied on the result from the convolution layer 17. We group the various convolution layers with different resolutions (R1 to R5). The final classification layer gives the resultant output that has the resolution of $1 \times 1 \times 3$ (three classes). The various resolutions are given in Table 1.

The image dataset that has been used in the proposed approach consists of 300 images of sports events that we divided into three classes with same number of images. The images have been taken from well-known social network Flicker dataset [16]. In this paper, the results generated by applying CNN on the raw pixels of social

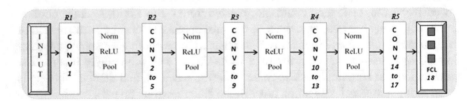

Fig. 1 The proposed CNN architecture for event detection in social network images

Table 1 Groupings as per the resolution of convolution layers

No.	Resolution	Convolution layers
R1	$112 \times 112 \times 64$	1
R2	$56 \times 56 \times 64$	2–5
R3	$28 \times 28 \times 128$	6–9
R4	$14 \times 14 \times 256$	10–13
R5	$7 \times 7 \times 512$	14–17
FCL	$1 \times 1 \times 3$	18

images show a decent amount of classification accuracy. But algorithm efficiency can get decreased because of the size and complexity of these images in the spatial domain. Input images of the size of $200 \times 200 \times 3$ are fed to the first convolution layer of the proposed model and resultant is a one-dimensional vector of length 3.

In the initialization phase, we do not use any other data except the images of the considered Flicker dataset and train the network from scratch. The resultant values for training are represented as sparse, binary vectors corresponding to the ground truth classes. The target (label vector) is in the length of number of classes chosen for classification. For the implementation purpose, we train our model with all combinations of training and testing set images with 10–90% portions from the dataset and test the accuracy of the proposed framework.

4 Experimental Procedure

In the following section, we present the details of implementation of the proposed methodology which is succeeded by the final results of the classification accuracy obtained.

4.1 Implementation Details

We start with rescaling all images to 200×200 and feeding them to our network. It follows the 17 convolution layers to process on the images. In the first convolution layer (Conv 1), 64 filter-maps are applied to the input image dataset. This step is followed by *ReLU* (Rectified Linear Operator), sum and a max pool layer. Next, we group four consecutive convolution layers with a common number of filter-maps as given in Table 1, and the same process is repeated till convolution layer 17 that contains 512 filter-maps. Using the ResNet-18 structure, a bridge is created between alternate convolution layers. The resultant data is fed to the Fully Connected Layer that gives the final output after mapping the data with particular classes, thus obtaining the final prediction accuracy.

Filter is applied to input images to obtain the features from convolution to give coarse resultant images, as shown in Fig. 2. The movement of the filter matrix is as per one unit for each element in both horizontal and vertical directions. Next, element-wise multiplication for the overlapping elements is performed. Max pooling is used for the size reduction of the convoluted features to simplify other computational requirements. The spatial resolution is lowered and zeros are padded on the boundaries of the obtained feature matrix to control its size. To obtain smaller feature vectors, both convolution and pooling layers are applied, iteratively on the dataset.

Fig. 2 Sample coarse images

Fig. 3 The classified Flicker images using ResNet-18

Table 2 Final results obtained

Sample size % (train/test)	Accuracy	Processing time (s)
10/90	94.1	58
20/80	93.7	49
30/70	96.4	52
40/60	97.4	50
50/50	**98.7**	49
60/40	98.4	48
70/30	97.9	46
80/20	96.8	**45**
90/10	93.7	46
Average	96.3	49

4.2 Simulation Results

We empirically study the performance of ResNet-18 in this section. Experimental results show that our method gives very good results in terms of accuracy. Because ResNet is a recently proposed model, the literature lacks enough works that compress this network, especially in case of social images. We report the performance of ResNet-18 on the social images from the Flicker dataset. The results have been calculated for a different number of images in the training and testing datasets, shown in Table 2. The results show that the coarse inputs shown in Fig. 2 are transferred to the various classifications of the social images as shown in Fig. 3. We compare the results and calculate the average accuracy obtained for the given set of social images. The average accuracy obtained is 96.3% with 98.7% accuracy being the highest. Results presented in Table 2 verify that the CNN is very appropriate for event classification task of social images.

5 Conclusion

In this paper, CNN is used for event detection from social network images. The sports events have been divided into three classes based on the various image features captured, including racing, field sports and water sports. The proposed framework consists of 17 convolution layers with 64, 128, 256 and 512 filter-maps, applied to the input image dataset alternated with ReLU, sum and max pool layers. The output is given to the Fully Connected Layer that gives the final output. The final result demonstrates the efficiency of the proposed approach for classifying the social images by giving the highest accuracy of 98.7% for event detection. In the future, we will extend the work for other social events and also increase the number of images for training the network to improve the efficiency further.

Acknowledgements This research work is funded by SERB, MHRD, under Grant **[EEQ/-2016/000413]** for Secure and Efficient Communication inside Partitioned Social Overlay Networks project, currently going on at National Institute of Technology Goa, Ponda, India.

References

1. Krizhevsky A, Sutskever I, Hinton GE (2012) Imagenet classification with deep convolutional neural networks. In: Advances in neural information processing systems. pp 1097–1105
2. Karayev S, Trentacoste M, Han H, Agarwala A, Darrell T, Hertzmann A, Winnemoeller H. Recognizing image style. arXiv:1311.3715
3. LeCun Y, Bottou L, Bengio Y, Haffner P (1998) Gradient-based learning applied to document recognition. Proc IEEE 86(11):2278–2324
4. Nielsen MA (2015) Neural networks and deep learning. Determination Press
5. Zhu L (2017) A novel social network measurement and perception pattern based on a multi-agent and convolutional neural network. Comput Electr Eng
6. Simonyan K Zisserman A (2014) Very deep convolutional networks for large-scale image recognition. arXiv:1409.1556
7. Girshick R, Donahue J, Darrell T, Malik J (2014) Rich feature hierarchies for accurate object detection and semantic segmentation. In: Proceedings of the IEEE conference on computer vision and pattern recognition. pp 580–587
8. Shen W, Wang X, Wang Y, Bai X, Zhang Z (2015) Deepcontour: a deep convolutional feature learned by positive-sharing loss for contour detection. In: Proceedings of the IEEE conference on computer vision and pattern recognition. pp 3982–3991
9. Tong C, Li J Zhu F (2017) A convolutional neural network based method for event classification in event-driven multi-sensor network. Comput Electr Eng
10. Xingyi R, Meina S, Haihong E, Junde S (2017) Social media mining and visualization for point-of-interest recommendation. J China Univ Posts Telecommun 24(1):67–86
11. Zeiler MD, Fergus R (2013) Stochastic pooling for regularization of deep convolutional neural networks. arXiv:1301.3557
12. Lee K, Park DC (2015) image classification using fast learning convolutional neural networks. Adv Sci Technol Lett 113:50–55
13. He K, Zhang X, Ren S, Sun J (2016) Deep residual learning for image recognition. In: Proceedings of the IEEE conference on computer vision and pattern recognition. pp 770–778
14. Alqhtani SM, Luo S, Regan B (2015) Fusing text and image for event detection in Twitter. arXiv:1503.03920
15. Keshavamurthy BN, Sharma M, Toshniwal D (2010) Privacy-preserving naive bayes classification using trusted third party and offset computation over distributed databases. In: Proceedings of the information and communication technologies. Springer, Berlin, Heidelberg. pp 529–534
16. Young P, Lai A, Hodosh M, Hockenmaier J (2014) From image descriptions to visual denotations: new similarity metrics for semantic inference over event descriptions. Trans Assoc Comput Linguist 2:67–78

vMeasure: Performance Modeling for Live VM Migration Measuring

Minal Patel, Sanjay Chaudhary and Sanjay Garg

Abstract The datacenter is a group of computing resources like networks, servers, storage, etc. These resources provide on-demand access to cloud computing. The multiple instances can run simultaneously using virtualization, and virtual machines (VMs) are being migrated for load balancing, energy optimization, and fault tolerance. When servers are heavily loaded or running large data, migration of a VM from one host to appropriate another host is a must. The performance of live migration is evaluated by getting optimal migration cost. The existing Xen-based migration is designed based on simple techniques such as LRU and compression. On the other hand, a number of techniques have been applied to predict dirty pages while migrating VM. On both techniques including Xen-based and prediction, the lacking of dirty pages monitoring is the key issue which does not handle VM migration properly. In our research work, we have applied exponential model to handle dirty pages efficiently. The proposed model is designed based on keeping maximum WWS on constant dirty rate. The state of the art is shown that migration time taken by number of iterations will be $(W_i)max/R$ (for memory intensive pages) otherwise $(W_i)avg/R$. The experimental results show that the vMeasure approach is able to give optimal downtime and migration time on three different workloads. The proposed model (called vMeasure approach) is able to reduce 13.94% downtime and 11.76% total migration time on an average.

M. Patel (✉)
Computer Engineering Department, A. D. Patel Institute of Technology,
Anand, Gujarat, India
e-mail: mppatel.adit@gmail.com

S. Chaudhary
School of Engineering and Applied Science, Ahmedabad University,
Ahmedabad, Gujarat, India
e-mail: sanjay.chaudhary@ahduni.edu.in

S. Garg
Institute of Technology, Nirma University, Ahmedabad, Gujarat, India
e-mail: gargsv@gmail.com

© Springer Nature Singapore Pte Ltd. 2019
M. L. Kolhe et al. (eds.), *Advances in Data and Information Sciences*, Lecture Notes
in Networks and Systems 39, https://doi.org/10.1007/978-981-13-0277-0_16

185

1   Introduction

Cloud computing is the framework of recent technology which can set the vision of computing utilities into a reality. The virtualization is the key technology to provide migration of virtual machines. Live migration can overcome the difficulties of process migration. Both have two basic methods to perform migration: (i) pre-copy and (ii) post-copy [1, 2]. The pre-copy algorithm [3] has five stages to complete migration between hosts: pre-migration, reservation, iterative process, commitment, and activation. Modified memory pages are recopied iteratively in pre-copy while post-copy does not allow to copy pages iteratively, which is a different approach than pre-copy [4, 5]. The objective of these approaches is to calculate performance metrics of live migration. These metrics are total migration time, downtime, and the number of pages transferred. During the live migration process, guests are transferred to another physical system with no noticeable downtime. Pre-copy is widely used approach compared to post-copy in live migration system [1].

2   Related Work

The main purpose of this section is to discuss a survey on live migration algorithms. Different methods of dirty page rate are also surveyed.

The primary work of live migration was performed by [1] which had been focused on two major dimensions. First, the paper has discussed issues of working model of Xen, including stop-and-copy condition, and second, it has evaluated basic workloads on a real platform as a benchmark study. In [6], two models of live migration are proposed: (i) performance model and (ii) energy model. The proposed work is able to produce more than 90% prediction accuracy with reductions in both migration and energy cost.

Aldhalaan and Menasc [7] are focused on analytic performance modeling for the uniform dirtying rate. The extension of this analytical model is based on live migration with hot pages. This proposed algorithm is developed to execute hot pages mainly at service downtime phase. Mann et al. [8] have discussed the issues of network performance at the time of live migration. The purpose of this paper is to design a system which cannot degrade network performance for VM migrations. This new concept is known as Remedy in which target hosts are given ratings for migration.

The iAware system is proposed in [9]. The iAware is a lightweight interference aware VM live migration approach. It is designed to analyze the important relationships between VM performance interference and key elements. The tests of benchmark workloads are maintained on a Xen-based cluster. In [10], the services of live VM migration progress management are utilized in the hybrid cloud. The pacer is the very first system able to effectively predict the migration time, synchronizing the migrations of multiple application elements to reduce the overall performance degradation.

Nathan et al. [11] have given a different kind of dirty and skip models. The comparative analysis of these models as well as proposed model has been discussed based on KVM and Xen. Existing models have limitations in three fundamental parameters: (i) the writable working set size, (ii) the number of pages eligible for the skip technique, and (iii) the relation of the number of skipped pages with the page dirty rate and the page transfer rate.

Our approach is dependent on performance modeling [6–9, 11]. The results are shown that evaluation of modified migration log-based dirty page model is able to perform optimally with existing algorithms.

3 Problem Overview

In this section, the Xen-based live migration mechanism is adopted which has been discussed in [3, 12, 13]. Domain0 is the host OS to run VMM program of Xen. It can manage other VMs and make the functioning of each VM for providing reliable computing environment. Xen live migration is measured by three bitmap structures [14]: to_send, to_skip, and to_fix. In Xen environment, a page at the time of migration can be categorized into either dirty page or non-dirty page. The VM is being migrated from host1 to host2 based on network storage. Each VM is configured to run workloads so during the migration process, this workload deals with a set of hot pages which are modified frequently called WWS. The network traffic can be reduced if proper mechanism of transferring dirty pages are applied and hence, system performance can be improved. In this paper, a base mathematical model of live migration is taken as a reference to collect set of WWS pages. The same model is enhanced to improve pre-copy algorithm that is designed with linear regression of both dirty rate and data rate parameters.

If contents get changed after copying page, then the page becomes a dirty page. If the state does not get changed, it is considered non-dirty page. If the page is dirtied in the previous round, then it is known as skip page, which is managed by to_skip bitmap structure [15]. A non-dirty page which is not dirtied in both previous and current iterations is being transferred immediately using to_send bitmap structure of Xen [3].

During the migration process, Xen can identify dirty page by shadow page table mechanism. This mechanism works based on set bits in the bitmap of the dirty log. If bitmap of the dirty log has any set bit, it will request to shadow-based control parameters for managing the page. The process of migration is continued until stop-and-copy condition given in [16].

The objectives are to improve live migration system. The vMeasure is evaluated and compared with existing models. The major contributions of this paper are given below:

- In this work, live migration system is proposed, and it calculates optimal migration costs. Three models are discussed along with proposed model which can facilitate the costs of migration based on monitoring dirty page techniques. The accurate evaluation of dirty pages during migration can improve the performance of live migration system. Following performance models are discussed: (i) migration log-based model [6], (ii) hot pages-based model [7], and (iii) proposed model (vMeasure approach).
- Three models are validated by performing experiments on different workloads. Following workloads are selected for experiments: (i) Kernel, (ii) RUBiS database, and (iii) OLTP (online transaction processing).

4   Performance Models for Live Migration

This section covers types of dirty models in detail. Dirty model is generated by taking the product of iteration period and the page dirty rate. The page dirty rate is recognized as the number of pages dirtied in a second [6].

It is represented by the following equation [16]:

$$page\ dirty\ rate\ =\ pages\ dirtied\ in\ the\ previous\ iteration/time\ taken\ by$$
$$all\ previous\ iterations$$

(1)

The dirty bitmap is generated by taking number of dirty pages during any iteration. Different methods are available to generate dirty bitmap in VM migration. These methods are (i) average dirty page rate method, (ii) exponential dirty page rate method, and (iii) probability dirty page rate method. First, average dirty page rate method is discussed with its current usage in VM migration. After that, exponential methods are discussed in detail with the working of its two models. The probability dirty page rate method is also briefly described in this section. At the end of this section, the proposed modified log-based model (vMeasure approach) is shown, and the theorem is also proposed based on it.

4.1   Average Dirty Page Rate Method

Average dirty page rate method [9, 16] is able to estimate the page dirty rate with an average amount of set bits per second in dirty bitmaps. Average dirty model is represented by the following equation:

$$Avgerage\ set\ bits\ in\ dirty\ bitmap\ per\ second\ = set\ bits\ in\ dirty\ bitmap$$
$$collected\ at\ the\ ith\ second/bitmap\ collection\ period\ = \Sigma\ d_i/p$$

$$(2)$$

It can also be represented by the product of dirty rate & time taken for each iteration divided by the total time taken for all iterations. Average dirty rate model is in use by migration activity of different hypervisors.

4.2 Exponential Dirty Page Rate Method

The exponential method is the second type of method to use in dirty page rate model. The exponential average method works by using an exponential moving average instead of simple average. The first model, migration log-based model is also known as the base model of live migration. The second model is hot pages-based model that has been designed to work with those pages which have considerably more dirtying rate than the other pages.

4.2.1 Migration Log-Based Model

The primary goal of this model [6] is to verify which VM should be migrated with the least amount of migration cost. An essential reason for cost-aware migration decision is how to accurately predict the performance of each VM migration in the datacenter. Parameters for this model are shown in Table 1.

Total pages transferred called network traffic is given by

$$V_{mi} = V_m\ (1 - \lambda^{n+1})/(1 - \lambda) \tag{3}$$

Total migration time is given by

$$T_{mi} = V_m/R\ *\ (1 - \lambda^{n+1})/(1 - \lambda) \tag{4}$$

where $\lambda = D/R$ is the ratio between dirtying rate and transmission rate.

Table 1 Parameters of migration log-based model

V_m: memory size of VM
V_{mi}: network traffic or total number of pages
T_{mi}: total migration time
T_{down}: downtime
R: transmission rate of memory
D: dirty rate of memory
V_{th}: threshold value transferred at the last iteration

During the stop-and-copy phase, the time is spent on resuming VM on destination host other than down time. This time is denoted as T_{resume}, which is constant and set it to 20 ms. The downtime is given by

$$T_{down} = T_n + T_{resume} \tag{5}$$

The number of pre-copying iterations is $log_\lambda[V_{th}/V_m]$ by taking inequality $V_n \leq V_{th}$. In [6], it is stated that efficient migration is performed with the smaller size of memory image and smaller λ because it is able to produce less network traffic. In Xen, frequently modified pages are handled by to_fix bitmap, and it is also observed that hot pages in each iteration are linearly dependent on the size of WWS.

During each iteration of live migration, the value of γ is calculated by $\gamma = aT_i + bD + c$, where a, b, and c are constants which are being learned by taking different values of these constants [6]. This value is formulated by taking multiple observations of model parameters which can estimate γ.

4.2.2 Hot Pages-Based Model

In [7], the analytic performance model is presented to perform better network utilization with proposing optimal α value. Hot pages are pages which have a higher rate of dirtying than the other pages. Parameters of hot pages-based model are given in Table 2.

In this model, it has been shown that a small number of pages have higher chances being modified than others. All pages are migrated in first round (including hot pages) and only hot pages are transferred when VM is down.

Downtime and pre-copy time are represented by the following equations:

$$T_{down} = P_s * \tau * \rho^{n+1} \tag{6}$$

$$T_{pre-copy} = P_s * \tau * [(1 - \rho^{n+1})/(1 - \rho)] \tag{7}$$

Total migration time is calculated based on $T_{precopy}$ and T_{down}.

In this model, the essential part is to find out fraction of α of the pages transferred during migration, and this phase will be ended till such fraction of α of the pages completed. When the value of α increases, downtime gets increased which can affect

Table 2 Parameters of hot pages-based model

P_s: *number of memory pages on VM s* $(0 \leq j \leq Ps, j \in N)$
τ: *time to transmit a page over the network* $\tau = S/B$
n: last iteration number
ρ: *network utilization* $\rho = D\tau$

the performance of a live migration system. It is also stated in [7] that lower α can give more network utilization at the time of migration. These issues are solved in this model by taking the problem of nonlinear optimization through which the optimal value for α can be found.

4.3 Probability Dirty Page Rate Method

In probability page dirty rate method [16], the dirty probability of each page is calculated based on the probability of each page getting dirty. It is the ratio of the dirty rate of recurrence of that page to the summation of page dirty frequency of all the pages. The very first disadvantage of these methods is that they are usually not able to acquire all page dirtying characteristics of applications due to a lot more estimation time. Due to this reason, these methods are not able to predict the number of unique pages dirtied. The second weakness is that it cannot find the same set of pages which are dirtied every second and number of new pages dirtied per second.

4.4 Proposed Dirty Page Rate Model

Three different types of dirty page rate models have been discussed in this section. Average dirty page rate method is useful for those workloads which have constant WWS. Exponential dirty page rate method is discussed for hot pages as shown in two models: (i) migration log-based and (ii) hot pages-based models. The proposed model known as vMeasure is the enhanced model of migration log-based model given in [6]. This model is developed by combining average and exponential-based methods. This model takes maximum WWS value of memory pages when the dirty rate is nonlinear and it takes average dirty page rate for pages of linear dirty rates. The writable working set is designed to measure the set of frequently changed dirty pages in memory at a given point of time. The proposed model is designed with the consideration of memory intensive applications.

Proposed dirty page rate algorithm (modified migration log-based model) is shown in Table 3:

The proposed theorem is based on vMeasure approach. The theorem and proof of the same are given below:

Theorem *The migration time taken by a number of iterations will be* $(W_i)max/R$ *(for memory intensive pages) otherwise* $(W_i)avg/R$.

Proof Let i denotes iteration number, V_{mem} represents memory size, D denotes dirty rate, and bandwidth is represented by R. From proposed dirty page algorithm, it is derived that the maximum model is applied for the following conditions [17]:

- Memory intensive pages and $V_{i+1} > V_i$.
- Memory intensive pages and $V_{i+1} \leq V_{thd}$.

Table 3 Proposed dirty page rate algorithm

Input: Vmem, Vthd, D, R and Output: Vmig, Tmig, Tdown

1. $i = 1$

2. $V_0 < -V_{mem}$

3. for $i = 2$ to max_iteration

4. $T_i < -V_i/R$

5. $\gamma < -aT_i + bD + c, W_i = \gamma T_i D$

6. if(high dirty page)

7. $(W_{i+1})max = max((W_i)max, \gamma T_i D)$

8. $V_{i+1} < -T_i D(W_{i+1})max, (W_{i+1})sum < -(W_i)sum + \gamma T_i D$

9. else

10. $(W_{i+1})sum < -(W_i)sum + \gamma T_i D$

11. $V_{i+1} < -T_i D((W_{i+1})sum/(i + 1)), (W_{i+1})max < -(W_i)max$

12. If (x) then

13. $V_{i+1} = max((W_{i+1})max, (W_{i+1})sum/(i + 1))$

14. $T_{i+1} < -V_{i+1}/R$

15. else

16. $V_{i+1} = min((W_{i+1})max, (W_{i+1})sum/(i + 1))$

17. $T_{i+1} < -V_{i+1}/R$

18. if (stop-and-copy condition) then

19. $T_{down} < -T_{i+1} + T_{resume}$.

Where $x = (V_{i+1} \leq V_{thd} || V_{i+1} > V_i)$ and memory intensive pages

---

And an average model is applied for the following conditions:

- Memory intensive pages and $V_{i+1} < V_i$.
- Not Memory intensive pages.

From the proposed dirty page algorithm (given above), optimal migration time will be $(W_i)max/R$ (for memory intensive pages) otherwise $(W_i)avg/R$ (for non-memory intensive pages). This can be proved by $T_i = V_i/R$ and V_i. V_i is represented by either $V_i = max((W_i)max, (W_i)sum/(i))$ or
$V_i = min((W_i)max, (W_i)sum/(i))$. [here, $(W_i)sum$ represents $(W_i)avg$]

In modified migration log-based model [17], higher memory dirty rate pages are identified by taking a maximum of $((W_{i+1})_{max}, (W_{i+1})_{avg})$ in the algorithm which can decide to take the maximum value of WWS. If the pages are not considered in high dirty rate mode, then the average model is applied by taking a minimum of $((W_{i+1})_{max}, (W_{i+1})_{avg})$ in the algorithm for available W_{i+1} values.

5 Evaluation of Dirty Page Rate Models

In this section, three different workloads (Kernel, RUBiS Database, and OLTP) are collected to evaluate the performance of existing models with the modified migration log model. The experiments are based on real data set [11] which have already given the detail of all three workloads.

As per the experimental results in Fig. 1a, modified migration log model has optimal downtime than other two models for all three workloads. However, the performance of log-based and hot pages models for RUBiS database and OLTP workloads is approximately same.

According to the results of experiment in Fig. 1b, modified migration log model has optimal total migration time than the other two models for all three workloads. However, the performance of log-based and hot pages models for all three workloads is approximately same.

Based on above experiments and results discussion, three observations are derived from proposed vMeasure approach:

- Network traffic: The network traffic for hot pages model and modified log-based model is almost identical for both Kernel compile and OLTP workloads. For Rubis–DB workload, the network traffic for log-based model and hot pages model is identical. The network traffic is generated based on number of pages transferred during iteration. Modified log base model is able to generate more number of unique pages dirtied so it outperforms better among all three workloads. Adaptive dirty rate is able to stop the migration process when page dirty rate is greater than threshold but such condition is not applicable with nonadaptive live migration. This is the reason for nonadaptive live migration in which more number of iterations is executed and hence, network traffic generated is also increased. Our model is able to give optimal performance with reduced network traffic.
- Downtime: The Rubis-DB workload generates more downtime compare to other two models due to its increasing dirty page rate criteria. The number of pages to be sent during stop-and-copy phase is also higher in Rubis-DB compare to other two workloads.

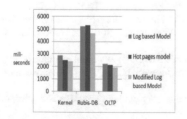

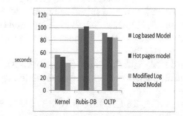

(a) Downtime for Different Workloads

(b) Total Migration Time for Different Workloads

Fig. 1 Performance parameters of live migration

- Total migration time: With adaptive dirty rate and adaptive data rate, the migration time for Rubis-DB worklaod is higher than the other two workloads for all three models. The reason is that page dirty rate increased significantly for Rubis-DB workload, so the migration time is also increased.

6 Conclusion and Future Work

This paper presented performance models to calculate performance metrics. Three models including proposed vMeasure have been compared.

- In this work, we design proposed model to estimate the performance of live migration system. Based on the analysis and experiments, we showed that three models have their migration costs based on network traffic and memory dirtying rate. The maximum throughput can be achieved by adaptive network capacity and adaptive dirtying rate.
- The model would facilitate the administrator of datacenters to explore the selection for choosing optimal migration decision. Results show that the proposed model is able to reduce 11.76% total migration time and 13.94% downtime on an average.
- In future work, skip model can be implemented to improve dirty page mechanism. Other dynamic workloads can also be used to evaluate detail comparative study of different models.

References

1. Clark C, Fraser K, Hand S, Hansen JG, Jul E, Limpach C, Pratt I, Warfield A (2005) Live migration of virtual machines. In: Proceedings of the 2nd conference on symposium on networked systems design and implementation, vol 2, May 2005. USENIX Association, pp 273–286
2. Patel M, Chaudhary, S (2014) Survey on a combined approach using prediction and compression to improve pre-copy for efficient live memory migration on Xen. In: 2014 International conference on parallel, distributed and grid computing (PDGC). IEEE, pp 445–450
3. Barham P, Dragovic B, Fraser K, Hand S, Harris T, Ho A, Neugebauer R, Pratt I, Warfield A (2003) Xen and the art of virtualization. ACM SIGOPS Oper Syst Rev 37(5):164–177
4. Shribman A, Hudzia B (2012) Pre-Copy and post-copy VM live migration for memory intensive applications. Euro-Par Parallel Processing Workshops August 2012. Springer, Berlin, Heidelberg, pp 539–547
5. Ahmad RW, Gani A, Hamid SHA, Shiraz M, Xia F, Madani SA (2015) Virtual machine migration in cloud data centers: a review, taxonomy, and open research issues. The Journal of Supercomputing 71(7):2473–2515
6. Liu H, Jin H, Xu CZ, Liao X (2013) Performance and energy modeling for live migration of virtual machines. Clust comput 16(2):249–264
7. Aldhalaan A, Menasc DA (2013) Analytic performance modeling and optimization of live VM migration. Computer Performance Engineering. Springer, Berlin, Heidelberg, pp 28–42
8. Mann V, Gupta A, Dutta P, Vishnoi A, Bhattacharya P, Poddar R, Iyer A (2012) Remedy: network-aware steady state vm management for data centers. 2012 Networking. Springer, Berlin, Heidelberg, pp 190–204

9. Xu F, Liu F, Liu L, Jin H, Li B, Li B (2014) iaware: Making live migration of virtual machines interference-aware in the cloud. IEEE Trans Comput 63(12):3012–3025
10. Zheng J, Ng TS, Sripanidkulchai K, Liu Z (2013) Pacer: a progress management system for live virtual machine migration in cloud computing. IEEE Trans Netw Serv Manag 10(4):369–382
11. Nathan S, Bellur U, Kulkarni P (2015) Towards a comprehensive performance model of virtual machine live migration. In: Proceedings of the 6th ACM symposium on cloud computing, August 2015. ACM, pp 288–301
12. Jin H, Deng L, Wu S, Shi X, Pan X (2009) Live virtual machine migration with adaptive memory compression. In: IEEE International Conference on Cluster Computing and Workshops CLUSTER'09, August 2009. IEEE, pp 1–10
13. Patel M, Chaudhary S, Garg S. Improved pre-copy algorithm using statistical prediction and compression model for efficient live memory migration. Int J High Perform Comput Netw
14. Cui W, Song, M (2010) Live memory migration with matrix bitmap algorithm. In: 2010 IEEE 2nd symposium on web society (SWS), August 2010. IEEE, pp 277–281
15. Patel M, Chaudhary S, Garg S (2016) Performance modeling of skip models for VM migration using Xen. In: 2016 International conference on computing, communication and automation (ICCCA). IEEE, pp 1256–1261
16. Nathan S, Kulkarni P, Bellur U (2013) Resource availability based performance benchmarking of virtual machine migrations. In: Proceedings of the 4th ACM/SPEC international conference on performance engineering. ACM, pp 387–398
17. Patel M, Chaudhary S, Garg S (2017) Performance modeling and optimization of live migration of virtual machines in cloud infrastructure. In: Research advances in cloud computing, November 2017. Springer, Singapore, ISBN: 978-981-10-5026-8

Latest Trends in Recommender Systems 2017

Poonam Singh, Sachin Ahuja and Shaily Jain

Abstract Recommender systems are in trend from last two decades. Most of the early recommender systems were made using content-based and collaborative filtering methods. Computational intelligence and knowledge base were used in mid-90s. Later recommender systems used social networking, group recommendations, context-aware, and hybrid systems. Today in the era of big data and e-commerce, massive amount of data from web and organizations provide new opportunities for recommender systems. Increased information obtained from high volume of data can be used in recommender systems to provide users with personalized product or service recommendations. From 2013, most of the social networking site like Facebook and Instagram and online shopping giants like Amazon, Flipkart, etc., started providing personalized recommendations to engage and attract users. Most of the review papers are full of applications of recommender systems which do not give the clear idea about methods, techniques, and shortcomings of these systems. Hence, this paper presents an analysis of methods and techniques in current (majorly from 2013) recommender systems. Current challenges have been identified to carry out the work in future.

Keywords Recommender system · Collaborative filtering · Cold start problem

1 Introduction

Recommender system can be defined as programs which attempt to recommend services or products or any suitable item to businesses individuals or any particular

P. Singh (✉) · S. Ahuja · S. Jain (✉)
Department of Computer Science and Engineering, Chitkara University Institute of Engineering and Technology, Chitkara University, Chandigarh, India
e-mail: Poonam.cse@chitkarauniversity.edu.in

S. Jain
e-mail: shaily.jain@chitkarauniversity.edu.in

S. Ahuja
e-mail: sachin.ahuja@chitkara.edu.in

© Springer Nature Singapore Pte Ltd. 2019
M. L. Kolhe et al. (eds.), *Advances in Data and Information Sciences*, Lecture Notes in Networks and Systems 39, https://doi.org/10.1007/978-981-13-0277-0_17

197

user [1]. The recommendations provided by these systems are the predictions on the basis of one or both of the following two criteria:

- Interest of user in an item (these interests are measured using related information about the item or by using information about the users).
- The interaction between user and items.

Purpose of recommender systems:
The target behind the development of recommender systems is to prospect the relevant information and services from massive data which can be used to provide patronized services.
Features of recommender systems:
The ability to guess the user's preference and interest is the most important feature of recommender system.

In the recent years, educational institutions are interested in recommendation system as they are storing massive amount of data. There is multipurpose involvement of recommendation systems in the field of technology-enhanced learning (TEL) as they help users (learners/institutes/teachers) in finding user-specific appropriate learning activities and/or learning objects in the process of teaching/learning, recommending learning pathways or educational scenarios, predicting ranks, and rates [2].

Use of recommender systems now a day is reasonably higher due to the high demand on interpreting data that is stored in educational institutions. Learning Analytics is currently the research field within TEL (Technology-Enhanced Learning) that focuses on understanding and supporting learners based on their data [3].

Recommender systems became extremely interesting for TEL that focuses on understanding and supporting learners based on their data. Research on personalization technologies is a strong topic with a large amount of national and international funded research grants [4].

Popescu found a problem in recommender system which is cold start in 2002. He defined cold start problem as recommendation for user or item which does not have any rating [5].

2 Existing Literature

2.1 Content-Based (CB) Recommendation Techniques

In 2002, Burk et al. gave a definition of content-based technique that machine learning provided a classifier system which is now termed as content-based recommender systems. These systems applied supervised machine learning to induce classifier that can differentiate likings and disliking of specific users for particular items.

In 2007, recommendation systems used CB recommendation on the basis of similarity of item with users choice in past [6].

2.2 Collaborative Filtering (CF)-Based Recommendation Techniques

Collaborating filtering became one of the most popular algorithms during 1994. Many variations of CF were proposed in 2010 in paper [7]. Recommender systems in paper [8] year 2011 used CF. It recommends items by recognizing users whose preferences match with the given user and then system recommend the item liked by these similar users.

Collaborative filtering can be applied to provide personalized recommendation with respect to item, content, and services to the user deal with massive information [9].

Item-based CF and user-based CF are two major classifications of collaborative filtering techniques [10]. Peoples are recommended on the basis of users showed interest in similar type of items; hence, similarity of opinion is considered [11].

Weighted CPC measure was calculated to compute the similitude of users. In weighted CPC, Jaccard metric was applied as weighting scheme with CPC [12]. Pearson correlation-based similarity can be used to measure the similarity between users or items [11].

2.3 Knowledge-Based Recommendation Techniques

Those recommender systems in which knowledge of users and items/products are used to recommend are classified as knowledge-based (KB) recommender system. The personal logic recommender system offers a dialog that effectively walks the user down a discrimination tree of product features [13].

Most common form of KB recommendation technique is reasoning on the basis of cases where case represents items. These recommendation systems recommend by matching the item to the prepared profile of users or on the basis of query [14].

Domain of knowledge can be prepared with help of ontology. Ontology is a method to represent Knowledge formally, it represents the view of domain and association between the perception of theses domains [15]. Further domain ontology can be used to determine the linguistic similarity between items [16].

2.4 Hybrid Recommendation Techniques

To overcome the disadvantages of basic recommendation techniques, two or more recommendation techniques were integrated, and the new technique formed after the integration is termed as hybrid recommendation technique. In hybrid recommendations, CF in integration with other techniques is most widely applied as a solution to ramp-up problem [6].

According to Burke [17], there are seven basic hybrid recommendation techniques. Five most widely used hybrid mechanisms are as follows: switching [18], weighted [19], mixed [20], feature combination, feature augmentation [21, 22].

To prevent the problems in basic recommendation technique like cold start, scalability, and/or sparseness, hybrid recommendation technique is applied which is mostly CF united with preexisting recommendation technique [23].

2.5 Computational Intelligence-Based Recommendation Techniques

In 1956, John McCarthy fabricated a technique which was flexible enough to absorb the change in the situations and named it computational intelligence (CI). Along with flexibility, the techniques have characteristics of generalization, association, discovering, and reasoning.

CI techniques are most applied solution for many problems of intent and implementation in the field of intelligent systems where preexisting formal models are not easy to apply [24].

Most of the model-based recommender systems were made on the basis of Bayesian classifiers [25]. In 2014, a method to spot and resolve the problem of gray-sheep users (users whose choice is different from other users of same profile) was proposed in [26]. In [27], to procure optimal similarity functions, a genetic algorithm method was introduced.

2.6 Social Network-Based Recommendation Techniques

Social network analysis technique was necessity in the world of Internet as social networking tools are expanding exponentially for the systems present on web. Web recommender systems are helping the users engaged in online activities like online social commenting, tagging, and making online friends as done on social networking sites like Facebook [2].

In 2010, recommendation for activities related to education was represented using social relationship of authors in which they considered the number of times same two authors have authored the papers [28].

2.7 Context Awareness-Based Recommendation Techniques

Context can be defined as derived knowledge from a system about any object which can be used to portray the scenario. The object can be any place, person, or situation which is related to the specified user and application (context includes the attributes of user and application) [29].

Wang et al. [30] found that rating function can be expressed with multiple dimensions. He derived that instead of two dimensions (user and item) of rating function it can be refined using context as additional dimension hence rating function from then became function with following dimensions:

User (domain of users),
Item (domain of items), and
Context (contextual information associated with the application).

Authors proposed a contextual-based method to recommend personalized recommendations for tourists [31]. In [4] authors used contextual information in technology-enhanced learning.

2.8 Group Recommendation Techniques

Group recommendation system is fit for those domains where users involve in groups not individually, for example, the cases of movies, music, and events [32]. Group recommendation techniques are applied where physical presence of user is not feasible or if by chance they are present physically they are able to express their preference clearly [33]. Latest in [34] a new model that is nonparametric Bayesian probabilistic latent factor model is introduced which explored the similarity of users within group hence able to handle variable group of size and similarity level.

Quijano-Sanchez et al. [35] used average strategy; Popescu [36] adopted the voting mechanism.

3 Latest Trends

Following methods and techniques were followed during 2013–2017 (chronologically arranged).

Authors proposed an implicit method of recommendation applicable in u-commerce for the real-time environment. This method is based on recency, frequency, and monetary; it overcomes the requirement of irksome responses from users. A method is proposed for u-commerce so it was implemented on cosmetic Internet shopping, and authors claimed that method recommended with higher accuracy [37].

A probabilistic voting scheme was prepared to deal with dynamicity of changing preferences of voters. The algorithm presented score of voters using probabilities hence helped users to state their preferences accurately and claimed that both static and dynamic cases contributed in calculating the preference for deterministic case [37].

A contextual-based method was proposed to recommend personalized recommendations for tourists. In this paper, they derived useful tourist's location from their geotagged photographs and applied collaborative filtering. Author claimed that this method has shown improvement especially in case of long and real tourist visits. The method was applied on Flickr dataset of various cities in China. Authors further concluded that inclusion of context improves the precision of prediction [31].

A hybrid preference model, namely, location-based social matrix factorization, was proposed. This approach is advanced version of matrix factorization in which both of the similarities that are venue similarity and social similarity of users were considered. Author verified the improved effect and efficiency of the method using two datasets deduced from location-based social networks [38].

A web-based tutoring system Protus [39] was implemented in which adaptive hypermedia techniques were applied to improve the quality of system. Results proved that there is improvement in the system when collaborative filtering is applied on the combination of the mining of frequent sequence in the weblogs and learners style recognition. Authors further concluded that the new system gained a noticeable effect on learners' self-learning. But this is a point to be noticed that the implementation of adaptive techniques was not performed locally for each individual recommendation.

A live prototype system, namely, TRecS, was introduced [40]. TRecS is a hybrid recommender framework with three new features:

- A novel hybrid approach integrating three sub-recommender systems,
- New visualization within user interface, and
- New feature was similarity matrix analysis.

Similarity matrix analysis is a thorough analysis of similarity matrices of elemental three single recommender systems which transformed their inter-matrix similarity matrix in a single number.

A fuzzy logic model was prepared to recommend career path. In this model, the data of the students was delineated to be a part of knowledge base. Career test was a part of this recommender system which needs to be filled by students. Fuzzy logic was applied on information collected from knowledge base and career test. System is applied on an institute to recommend the complete path of career to students [41].

An analysis in [42] concluded that job transition model exceeds the performance of cosine similarity, where job transition model was a model with job transitions trained on the career progressions of a set of users. Further performed an analysis on 2,400 LinkedIn users and proved that fusion of career transition with cosine similarity produces recommendations of almost same quality.

A new algorithm CCD++ was introduced which revise the factor of rank—one by one. Author analytically proved that this new algorithm can be easily parallelized on both multi-core and distributed systems. To analyze the performance, CCD++ was performed on a synthetic dataset containing 14.6 billion ratings, on a distributed memory cluster with 64 processors. Results proved that CCD++ is 49 times faster than SGD (stochastic gradient descent) and 20 times faster than ALS (alternating least squares) [43].

A location-aware recommender system LARS can tackle the problem of three types of location-based ratings, and these are spatial ratings for nonspatial items, nonspatial ratings for spatial items, and spatial ratings for spatial items. LARS implemented two techniques user partitioning applicable on spatial ratings and travel penalty applicable on spatial items. They further claimed that this system is scalable, efficient, and hence, this is an improved recommender [44].

A hybrid recommendation algorithm was evaluated to solve gray-sheep users. In the paper, it was proved that if collaborative filtering-based algorithm is applied in presence gray-sheep users, the results are more error prone and inaccurate. In the proposed algorithm, gray-sheep users were searched using clustering, similarity threshold isolated them from rest of the data. Further, it was proved that content-based profile of gray-sheep users can improve the accuracy [26].

In the year 2015, [45] prepared a universal framework for common explanation of recommender system for TEL. In the work, the authors not only prepared framework but also compared their framework with existing frameworks of same area that is TEL. There is scope to work in area of personalized e-learning.

Recommendation system using fuzzy linguistic model was introduced [46]. In this model instead of singular criteria, multiple criteria were applied. Further comparative analysis of this paper proved that performance of fuzzy multi-criteria recommendation is better than multi-criteria recommendation in terms of accuracy and precision. Further authors suggested that accuracy of recommender system can be improved using hybrid fuzzy approaches.

In framework introduced in paper [47] mining of selected items and user was performed these item and users are those which can reflect preferences in case of users and characteristics in case of items. In this framework, one unified model was prepared using these two selection tasks. Further performance is optimized by deriving alternate minimization algorithm.

A novel user similarity measure [48] is proposed and concluded that it can be a good solution for cold start condition. In the work, an asymmetric user similarity method was applied to measure the influence of user on his neighbor and influence of neighbors on user. Further matrix factorization was applied to detect the closeness in users who rated the items.

An approach [49] was proposed to improve stability in recommendation. In the proposed approach, two scalable, general-purpose approaches which are bagging and iterative smoothing were united with different traditional algorithms. Further experiments performed on real-world rating data proved that proposed approach enhanced the stability of recommendation without compromising predictive accuracy.

A linear regression-based algorithm [50] was evaluated to enhance the service recommender systems. A stochastic gradient descent learning procedure is used to integrate both context-dependent preferences (obtained after analyzing three variations of contextual weighting methods) and context-independent preferences (obtained after analyzing two regression models). It was concluded that finding correlation of users' opinions including contextual factors is important, the accuracy can be upgraded by integrating context-dependent and context-independent preferences, and further aspect-related terms can be applied to discriminate users aspect level for different contexts.

A recommendation algorithm based on divide and conquer technique was proposed. The experiment divided the space of users into smaller spaces using multiplicatively weighted Voronoi diagrams and then applied recommendation algorithm to these regions. Spatial Autocorrelation indices are calculated for the smaller region (formed using Voronoi decomposition). The algorithm is verified on Movielens and Book-Crossing datasets and proved that it is scalable and trim down the running time [51].

A recommender system for e-commerce [52] was made to absorb changes because of the diversity of e-commerce and change in purchase patterns of customers. The recommender system is based on periodicity analysis. In order to recommend service to customers instead of using profiles, mining of sequential patterns was performed, and clusters of categories with weight using FRAT analysis were made.

An ontology using the fuzzy linguistic modeling was developed to distinguish the trust between users. As ontology can help in profile building during the process of recommendation to portray the users; hence, the ontology was presented in this paper. Further, a method is provided to update the user profile by accumulated trust information that was collected in trust ontology [53].

A method for recommendations which outperformed other recommendation systems in presence of sparse data was proposed. The method provided more personalized recommendation with the help of user-specific biclusters. In the experiment, a rank score was created for candidate items; this score was a combination of local similarity of biclusters and global similarity. There is no need for tuning parameters as the method is parameter free [54].

An approach in which filters were applied on context-aware recommendation is the matrix factorization. Matrix factorization is applied on the user's ratings (similar to the target) where similar users can be found using contextual information. Authors further suggested that clustering approach can enhance this accuracy of the present approach [55].

A framework [56] was introduced to deal with the problems of collaborative filtering; those problems are data sparsity, first rater, cold start, and scalability jointly. But the work in the paper used ontology of movies, in other words, recommendation system framework specifically made for movies.

A design for persuasive recommendation system [57] is specifically made for Ecuador's students to solve the problem of students to compare and analyze option of courses available on single URL address. Research work covered implementation of automatic filtering and analysis on collection of data here data was determined using characteristics from profiles of candidates and programs (offered by universities). The recommendations can be supported by a set of rules used for matching user's profile (profile of students) with items (courses available case).

A model, namely, alternating direction method (ADM) based nonnegative latent factor (ANLF) [58] can be applied on incomplete matrices in collaborative filtering. The model overcomes the disadvantages of NMF (nonnegative matrix factorization) based CF recommenders that is low convergence rate and high complexity. ANLF can be efficiently applied on sparse data as complexity in ANLF is linear with size of data.

Authors proposed a new factorization model that applied SVD (singular-value decomposition) with rating elicitation for approximation of matrix having missing values hence named it enhanced SVD. As all ratings were elicited simultaneously, it reduced the computational cost. As the model is trained by very high quality elicited ratings obtained from very dense matrix, it can enhance the performance when integrated with other SVD-based algorithms [59].

Sparsity is one of the major problems in traveling system specifically when user wants to visit a new place; hence, a recommender system [60] was prepared in which they considered three variables; these are time, user, and category. Missing values were also treated using context awareness.

In the field of photography selection of site is one of the prime tasks hence a context-aware recommendation [61] was introduced. This system can recommend real-time sites for photography. The approach utilized the metadata of images from camera to fetch basic information which can be used to make context-aware recommendations. They considered three dimensions, viz., camera, location, and time, based on features, fame, and oneness.

Solution for cold start problem [62] was evaluated by classifying them into two types: items without ratings and items with few ratings. This is a model which combined the neural network's deep learning with collaborative filtering. Where neural network is used to eradicate the item theme properties and cold start problem was resolved using SVD++.

A context- and social-aware real-time recommender system for mobile users was proposed. This recommender system is specially prepared for tourists [63].

4 Future Challenges

- Current recommender systems being used in the field of TEL are based on predefined strategy and lacks in portability and adaptability [36].

- Parveen et al. suggested that TEL recommender systems can be analyzed and compared more accurately and efficiently if we design and develop a framework which is well defined, comprehensive, and unified [46].
- Cold start problems can be tackled using rating elicitation in which profiles of user and items can be prepared with the help of initial interviews. Elicitation of most useful rating is still a challenge [47].
- Collaborative filtering still suffers from several major limitations like data sparsity and scalability [48].
- There is need of new methods for measuring of influence of those contextual conditions which are independent of explicit and baseline rating predictors [49].
- Web recommender systems suffer from several shortcomings, such as scalability, sparsity, first rater, and cold start problems [47, 48, 56].

5 Conclusions

In the analysis of latest recommender systems, we found that from year 2013, most of the recommender systems are just the applications of previous methods in different domains. Enhanced techniques majorly revolved around the problems of sparsity, scalability, and accuracy. Moreover, computation cost was high due to the presence of high volume of data; we can use parallelization of methods to improve the speed.

As summarized in Table 1 most of the recommender systems followed one of the following techniques: hybrid technique (fusion of two or more than two techniques), advanced matrix factorization, fuzzy logics, context-aware, and location-aware systems. Out of these techniques in the recent recommendation systems, context-aware technique is the most widely applied technique. During these years, recommender systems solved the following problems completely: low convergence, high complexity, context-aware mobile recommendation, gray-sheep users' problem, and spatial and nonspatial ratings.

Despite the presence of many methods and technique, the accuracy of recommendation is still an open problem. Following are some other problems which are still not resolved completely: dynamic choice of users in different contexts, personalized recommendation, cold start problem, sparsity, and scalability.

Table 1 Summary of latest trends

Approach	Method	Advantage	Future scope
Probabilistic	Presented score of user using probabilities	Deal with the dynamicity of changing preference [36]	Recommendation for group was suggested but did not consider group dynamics which can be a scope for future work Applied generalized plurality rule which is not truthful [31] Can be applied in other fields where users change their choices dynamically
Adaptive hypermedia	Collaborative filtering is applied on the combination of the mining of frequent sequence in the weblogs and learners style recognition [39]	Increased efficiency	Implementation of adaptive techniques was not performed locally for each individual recommendation
Advanced matrix factorization	Revise the factor of rank—one by one performed on a synthetic dataset containing 14.6 billion ratings, on a distributed memory cluster with 64 processors [43]	Can be easily parallelized on both multi-core and distributed systems CCD++ is 49 times faster than SGD (stochastic gradient descent) and 20 times faster than ALS (alternating least squares)	Results were compared with SGD and ALS only further all results were based on only one dataset movie lens
Fuzzy	Instead of singular criteria, multiple criteria were used [46]	Fuzzy multi-criteria recommendation is better than multi-criteria recommendation in terms of accuracy and precision	Accuracy of recommender system can be improved using hybrid fuzzy approaches
Matrix factorization	Matrix factorization is applied on the user's ratings (similar to the target) where similar users can be found using contextual information [55]	Deal data sparsity	Ontology-based similarity method can be applied to improve the accuracy in case of highly sparse data and small trained data
Advanced collaborative filtering	Alternating direction method (ADM) based nonnegative latent factor (ANLF) [58]	Deal low convergence rate and high complexity Can deal extreme sparse data Handle incomplete matrix in collaborative filtering	Can be useful in learning analytics having extreme sparsity under nonnegative constraints
Enhanced SVD	Rating elicitation for approximation of matrix having missing values [59]	Reduced the computational cost	Cannot deal cold start problem
Hybrid (context aware)	Context-aware tensor decomposition is used to model preferences of users and for rating weighted hypertext induced topic search point of interest is used [60]	Deal sparsity Missing values dealt using context awareness	Cannot deal cold start problem
	Combined the neural network's deep learning with collaborative filtering [62]	Deal cold start problem	Storage and computation cost is high

References

1. Bobadilla J, Ortega F, Hernando A, Gutiérrez A (2013) Recommender systems survey. Knowl-Based Syst 46:109–132
2. Lu J, Dianshuang W, Mao M, Wang W, Zhang G (2015) Recommender system application developments: a survey. Decis Support Syst 74:12–32
3. Drachsler H, Verbert K, Santos OC, Manouselis N (2015) Panorama of recommender systems to support learning. In: Recommender systems handbook. Springer, US, pp 421–451
4. Verbert K, Manouselis N, Ochoa X, Wolpers M, Drachsler H, Bosnic I, Duval E (2012) Context-aware recommender systems for learning: a survey and future challenges. IEEE Trans Learn Technol 5(4):318–335
5. Schein AI, Popescul A, Ungar LH, Pennock DM (2002) Methods and metrics for cold-start recommendations. In: Proceedings of the 25th annual international ACM SIGIR conference on research and development in information retrieval. ACM, pp 253–260
6. Pazzani MJ, Billsus D (2007) Content-based recommendation systems. In: The adaptive web. Springer, Berlin, Heidelberg, pp 325–341
7. He J, Chu WW (2010) A social network-based recommender system (SNRS). In: Data mining for social network data. Springer, US, pp 47–74
8. Lops P, Gemmis MD, Semeraro G (2011) Content-based recommender systems: state of the art and trends. In: Recommender systems handbook. Springer, US, pp 73–105
9. Adomavicius G, Tuzhilin A (2005) Toward the next generation of recommender systems: a survey of the state-of-the-art and possible extensions. IEEE Trans Knowl Data Eng 17(6):734–749
10. Sarwar B, Karypis G, Konstan J, Riedl J (2001) Item-based collaborative filtering recommendation algorithms. In: Proceedings of the 10th international conference on World Wide Web. ACM, pp 285–295
11. Deshpande M, Karypis G (2004) Item-based top-n recommendation algorithms. ACM Trans Inf Syst (TOIS) 22(1):143–177
12. Shambour Q, Jie L (2011) A hybrid trust-enhanced collaborative filtering recommendation approach for personalized government-to-business e-services. Int J Intel Syst 26(9):814–843
13. Trewin S (2000) Knowledge-based recommender systems. Encycl Lib Inf Sci 69(Supplement 32):180
14. Smyth B (2007) Case-based recommendation. In: The adaptive web, pp 342–376
15. Middleton SE, Roure DD, Shadbolt R (2009) Ontology-based recommender systems. In: Handbook on ontologies. Springer, Berlin, Heidelberg, pp 779–796
16. Cantador I, Bellogín A, Castells P (2008) A multilayer ontology-based hybrid recommendation model. Ai Commun 21(2–3):203–210
17. Burke R (2007) The adaptive web. Chapter hybrid web recommender systems, p 377
18. Billsus D, Pazzani MJ (2000) User modeling for adaptive news access. User Model User-Adap Inter 10(2–3):147–180
19. Mobasher B, Jin X, Zhou Y (2004) Semantically enhanced collaborative filtering on the web. In: Web mining: from web to semantic web. Springer, Berlin, Heidelberg, pp 57–76
20. Smyth B, Cotter P (2000) A personalised TV listings service for the digital TV age. Knowl-Based Syst 13(2):53–59
21. Wilson DC, Smyth B, Sullivan Derry O (2003) Sparsity reduction in collaborative recommendation: a case-based approach. Int J Pattern Recognit Artif Intell 17(05):863–884
22. Sullivan DO, Smyth B, Wilson D (2004) Preserving recommender accuracy and diversity in sparse datasets. Int J Artif Intel Tools 13(01):219–235
23. Bellogín A, Cantador I, Díez F, Castells P, Chavarriaga E (2013) An empirical comparison of social, collaborative filtering, and hybrid recommenders. ACM Trans Intel Syst Technol (TIST) 4(1):14
24. Abbas A, Zhang L, Khan SU (2015) A survey on context-aware recommender systems based on computational intelligence techniques. Computing 97(7):667–690
25. Amatriain X, Pujol JM (2015) Data mining methods for recommender systems. In: Recommender systems handbook. Springer, US, pp 227–262

26. Ghazanfar MA, Prügel-Bennett A (2014) Leveraging clustering approaches to solve the gray-sheep users problem in recommender systems. Expert Syst Appl 41(7):3261–3275
27. Bobadilla J, Ortega F, Hernando A, Alcalá J (2011) Improving collaborative filtering recommender system results and performance using genetic algorithms. Knowl-Based Syst 24(8):1310–1316
28. Hwang S-Y, Wei C-P, Liao Y-F (2010) Coauthorship networks and academic literature recommendation. Electron Commer Res Appl 9(4):323–334
29. Dey AK, Abowd GD, Salber D (2001) A conceptual framework and a toolkit for supporting the rapid prototyping of context-aware applications. Human Comput Interact 16(2):97–166
30. Wang W, Zhang G, Jie L (2016) Member contribution-based group recommender system. Decis Support Syst 87:80–93
31. Majid A, Chen L, Chen G, Mirza HT, Hussain I, Woodward J (2013) A context-aware personalized travel recommendation system based on geotagged social media data mining. Int J Geogr Inf Sci 27(4):662–684
32. O'connor M, Cosley D, Konstan JA, Riedl J (2001) PolyLens: a recommender system for groups of users. In: ECSCW 2001. Springer, Netherlands, pp 199–218
33. Jameson A, Smyth B (2007) Recommendation to groups. In: The adaptive web, pp 596–627
34. Chowdhury N, Cai X (2016) Nonparametric Bayesian probabilistic latent factor model for group recommender systems. In: International conference on web information systems engineering. Springer International Publishing, pp 61–76
35. Quijano-Sanchez L, Recio-Garcia JA, Diaz-Agudo B, Jimenez-Diaz G (2013) Social factors in group recommender systems. ACM Trans Intel Syst Technol (TIST) 4(1):8
36. Popescu G (2013) Group recommender systems as a voting problem. In: International conference on online communities and social computing. Springer, Berlin, Heidelberg, pp 412–421
37. Cho YS, Moon SC, Jeong S-P, Oh I-B, Ryu KH (2013) Clustering method using item preference based on RFM for recommendation system in u-commerce. In: Ubiquitous information technologies and applications. Springer, Dordrecht, pp 353–362
38. Yang D, Zhang D, Yu Z, Wang Z (2013) A sentiment-enhanced personalized location recommendation system. In: Proceedings of the 24th ACM conference on hypertext and social media.ACM, pp 119–128
39. Vesin B, Klašnja-Milićević A, Ivanović M, Budimac Z (2013) Applying recommender systems and adaptive hypermedia for e-learning personalizatio. Comput Inform 32(3):629–659
40. Hornung T, Ziegler C-N, Franz S, Przyjaciel-Zablocki M, Schätzle A, Lausen G (2013) Evaluating hybrid music recommender systems. In: Proceedings of the 2013 IEEE/WIC/ACM international joint conferences on web intelligence (WI) and intelligent agent technologies (IAT), vol 01. IEEE Computer Society, pp 57–64
41. Razak TR, Hashim MA, Noor NM, Halim IHA, Shamsul NFF (2014) Career path recommendation system for UiTM Perlis students using fuzzy logic. In: 2014 5th international conference on intelligent and advanced systems (ICIAS). IEEE, pp 1–5
42. Heap B, Krzywicki A, Wobcke W, Bain M, Compton P (2014) Combining career progression and profile matching in a job recommender system. In: Pacific Rim international conference on artificial intelligence. Springer, Cham, pp 396–408
43. Yu H-F, Hsieh C-J, Si S, Dhillon IS (2014) Parallel matrix factorization for recommender systems. Knowl Inf Syst 41(3):793–819
44. Sarwat M, Levandoski JJ, Eldawy A, Mokbel MF (2014) LARS*: an efficient and scalable location-aware recommender system. IEEE Trans Knowl Data Eng 26(6):1384–1399
45. Khribi MK, Jemni M, Nasraoui O (2015) Recommendation systems for personalized technology-enhanced learning. In: Ubiquitous learning environments and technologies. Springer, Berlin, Heidelberg, pp 159–180
46. Parveen H, Ashraf M, Parveen R (2015) Improving the performance of multi-criteria recommendation system using fuzzy integrated meta heuristic. In: 2015 international conference on computing, communication and automation (ICCCA). IEEE, pp 304–308
47. Zhang X, Cheng J, Qiu S, Zhu G, Hanqing L (2015) Dualds: a dual discriminative rating elicitation framework for cold start recommendation. Knowl-Based Syst 73:161–172

48. Pirasteh P, Hwang D, Jung JJ (2015) Exploiting matrix factorization to asymmetric user similarities in recommendation systems. Knowl-Based Syst 83:51–57
49. Adomavicius G, Zhang J (2015) Improving stability of recommender systems: a meta-algorithmic approach. IEEE Trans Knowl Data Eng 27(6):1573–1587
50. Chen G, Chen L (2015) Augmenting service recommender systems by incorporating contextual opinions from user reviews. User Model User-Adap Inter 25(3):295–329
51. Das J, Majumder S, Dutta D, Gupta P (2015) Iterative use of weighted voronoi diagrams to improve scalability in recommender systems. In: Pacific-Asia conference on knowledge discovery and data mining. Springer, Cham, pp 605–617
52. Cho YS, Moon SC (2015) Recommender system using periodicity analysis via mining sequential patterns with time-series and FRAT analysis. JoC 6(1):9–17
53. Martinez-Cruz C, Porcel C, Bernabé-Moreno J, Herrera-Viedma E (2015) A model to represent users trust in recommender systems using ontologies and fuzzy linguistic modeling. Inf Sci 311:102–118
54. Alqadah F, Reddy CK, Junling H, Alqadah HF (2015) Biclustering neighborhood-based collaborative filtering method for top-n recommender systems. Knowl Inf Syst 44(2):475–491
55. Codina V, Ricci F, Ceccaroni L (2016) Distributional semantic pre-filtering in context-aware recommender systems. User Model User-Adap Inter 26(1):1–32
56. Moreno MN, Saddys S, Vivian FL, María DM, Sánchez AL (2016) Web mining based framework for solving usual problems in recommender systems. A case study for movies recommendation. Neurocomputing 176:72–80
57. Pinto FM, Estefania M, Cerón N, Andrade R, Campaña M (2016) iRecomendYou: a design proposal for the development of a pervasive recommendation system based on student's profile for ecuador's students' candidature to a scholarship. In: New advances in information systems and technologies. Springer, Cham, pp 537–546
58. Luo X, Zhou MC, Li S, You Z, Xia Y, Zhu Q (2016) A nonnegative latent factor model for large-scale sparse matrices in recommender systems via alternating direction method. IEEE Trans Neural Netw Learn Syst 27(3):579–592
59. Guan X, Li C-T, Guan Y (2016) Enhanced SVD for collaborative filtering. In: Pacific-Asia conference on knowledge discovery and data mining. Springer International Publishing, pp 503–514
60. Ying Y, Chen L, Chen G (2017) A temporal-aware POI recommendation system using context-aware tensor decomposition and weighted HITS. Neurocomputing 242:195–205
61. Rawat YS, Kankanhalli MS (2017) ClickSmart: a context-aware viewpoint recommendation system for mobile photography. IEEE Trans Circuits Syst Video Technol 27(1):149–158
62. Wei J, He J, Chen K, Zhou Y, Tang Z (2017) Collaborative filtering and deep learning based recommendation system for cold start items. Expert Syst Appl 69:29–39
63. Colomo-Palacios R, García-Peñalvo FJ, Stantchev V, Misra S (2017) Towards a social and context-aware mobile recommendation system for tourism. Pervasive Mob Comput 38:505–515

Swarm Intelligence for Feature Selection: A Review of Literature and Reflection on Future Challenges

Nandini Nayar, Sachin Ahuja and Shaily Jain

Abstract Feature subset selection is considered to be a significant task in data mining. There is a need to develop optimal solutions with higher order of computational efficiency. In this paper, we reviewed the problems encountered during the process of feature selection and how swarm intelligence has been used for extraction of optimal set of features. It also gives a concise overview of various swarm intelligence algorithms like particle swarm optimization, ant colony optimization, bacteria foraging algorithms, bees algorithm, BAT algorithms and the various hybrid approaches that have been discovered using these approaches.

Keywords Swarm intelligence · Data mining · Optimization

1 Introduction

The dataset with a large number of features (attributes) is known as high-dimensional dataset. The accuracy of a model can be enhanced by using a wisely selected subset of features, rather than using every available feature. Researchers have proved that feature subsets give superior results (in terms of classification accuracy) as compared to entire set of features.

N. Nayar (✉) · S. Ahuja · S. Jain
Department of Computer Science and Engineering, Chitkara University Institute of Engineering and Technology, Chitkara University, Chandigarh, India
e-mail: nandini.nayar@chitkarauniversity.edu.in

S. Ahuja
e-mail: sachin.ahuja@chitkara.edu.in

S. Jain
e-mail: shaily.jain@chitkarauniversity.edu.in

© Springer Nature Singapore Pte Ltd. 2019
M. L. Kolhe et al. (eds.), *Advances in Data and Information Sciences*, Lecture Notes in Networks and Systems 39, https://doi.org/10.1007/978-981-13-0277-0_18

211

1.1 Feature Selection

Databases that contain a huge amount of data with high-dimensionality are exceptionally common. Feature selection may be referred as the process of selecting the optimal subset of features based on certain criterion [1]. Feature selection and variable selection are essential for knowledge discovery from huge amount of data. Traditional variable selection methods (e.g. Cp, AIC and BIC) involve a combinatorial optimization problem that is NP-hard. Their computational time increases with the dimensionality. Due to this expensive computational cost, these traditional procedures become fairly infeasible for high-dimensional data analysis. Thus, more innovative variable selection procedures are essential to deal with high-dimensionality [2]. Feature selection has a vital role in data preprocessing techniques that are used for data mining [3].

1.2 Classification and Clustering

Classification is used to classify data into various classes according to some criteria that can be used to envisage the group membership for various data instances. The major classification methods [4] include decision trees, Bayesian networks, k-nearest neighbour (KNN), genetic algorithms, case-based reasoning and the fuzzy logic techniques.

Feature selection methods can be wrapper-based, filter-based or embedded. The classification models that rely on filter-based approach perform the feature selection during the preprocessing stage but ignore the learning algorithm. The wrapper-based models perform the process of feature selection by considering every possible subset, and the subsets of features are ranked according to their predictive power [5]. However, the embedded model [6], selects features while considering the design of classifier. Data clustering refers to the method of assembling the data into various clusters, in such a way that each cluster contains data of massive similarity among each other and a high level of distinction with the data of other clusters [7].

1.3 Swarm Intelligence (SI)

The term swarm is used for a group of animals like birds, fishes, insects (such as ants, termites and bees) that exhibit a cooperative behaviour. The intellect of the swarm basically lies in the set of interactions among these individuals and their environment [8]. Moreover, in combined efforts, they are habitually able to solve a variety of complex tasks that demonstrate their group intelligence [9]. As a result, many SI-based systems are being developed by observing the efficacy and successful behaviour of these swarms in nature that allows them to solve complex problems

[10]. Various approaches have been developed, based on collective performance of population of these individuals with restricted abilities, devoid of any central control that are being used in diverse fields [11]. The advantages of SI approaches over the conventional techniques include their 'robustness' and 'flexibility' [12]. SI-based algorithms are being widely deployed in diverse real-life problems [13].

1.4 Importance of Swarm Search in Feature Selection

- A dataset may contain redundant or inappropriate features. These features must be removed as it would have a negative influence on performance of classifier model. Moreover, it becomes a typically difficult task to determine the usefulness of each feature. Therefore, SI may be used to overcome these problems and may lead to more accuracy [14].
- It is not necessary for the user to explicitly limit the size of feature set, because swarm search can provide the best possible length of the feature set [15].
- The classification problem has become very complex, especially when the number of possible different combinations of variables is large. For that reason, SI algorithms are found to be more apt to solve these complicated problems. Moreover, they are also capable of finding a global optimal solution [16].

2 Literature Studied

To develop consistent and reliable models for solving data mining problems, it is anticipated that the features must have useful information, and the number of features must be as small as possible. If the right subset of features is chosen, the accuracy of model is improved [17]. Mostly, it is hard to identify which features are important and which are not. Sometimes, eliminating an attribute that has a less amount of correlation with the target variable may have a considerable influence in composite dependencies with some other attributes, which may deteriorate the performance [18, 19]. Dimensionality reduction refers to the task of reducing the dimensionality of patterns while preserving vital information. If the significance of particular features is recognized in advance, one can considerably reduce the data space size using feature selection. But if it is difficult to judge the significance of features due to large number of dimensions, then we can use dimensionality reduction methods [20]. This reduced subset can be fed as input to various swarm intelligence algorithms for extraction of classification rules [21].

Swarm intelligence (SI) paradigm consists of two sub-fields: ACO (ant colony optimization) and PSO (particle swarm optimization). SI algorithms are considered to be adaptive strategies [22] and can be applied to search and optimization domains. Ant colony optimization (ACO) is among the oldest techniques for optimization that was stimulated by swarm intelligence [23]. Particle swarm optimization (PSO) is an

Table 1 Applications of SI algorithms

Algorithm	Applications
Particle swarm optimization (PSO)	Bioinformatics, task assignment, flowshop scheduling, industrial applications, biomedical image registration, constraint handling, neural network, electromagnetic applications, multi-objective problems and power systems [12]
Ant colony optimization (ACO)	Neural network training, rule extraction, weight optimization, Bayesian network structure learning [10], bioinformatics, biomedical and multi-objective problems [12]
BAT algorithm	Continuous optimization, combinatorial optimization and scheduling, inverse problems and parameter estimation, image processing, fuzzy logic and data mining [34]
Firefly algorithm	Multi-modal optimization, industrial optimization, wireless sensor networks, antenna design, robotics and semantic web [27]
Cuckoo algorithm	Business optimization applications [30], design optimization problems [31], improved Cuckoo algorithm used in unconstrained optimization problems [32]

eminent SI algorithm, based on social behaviour of birds or fish [24]. The concept of BAT-inspired algorithm is derived from echolocation behaviour of the microbats that use echolocation to find their prey and differentiate the diverse insects even in utter darkness [25]. Bees algorithm (BA) is based on the food-searching behaviour of a group of honey bees [26]. Firefly algorithm is based on collective decisions that are connected with the flashing light behaviour of fireflies [27] and is an area of continuous research [28, 29]. Cuckoo search algorithm is also gaining popularity for solving various optimization problems [30–32]. In [33], fruit fly optimization algorithm is used for optimization of artificial neural network model.

Some broad applications of various swarm intelligence algorithms have been discussed in Table 1.

The SI algorithms are studied in five major categories that include PSO, ACO, ABS (ant-based sorting), hybrid approaches and other SI-based approaches. SI-based approaches can be combined with some other meta-heuristics or AI methods [35]. A novel computational technique has been proposed in [36] that uses artificial fish swarm algorithm (AFSA) as a tool for rule classification. In [37], the dataset comprises microarray cancer data, and experimental comparison is performed on five methods: Random Forest, C4.5, Bagging C4.5, AdaBoosting C4.5 and LibSVM's. The results clearly demonstrate that all these methods are benefitted by data preprocessing. In [38], SI-based techniques for rule-mining in medical domain (from STULONG dataset) have been presented. An algorithm based on ACO/PSO with quality measure has been found to perform best in terms of accuracy (87.43%).

CBPSO (chaotic binary particle swarm optimization) method has been proposed for feature selection [39] that could achieve superior classification accuracy for six UCI datasets that include Wine, Vehicle, WDBC, Ionosphere, Satellite and SONAR data. CBPSO(2) has provided superior results for classification accuracy, i.e. Wine (99.44%), Vehicle (75.06%), WDBC (97.54%), Ionosphere (96.02%), Satellite (91.45%) and Sonar (95.75%). In [40], an algorithm has been proposed that is based on PSO with fuzzy-fitness function for selecting optimal feature subset among various available features.

In [41], a method has been proposed for feature selection, based on ant colony optimization and rough set theory, and performance of this method is compared with C4.5, that leads to the conclusion that proposed method is flexible. A hybrid algorithm based on PSO and ACO has been proposed for performing hierarchical classification. The proposed algorithm has been assessed on a biological dataset and the results demonstrate that this hybrid algorithm has provided superior results for classification accuracy and comprehensibility [42]. A neural network algorithm has been proposed in [43]. The dataset, which is derived from medical records, is used to study regularities in readmission of patients. The Random Forest (RF) and the Support Vector Machine (SVM) classifiers are used to model patients' characteristics. The results specify that the proposed model achieves 78.4% on prediction accuracy and 97.3% on sensitivity.

Data streams were represented by five datasets of various domains that had large amount of features from UCI archive. A new version of accelerated PSO, i.e. FS-APSO has been proposed. Results demonstrate that the highest average gain obtained with the group of incremental algorithms, coupled with FS-APSO was 0.7906 [44]. The modified PSO and its hybridization with some other algorithms have been efficiently used to tackle the problems related to clustering, especially in case of high-dimensional datasets [45], that leads to superior cluster formation, that eventually leads to the improved result of predictions and better analysis of data.

In [46], a hybrid approach has been proposed using integrated statistical method and discrete particle swarm optimization. It has been applied on breast cancer dataset, which has helped to improve accuracy (98.71%). In [47], two datasets were used: Enzyme-Prosite and GPCR-Prosite and the proposed algorithm (PSO-GA) that is helpful in selecting smaller number of features. In [48], a postsynaptic dataset is used. There were 443 original attributes. The discrete PSO algorithm could find smaller subset of attributes. With the help of fewer attributes, the binary PSO and DPSO could obtain better predictive accuracy as compared to the classification that involved all the 443 attributes. A PSO-based approach (PSO+SVM) has been developed in [49] for parameter determination and selecting features in SVM. The average accuracy rate with feature selection has more average accuracy rate that has been observed on 17 datasets from UCI repository. In [50], real-time data of earthquake was used for training and testing. The results demonstrate that artificial bee colony algorithm has shown 99.89% accuracy. In [51], thyroid gland dataset has been used, and improved simplified swarm optimization (SSO) classifier design method has shown to provide improved accuracy (99.06%) after implementing the elite concept.

A hybrid intrusion detection system has been proposed in [52] named as IDS-RS (intelligent dynamic swarm-based rough set) for classification of intrusion data that has achieved higher classification accuracy (93.3%). In [53], a churn dataset has

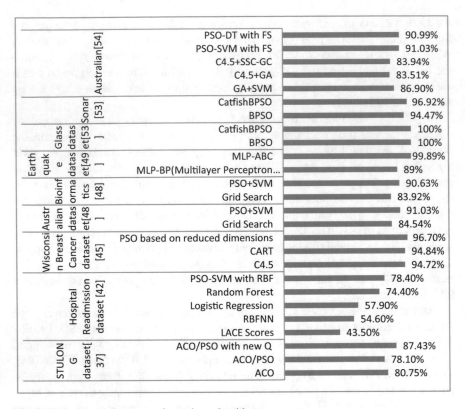

Fig. 1 Comparison of accuracy for various algorithms

been used, and the results show that Antminer+ has included domain knowledge, leading to models that are intuitively correct and more accurate. In [54], an optimization algorithm, known as catfish binary particle swarm optimization algorithm (CatfishBPSO), has been presented that has increased the classification accuracy for Wine dataset (99.44%), Segmentation dataset (97.92%) and WDBC (98.17%). In [55], the data of various banks of Taiwan has been collected and the results show that the proposed models (PSO-SVM and PSO-DT) were quite efficient in searching useful subset of features, with high accuracy rates.

The value of accuracy achieved by SI algorithms or hybrid approaches of SI algorithms has been summarized in the table below. This demonstrates that in almost all cases, the SI algorithms have achieved superior accuracy values as compared to other algorithms as shown in Table 2. Moreover, the cases where a hybrid approach has been followed, it has been observed that the hybrid approaches provide greater level of accuracy as compared to non-hybrid algorithms, as shown in Fig. 1.

Table 2 Comparison of classification accuracies for test datasets

Dataset	Algorithm	Accuracy (%)
STULONG dataset [38]	ACO	80.75
	ACO/PSO	78.1
	ACO/PSO with new Q	87.43
Hospital readmission dataset [43]	LACE scores	43.5
	RBFNN	54.6
	Logistic regression	57.9
	Random Forest	74.4
	PSO-SVM with RBF	78.4
Wisconsin breast cancer dataset [46]	C4.5	94.72
	CART	94.84
	PSO based on reduced dimensions	96.7
Australian dataset [49]	Grid Search	84.54
	PSO+SVM	91.03
Bioinformatics [49]	Grid Search	83.92
	PSO+SVM	90.63
Earthquake dataset [50]	MLP-BP (multilayer perceptron backpropagation)	89
	MLP-ABC	99.89
Glass dataset [54]	BPSO	100
	CatfishBPSO	100
Sonar [54]	BPSO	94.47
	CatfishBPSO	96.92
Australian [55]	GA+SVM	86.90
	C4.5+GA	83.51
	C4.5+SSC-GC	83.94
	PSO-SVM with FS	91.03
	PSO-DT with FS	90.99

3 Future Challenges

In case of datasets with higher dimensionality, selecting an optimal subset of features among the available features is a challenging problem. Although lots of research efforts are going on to promote various feature selection methods, there is no feature selection algorithm that yields better results across datasets [15]. An adaptive intelligent method can be developed, which is able to automatically identify the dynamic alterations of the characteristics of dataset that may improve the classification algorithm [56] in a self-directed manner. Moreover, context-aware selection of features is a useful field for future research. Many researchers have studied the heuristics to deal with the problems related to feature selection. However, the designs of these proposed methods have been restricted by some constraints, e.g. size of resulting feature set may be fixed or feature set becomes negligible. Though a comprehensive method [57] can be used to predict the most valuable features, making the use of brute-force method, but this may not be realistic for high-dimensional biomedical data. There exist many other algorithms, e.g. glow-worm, monkey-based, lion-based and wolf-based algorithms. These algorithms are not widely used in application-oriented domains like medical domain, social science and business studies [58].

Exhaustive search methods are available for finding the most suitable feature subsets. However, these methods become quite impractical for data streams which are typically very high in dimensions and if the amount of dimensions reaches to infinity [44].

Multi-label learning is emerging as one of the major research areas in the field of data mining. It is projected to solve the ambiguity problem in text-based classification. Moreover, to mine useful information from massive amount of data, the implication of multi-label learning is becoming more noticeable [59, 60]. Another vital area of interest can be parameter tuning and parameter control. A detailed study of parameters must be carried out to find the best parameter settings and a method must exist that can tune the parameters automatically that helps to achieve optimal performance. Another challenging question according to [34] is the issue of speeding up the convergence of an algorithm.

4   Conclusion

In this paper, various problems faced in feature selection for datasets with higher dimensionality are discussed. Hence, the role of swarm intelligence in selecting the relevant features in high-dimensional data has been studied. This proves swarm intelligence as an effective framework for those algorithms that are concerned with unorganized high-dimensional data. We have studied various SI algorithms which, in general, can be contemplated to be a proficient tool which solves optimization problems that are sophisticated in terms of computational ease. Thus, proving SI to be superior to other traditional feature selection techniques. A review of research in the area of data mining and swarm intelligence has been presented and the results summarized in Table 2 clearly demonstrate that the SI approaches and their hybrid approaches have achieved superior accuracy. The paper also includes the future challenges in feature selection.

References

1. Coello CC, Dehuri S, Ghosh S (eds) (2009) Swarm intelligence for multi-objective problems in data mining, vol 242. Springer

2. Fan J, Li R (2006) Statistical challenges with high dimensionality: feature selection in knowledge discovery. Proc Int Congr Math 1–27
3. Han J, Pei J, Kamber M (2011) Data mining: concepts and techniques. Elsevier
4. Phyu TN (2009) Survey of classification techniques in data mining. In: Proceedings of the international multiconference of engineers and computer scientists, vol 1, pp 18–20
5. Ruiz R, Riquelme Santos JC, Aguilar-Ruiz JS (2005) Heuristic search over a ranking for feature selection. In: IWANN, pp 742–749
6. Lal T, Chapelle O, Weston J, Elisseeff A (2006) Embedded methods. In: Feature extraction, pp 137–165
7. Rana S, Jasola S, Kumar R (2011) A review on particle swarm optimization algorithms and their applications to data clustering. Artif Intell Rev 35(3):211–222
8. Karaboga D, Gorkemli B, Ozturk C, Karaboga N (2014) A comprehensive survey: artificial bee colony (ABC) algorithm and applications. Artif Intell Rev 42(1):21–57
9. Franks NR, Pratt SC, Mallon EB, Britton NF, Sumpter DJT (2002) Information flow, opinion polling and collective intelligence in house–hunting social insects. Philos Trans R S Lond B Biol Sci 357(1427):1567–1583
10. Martens D, Baesens B, Fawcett T (2011) Editorial survey: swarm intelligence for data mining. Mach Learn 82(1):1–42
11. Holzinger K, Palade V, Rabadan R, Holzinger A (2014) Darwin or lamarck? Future challenges in evolutionary algorithms for knowledge discovery and data mining. In: Interactive knowledge discovery and data mining in biomedical informatics. Springer, Berlin, pp 35–56
12. Blum C, Li X (2008) Swarm intelligence in optimization. In: Swarm intelligence. Springer, Berlin, pp 43–85
13. Kumar D, Mishra KK (2017) Artificial bee colony as a frontier in evolutionary optimization: a survey. In: Advances in computer and computational sciences. Springer, Singapore, pp 541–548
14. Husain W, Yng SH, Jothi N (2016) Prediction of generalized anxiety disorder using particle swarm optimization. In: International conference on advances in information and communication technology. Springer International Publishing, pp 480–489
15. Fong S, Zhuang Y, Tang R, Yang X-S, Deb S (2013) Selecting optimal feature set in high-dimensional data by swarm search. J Appl Math 2013
16. Abraham A, Grosan C, Ramos V (eds) (2007) Swarm intelligence in data mining, vol 34. Springer
17. https://www.analyticsvidhya.com/blog/2016/12/introduction-to-feature-selection-methods-with-an-example-or-how-to-select-the-right-variables
18. Gu S, Cheng R, Jin Y (2016) Feature selection for high-dimensional classification using a competitive swarm optimizer. Soft Comput 1–12
19. Chapelle O, Scholkopf B, Zien A (2009) Semi-supervised learning (Chapelle, O. et al., Eds.; 2006) [Book reviews]. IEEE Trans Neural Netw 20(3):542–542
20. Kramer O (2013) Dimensionality reduction with unsupervised nearest neighbors. Springer
21. Lakshmi PV, Zhou W, Satheesh P (eds) (2016) Computational intelligence techniques in health care. Springer, Singapore
22. http://www.cleveralgorithms.com/nature-inspired/swarm.html
23. Dorigo M, Stutzle T (2004) Ant colony optimization. MIT Press, Cambridge, MA
24. Shi Y (2001) Particle swarm optimization: developments, applications and resources. In: Proceedings of the 2001 congress on evolutionary computation, vol 1. IEEE, pp 81–86
25. Yang X-S (2010) A new metaheuristic bat-inspired algorithm. In: Nature inspired cooperative strategies for optimization (NICSO 2010), pp 65–74
26. Pham DT, Ghanbarzadeh A, Koc E, Otri S, Rahim S, Zaidi M (2011) The bees algorithm-A novel tool for complex optimisation. In: Intelligent production machines and systems-2nd I* PROMS virtual international conference, 3–14 July 2006. SN, 2011
27. Fister I, Fister I Jr, Yang X-S, Brest J (2013) A comprehensive review of firefly algorithms. Swarm Evolut Comput 13:34–46
28. Lukasik S, Zak S (2009) Firefly algorithm for continuous constrained optimization tasks. In: ICCCI, pp 97–106

29. Gandomi AH, Yang X-S, Talatahari S, Alavi AH (2013) Firefly algorithm with chaos. Commun Nonlinear Sci Numer Simul 18(1):89–98
30. Yang X-S, Deb S, Karamanoglu M, He X (2012) Cuckoo search for business optimization applications. In: 2012 national conference on computing and communication systems (NCCCS). IEEE, pp 1–5
31. Yang X-S, Deb S (2010) Engineering optimisation by cuckoo search. Int J Math Modell Numer Optim 1(4):330–343
32. Valian E, Mohanna S, Tavakoli S (2011) Improved cuckoo search algorithm for global optimization. Int J Commun Inf Technol 1(1):31–44
33. Lin S-M (2013) Analysis of service satisfaction in web auction logistics service using a combination of fruit fly optimization algorithm and general regression neural network. Neural Comput Appl 22(3–4):783–791
34. Yang X-S, He X (2013) Bat algorithm: literature review and applications. Int J Bio-Inspired Comput 5(3):141–149
35. Inkaya T, Kayalıgil S, Ozdemirel NE (2016) Swarm intelligence-based clustering algorithms: a survey. In: Unsupervised learning algorithms. Springer International Publishing, pp 303–341
36. Zhang M, Shao C, Li M, Sun J (2006) Mining classification rule with artificial fish swarm. In: 2006 the sixth world congress on intelligent control and automation, WCICA 2006, vol 2. IEEE, pp 5877–5881
37. Hu H, Li J, Plank A, Wang H, Daggard G (2006) A comparative study of classification methods for microarray data analysis. In: Proceedings of the fifth Australasian conference on data mining and analytics, vol 61. Australian Computer Society, Inc., pp 33–37
38. Mangat V (2010) Swarm intelligence based technique for rule mining in the medical domain. Int J Comput Appl 4(1):19–24
39. Yang C-S, Chuang L-Y, Li J-C, Yang C-H (2008) Chaotic maps in binary particle swarm optimization for feature selection. In: 2008 IEEE conference on soft computing in industrial applications, SMCia'08. IEEE, pp 107–112
40. Chakraborty B (2008) Feature subset selection by particle swarm optimization with fuzzy fitness function. In: 2008 3rd international conference on intelligent system and knowledge engineering, ISKE 2008, vol 1. IEEE, pp 1038–1042
41. Ming H (2008) A rough set based hybrid method to feature selection. In: 2008 international symposium on knowledge acquisition and modeling, KAM'08. IEEE, pp 585–588
42. Holden N, Freitas AA (2005) A hybrid particle swarm/ant colony algorithm for the classification of hierarchical biological data. In: Proceedings 2005 IEEE swarm intelligence symposium, SIS 2005. IEEE, pp 100–107
43. Zheng B, Zhang J, Yoon SW, Lam SS, Khasawneh M, Poranki S (2015) Predictive modeling of hospital readmissions using metaheuristics and data mining. Expert Syst Appl 42(20):7110–7120
44. Fong S, Wong R, Vasilakos AV (2016) Accelerated PSO swarm search feature selection for data stream mining big data. IEEE Trans Serv Comput 9(1):33–45
45. Esmin AAA, Coelho RA, Matwin S (2015) A review on particle swarm optimization algorithm and its variants to clustering high-dimensional data. Artif Intell Rev 44(1):23–45
46. Yeh, W-C, Chang W-W, Chung YY (2009) A new hybrid approach for mining breast cancer pattern using discrete particle swarm optimization and statistical method. Expert Syst Appl 36(4):8204–8211
47. Nemati S, Basiri ME, Ghasem-Aghaee N, Aghdam MH (2009) A novel ACO–GA hybrid algorithm for feature selection in protein function prediction. Expert Syst Appl 36(10):12086–12094
48. Correa ES, Freitas AA, Johnson CG (2006) A new discrete particle swarm algorithm applied to attribute selection in a bioinformatics data set. In: Proceedings of the 8th annual conference on genetic and evolutionary computation. ACM, pp 35–42
49. Lin S-W, Ying K-C, Chen S-C, Lee Z-J (2008) Particle swarm optimization for parameter determination and feature selection of support vector machines. Expert Syst Appl 35(4):1817–1824
50. Shah H, Ghazali R, Nawi NM (2011) Using artificial bee colony algorithm for MLP training on earthquake time series data prediction. arXiv preprint arXiv:1112.4628

51. Yeh W-C (2012) Novel swarm optimization for mining classification rules on thyroid gland data. Inf Sci 197:65–76
52. Chung YY, Wahid N (2012) A hybrid network intrusion detection system using simplified swarm optimization (SSO). Appl Soft Comput 12(9):3014–3022
53. Verbeke W, Martens D, Mues C, Baesens B (2011) Building comprehensible customer churn prediction models with advanced rule induction techniques. Expert Syst Appl 38(3):2354–2364
54. Chuang L-Y, Tsai S-W, Yang C-H (2011) Improved binary particle swarm optimization using catfish effect for feature selection. Expert Syst Appl 38(10):12699–12707
55. Lin S-W, Shiue Y-R, Chen S-C, Cheng H-M (2009) Applying enhanced data mining approaches in predicting bank performance: a case of Taiwanese commercial banks. Expert Syst Appl 36(9):11543–11551
56. Kwon O, Sim JM (2013) Effects of data set features on the performances of classification algorithms. Expert Syst Appl 40(5):1847–1857
57. Fong S, Deb S, Yang X-S, Li J (2014) Feature selection in life science classification: meta-heuristic swarm search. IT Prof 16(4):24–29
58. Chakraborty A, Kar AK (2017) Swarm intelligence: a review of algorithms. In: Nature-inspired computing and optimization. Springer International Publishing, pp 475–494
59. Wu Q, Liu H, Yan X (2016) Multi-label classification algorithm research based on swarm intelligence. Clust Comput 19(4):2075–2085
60. Zhang M-L, Zhou Z-H (2014) A review on multi-label learning algorithms. IEEE Trans Knowl Data Eng 26(8):1819–1837

Identification and Analysis of Customer's Requirements from Refurbished Electronics in Order to Create Customer Value

Aparna Deshpande, Priyansha Chouksey and A. Subash Babu

Abstract The endeavor of the study is to add to the literature by categorizing and scrutinizing a range of customer needs in the refurbished electronics business, by the help of that we will generate the result in the formation of end user value in the reverse supply chain. Throughout developing and developed the world, negative customer's perception has prevented the refurbishment process to realize its full potential in the business environment. Though the practice has benefits and is eco-friendly, but lack of awareness and ignorance toward environmental hazard has resulted in this study. The study aims to create a positive perception of the consumer by creating end user significance. The purpose of this research paper focuses on that the demands of end user will increase day by day, and how an organization can create more importance to end user by administrating the dynamics distressing the fulfillment of desirable customer needs for refurbished electronics. The work proposed in this paper can help refurbishers to develop organizational strategies and promotion approaches.

Keywords Reverse supply chain · Electronic waste · Fault tree · Sensitivity analysis · Customer value

1 Introduction

In topical years, reverse supply chain has gained wide consideration through researchers and industrialist. This practice has numerous advantages, including conserving resources, reducing electronic waste, growing other environmental concerns, and lowering prices. The reverse supply chain is the successions of actions which

A. Deshpande · P. Chouksey
Department of Industrial and Production Engineering, S.G.S.I.T.S, Indore, India
e-mail: aparnadeshpande129@gmail.com

P. Chouksey
e-mail: priyanshachouksey2@gmail.com

A. Subash Babu (✉)
Department of Mechanical Engineering, IIT-Bombay, Mumbai, India
e-mail: subash@iitb.ac.in

© Springer Nature Singapore Pte Ltd. 2019
M. L. Kolhe et al. (eds.), *Advances in Data and Information Sciences*, Lecture Notes in Networks and Systems 39, https://doi.org/10.1007/978-981-13-0277-0_19

take place to repossess used products from consumers, returned and shipped back to manufacturers where different disposition actions take place like reuse, remanufacturing, refurbishing, and recycling. Then, these are sold in resulting marketplaces for supplementary profits [13]. There is an increasing need for the reverse supply chain because of rapid advancement in technology and inventions in electronics products. In this, amusement of electronics and good products from a landfill is an essential concern nowadays. Due to the presence of various perilous rudiments like mercury, lead, etc., whose inappropriate disposition can cause a severe threat to human life. Rapidly rising volume and enhanced rates of obsolescence of the returned electronic gadgets serve in enhancing electronics waste. These wastes include discarded electrical and electronic devices. Used electronics which are classified for resale, reuse, recycling, salvage, or disposal are also termed as e-waste. The study in this project was mainly done from a customer's perspective. Specifically mentioning from an Indian customer's view, the concept of "Reverse Supply Chain" is totally alien to them. The Indian electronic industry is advancing at a very high pace, but it was surprising to observe that still, reverse supply chain in electronics and refurbishment of electronics is not a jargon here. The idea of using a refurbished cell phone is still not quite welcomed among the Indian customers. Hence, the study deals with the process of refurbishment at a breath, but on how the process can lead to enhancement of customer value.

2 Literature Review

2.1 Reverse Supply Chain and Refurbishment

According to the state-of-the-art paper by Samir K. Srivastava, there were distinct units in an organization which took the responsibility of ensuring environmental eminence various elements of the entire cycle, for example, in product development, operations, process design, logistics, and waste management. Revolutions in 1980s and 1990s in the field of quality and supply chain, respectively, proved that integration of ongoing operations with environmental management is the best practice. The deterioration of environment in various ways like overflowing and unattended waste sites, decreasing raw material, and hiked pollution levels have motivated the use of environment friendly supply chain. Along with protection of nature, it has also resulted in growth in business. It has become more of a business value driver.

According to Nagurney and Toyasaki [6], various countries like U.S., Japan, and Europe for the management of their electronic waste have considered the option of recycling. This effort is the result of a spreading awareness about deleterious effects of e-waste on the environment. E-waste cannot be recycled in the same way as the household recycling because unlike household waste, e-waste is an amalgamation of various elements like some nonbiodegradable materials, some reusable, and some precious metals. Therefore, processing of e-waste is the most feasible solution. Proper analysis of their process including procurement of used products and refurbishing of received products can help create a secondary market for these products.

For legislation and policymaker, this is advantageous in reducing waste for protecting the environment; for a company, it reduces scrap and save material and reduces total production cost; and for a consumer, they get a variety of choices at lower prices [14]. Reverse supply chain structure varies across firms like the manufacturer of electronic devices such as Apple, HP, and Cisco, who operate their own refurbishing facilities. While some retailers outsource refurbishment of items to other firms/manufacturers or contracted partners, there are retailers in the market such as BestBuy who are running their own refurbishing facilities for the used electronic devices [14]. This creates high level of uncertainty in quality, quantity as well as the timing of returned items, short selling period, technologically obsolescence of electronics. Usually, in cell phone industry, it may be necessary to improve the processes of manufacturing or supplier is at the risk to be changed. This results in the fast decline in the monetary value of cell phones that are stored in inventory for long periods of time. This makes refurbishment process and their maintenance difficult. In the recovering of used products, improving the rate of collection of used cell phones depends on how the end users perceive the stake of used phone and various benefits of bringing it back to the manufacturer. Consumer's perception of product obsolescence and the re-marketability of refurbished phones impact the decisions made by both stakeholders party.

The refurbishing cost is highly dependent upon the condition of used devices and ability to resell which decreases as technology progresses. In short, consumers do not have positive attitudes toward purchasing relatively old refurbished products [7]. Generally, refurbisher's and consumer's interests are not inline with one another and are sometimes even in conflict [9].

2.2   Electronic Waste

The final destination of discarded electrical and electronics equipment (EEE) is frequently not in the same country or even on the same continent where the equipment is stationed. Transshipping electronic waste from developed to developing countries is a normal trend. On estimation, 23% of electronic waste produced in advanced countries is exported to growing countries [2]. Reprocessing of E-waste can be considered as part of the either "formal" or "informal" economic sector. Formal electronic waste recycling entails specifically built facilities with up to date equipment that promotes safe distillation of the retrievable materials. These facilities, for the most part, ensure safe working environment. Not surprisingly, these facilities are not found in developing countries due to the high cost of infrastructure, safety standards, and maintenance [11]. According to reports, around 1.5 lakh tones of e-waste were produced in Indian subcontinent in 2005, and in 2013, the quantity reached 8.5 lakh tons. India's maximum e-waste is generated by major cities—major being Delhi, Bangalore, Chennai, Kolkata, Ahmedabad, and Pune. There is no authentic and systematic mechanism to collect e-waste from users in India; the major portion of the e-waste lands in municipal bins. Retailers have provision for collecting mobile phones in general but not so much in use in day to day practices. Electronic waste management and handling rules were made in 2011, but came into effect on May 2012 in Indian subcontinent. Here, all the producers had to take responsibility for the complete product life cycle and as per the e-waste mandate, original equipment producer had to set up electronic waste collection stations to collect waste. But till now, majorly all the companies have failed to set up these facilities [5]. In India, mostly recycling sites are present in Bengaluru and Delhi [4].

2.3 Customer Value

Previous researchers were mainly focused on operational aspects focusing on minimizing costs and complying with legislation. Reverse supply chain was perceived as a cost of doing business. Over the last years, researchers have deviated toward analyzing entire reverse and closed-loop supply chain (CLSC) processes and strategic issues using survey or case-based methodologies, system dynamics mathematical and economic (e.g., game-theoretic) modeling. Most studies that explicitly deal with the idea of "value creation" have focused on forward supply chains with more firms showing an interest in reverse supply chains and CLSC; it is time for research make its way from technical and operational issues toward the strategic impact of reverse supply chain as a means to create value. Although research interests are slowly moving from minimizing costs to creating value, the existing literature on manifestations of value creation in CLSC is still scattered across topics [10]. The value in business markets is the perceived worth of monetary units of the set of economic, technical, service, and social benefits received by a customer firm in exchange for the price paid for a product, taking into consideration the available supplier's offerings and prices [1]. According to the anonymous author, four different types of values, namely, product value, value in use, possession value, and overall value, are linked together in consumer's evaluation process, which ultimately creates customer value. It is defined that customer value basically is perceived preference given by customer when evaluating product features, features performances, and output from use that facilitates or hinders customer from achieving the goals and purposes in use situations.

According to Wolfgang Ulaga, customer value is thus a subject of growing interest to professionals and researchers in marketing. According to industrial marketing management, there are three different perspectives. Traditionally, the subject of research was value in business market focused mainly on the evaluation of how any manufacturer creates value for customers and customer perception of superior value in this offering compared to competitors. But the current trend has increased concern toward the requirement to consider customers as a key asset of the firm.

By engaging and retaining customers, the customer equity management can therefore be considered as a second important aspect of customer value (the seller's perspective). "The Initiators of Changes in Customer's Desired Value: Results from a Theory Building Study" article written by Daniel J. Flint and Robert B. Woodruff identifies stress taken by customer is a major phenomenon leading to amend in customer's desired value from suppliers. This tension has three underlying dimensions—affective strength, perceived extensiveness, and temporal dynamism. In their study, they suggested there are five drivers for environmental change (altering customer's demands, demands internal to organizations, competitors moves, changes in supplier demands and performances, and altering macro-environment) and three drivers for current capability (today's knowledge levels, performance levels, and control levels) which all create tension at the customer level.

This high disruption of the environment and companies must focus on those strategies and management policies with a greater probability of achieving success and helping them get sustainable competitive advantages over time. In order to change consumer perception for refurbished cell phones, an organization should understand and determine ways to offer greater value to the customers.

Any organization that practices refurbishment process should ensure ways to consider their customer first and then introduce products and services which provide a certain value.

Creation and management of customer value are recognized for a long time as important elements of the company's strategies. Recently, growing competition ask for consistent value creation which results in growing interest in delivering higher customer value [3].

According to the recent survey of Dutch mobile phone owners, Van Weelden et al. [12] have identified the reasons behind the preference of consumer while buying. Consumer's perception for refurbished phones to be inferior is a hindrance to their purchase. In their study conducted with Dutch mobile phone users generated certain barriers to consumer in giving preference to refurbished phones which were cell phone users are oblivious about the existence of refurbished phones. They blindly refer refurbished phone as secondhand phone which may have a risk of being damaged, not fully functional. Their supply in retail channel is scarce, and getting refurbished phone is seen as a "hassle".

Environmental benefits were a minor consideration for consumers, with the majority being unaware of them. Other risks included returning the phone for maintenance and it becoming obsolete quickly. Another factor that influenced consumers was the knowledge about this practice. This study undertakes a systematic literature review about reverse supply chain specifically in refurbishment process followed by existing of customers' requirement based on human psychology. Customer requirements identified by Rebentischa et al. [8] were product quality, supplier flexibility, customer focus, innovation capacity, and customer expertise; this study penetrated deeper into the customer requirements and extended up to 40 customer requirements. These customer requirements were the exhaustive list covering all aspects of human wants.

The survey was conducted in groups based on age group (15–25 and 25–40+) and gender (female and male). This paper aims at answering the question: What is the important customer requirement? And what is the best combination of factors affecting the customer requirement that could create more customer value?

Furthermore, this systematic approach to understand customer need is then validated in a dynamic environment.

3   Research Methodology

In this section, the procedure followed in research is represented. The descriptions of following steps are given below.

3.1   Identifying Customer Requirements and Prioritizing Them

The approach presented in this work was initiated by enlisting the basic end user needs. End user needs are scrupulous personality and specifications of a good or service as determined by an end user. There are 40 customer requirements determined to cover all aspects of human needs. These requirements were derived from the literature review.

Following Table 1 contains the list of customer requirements.

The survey was conducted among the consumers categorized based on age (15–25 and 25–40+) and gender (male and female). A sample size of the survey was 100 in which 70 people responded in total. Their responses were analyzed and 15 top priority customer requirement as per the consumers were selected. (RIDs are simply number listed in above list to refer each customer requirement.) These top 15 customer requirements with RIDs are listed in Table 2.

The interview was conducted among professionals including college professors and industrialist. This was conducted to reach the very important customer requirements which are essential for the customer value creation. Requirements were rated on the basis of economically feasible aspect, maximum customer satisfaction aspects, and environmental sustainable aspects. The selected customer requirements are listed below (Table 3).

3.2   Analysis

This study used AHP tool to prioritize the very important customer requirements. AHP is analytical chain of process which is an ordered procedure for systematizing and investigating complex decisions, based on mathematics and psychology. It was given by Thomas L. Saaty in the 1970s and had been comprehensively studied and

Table 1 Customer requirements

• Less vulnerability of refurbished phones due to operational and disruption risk
• Effective performance in terms of economic sustainability, environmental sustainability, and operational performance
• Timely deliveries
• Short refurbishing lead time
• Innovation and adaptable to upgrade
• Promoting information transparency
• Supplier flexibility
• Optimum quality standards
• Reduced failure cases
• Customer service improvement
• Secured spare part supply or improved spare part service
• Return practices and services
• Return policies
• Accessibility
• Good ownership experience
• Lower customer disposal costs
• Product quality
• Brand image
• Communication and healthy customer relationship
• Feedback system
• Improved product design/quality/safety
• Fulfilling personal customer wishes
• Good promotion
• Economic feasibility
• Green design features and material used
• The distance to be traveled for product disposition should be minimum
• The time duration between product return and new product delivery should be less than a week
• 100% data retention is provided, i.e., no sort of mobile phone data is lost
• It should provide all function same as the original phone
• If the refurbished phone has an updated version there should be a scope to go back to the older version
• Transparency of process is maintained by making the customer well aware of the parts and processes used
• The price paid for refurbished phones should be at least 10% less than the original cost
• They should get a certain amount of resolution price at the time of old product return
• In the mind of the customer, the perceived value should be more than the perceived risk
• Higher warranty for the refurbished product than the new one

(continued)

Table 1 (continued)

• Service centers should be dedicated specifically for refurbished phones
• Refurbished phones should not slow down their functionality. Reduce efficiency and upgradations
• Refurbished phones should satisfy and maintain the status quotient
• RF phones should be equally marketed as the new phones so as to satisfy the mind of the customer that he is buying a good which seems to be highly demanded and showed off
• A customer buying RF product expects to be treated more exclusively (special) than the normal customer because he has this sense of admiration for himself because, in a way he has contributed in protecting the environment

Table 2 Important customer requirements as per surveyed by consumers

• Effective performance in stipulations of monetary sustainability, environmental sustainability, and operational performance **R2**
• Timely deliveries **R3**
• Innovation and adaptable to upgrade **R5**
• Optimum quality standards **R8**
• Reduced failure cases **R9**
• Customer service improvement **R10**
• Return policies **R13**
• Good ownership experiences **R15**
• Product quality **R17**
• Brand image **R18**
• Complete data retention and information transparency **R28**
• Same functionality as original **R29**
• Reduced price lesser than original **R32**
• Higher warranty for refurbished cell phones than new one **R35**
• Enhancing and maintaining status quotient **R38**

Table 3 Very important customer requirements selected by professionals

• Good ownership experiences **R15**
• Product quality **R17**
• Brand image **R18**
• Same functionality as original **R29**
• Reduced price lesser than original **R32**
• High warranty for refurbished cell phones than new one **R35**
• Enhancing and maintaining status quotient **R38**

Table 4 Pairwise comparison of all customer requirements

	R15	R17	R18	R29	R32	R35	R38
R15	1	5	10	5	0.1	0.2	5
R17	0.2	1	5	5	0.1	5	10
R18	0.1	0.2	1	0.2	0.1	0.2	1
R29	0.2	0.2	5	1	0.1	0.1	5
R32	10	10	10	10	1	5	5
R35	5	0.2	5	10	0.2	1	10
R38	0.2	0.1	1	0.2	0.2	0.1	1
Total	16.7	16.7	37	31.4	1.8	11.6	37

polished since then. It has scrupulous purpose while producing a decision in group. Users who use the AHP first crumble their decision problem into a hierarchy of more easily comprehended subproblems, each of which can be scrutinized separately. The elements of the hierarchy can relate to any aspect of the decision problem—substantial or insubstantial, vigilantly deliberated or approximately anticipated, sound or feebly understood—anything at the end for producing decision.

Refer Table 4, numeric weighting is given to every customer requirement in the matrix. For weighing, scale is assumed whose description is given below:

1.0 = Requirement is considered significant and preferred when moderated against the decisive factor company is evaluating it.
5.0 = The requirement is considered significantly more important or more preferred.
10.0 = The decisive factor enormously additional or more preferred.
0.2 = It is radically less important or preferred.
0.1 = It is enormously less important or preferred.

On the basis of above scale, weighs are assigned to all requirements, and sum of each column is obtained.

Then, standardized matrix is obtained after obtaining the sum (refer to Table 5). Each element weigh in Table 4 is divided by the sum of that column, and then, the value obtained is entered in the corresponding cell of the standardized matrix refer to Table 5.

Example: For "R15", weigh given is 1 and the sum of column 1 is 16.7. Therefore, in standardized matrix in cell (1, 1),

Value entered = 1/16.7 = 0.031746032.

Similarly, values of all the cells are calculated, and standardized matrix is obtained. After this, mean weight is calculated by taking an average of each row.

Like for row 1, sum of {cell (1, 1): cell (8, 8)}/8.

In this way, weightage of each requirement can be calculated.

Table 5 Standardized matrix with ranking of customer requirements

	Standardized matrix							Average	Ranking in %
	R15	R17	R18	R29	R32	R35	R38		
R15	0.05988024	0.2994012	0.27027027	0.1592357	0.055556	0.017241	0.135135	0.142388	14.23885
R17	0.01197605	0.05988024	0.13513514	0.1592357	0.055556	0.431034	0.27027	0.160441	16.04411
R18	0.00598802	0.01197605	0.02702703	0.0063694	0.055556	0.017241	0.027027	0.021598	2.159778
R29	0.01197605	0.01197605	0.13513514	0.0318471	0.055556	0.008621	0.135135	0.055749	5.574939
R32	**0.5988024**	**0.5988024**	**0.27027027**	**0.3184713**	**0.555556**	**0.431034**	**0.135135**	**0.415439**	**41.54388**
R35	0.2994012	0.01197605	0.13513514	0.3184713	0.111111	0.086207	0.27027	0.176082	17.60817
R38	0.01197605	0.00598802	0.02702703	0.0063694	0.111111	0.008621	0.027027	0.028303	2.830276

3.3 Fault Tree Analysis

According to the analysis done in the previous step, it was observed that "reduced price of refurbished electronics than original" is the prerequisite customer requirement for enhancing the customer value of refurbished electronics. In order to determine any company's probability to satisfy this customer requirement, fault tree analysis was conducted. In this analysis, reduced price of refurbished electronics than original was perceived as a desirable event. Factors critical to failure within diverse business environments were determined. The fault tree was thereby used to determine the failure of the event "Reduced price of the refurbished cell phone than the original price of a brand new phone," due to the effect of the sub crucial factors determined here. The interaction of these factors has to be considered for the successful implementation of the event. The objective of this fault tree is to explore the interaction among various factors responsible for the reduction of the price. The results of the survey conveyed that the customer would give preference to a refurbished cell phone over a brand new cell phone, only if they were provided at a price fairly lower than the branded new cell phone. Here, the event's failure studies typically deal with the factors whose various combinations and degree of interaction determine the failure of the event. The key basic events or the primary factors affecting the failure of the main event has been determined through literature study conducted. An attempt has been made to include all the details into a fault tree. Most of the factors can be classified as internal and external factors, and these factors cover the major parties involved in the reduction of price by a firm.

The internal factors which were identified here were those which fall under the control of the firm. The minimum total cost incurred by the firm from the initial stage of production acquisition to the sale in the secondary market constitutes all the major costs. The cost sustained by a firm depends on the efficient allocation of resources by the firm. So if the no-productive costs were determined and eliminated, it will surely end up reducing the price of the final product. Product existence phases are the journey of the product commencing manufacturing on the way to its end of life state. Here, time and cost both play an important role. For instance, here we have considered depreciation cost, so the minimization of this cost will also reduce the price. If the cost of depreciation of a good is minimum, then ultimately it will reduce the work of repairing and refurbishing. Also, older products have much lesser value than comparatively newer version and also less preferred by the consumer. The objectives of the organization are imperative in the pricing process. For example, if a firm has the policy to skim the market, it will not reduce the cost just for the sake of attracting more customers, whereas on the other hand, the firm aiming at market penetration will have a reduction in price as a primary motive. Brand name also has a great impact on the perspective of the customer, it entrusts the customer with the quality of goods received. It provides the customer with a sense of "value for money". Service support can be described as the backbone for the sustenance of the firm in the industry. The firm tends to create the highly satisfactory experience through various means. For an instance, the warranty can be termed as a kind of assurance that a

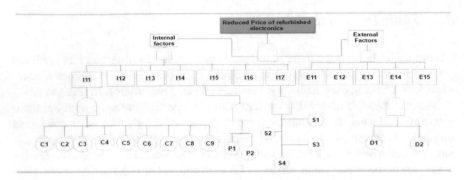

Fig. 1 Fault tree for "reduced price of refurbished electronics than original"

firm provides to its customer and agrees to take the complete responsibility of its good for a stipulated period of time. This is done to enhance customer's satisfaction. Customization refers to an attempt to incorporate each and every basic requirement of a customer in the product. This is a costly system, so to reduce the price, the degree of customization is reduced. To reduce the price of the refurbished cell phone, the company has allocated resources so that they can be utilized to their optimum capacity.

The external factors included in the study are those, which are beyond the control of the firm. They can be either static for a period of time but are mostly dynamic in nature. Consumer out of all is the most dynamic in nature. For a firm to decide its final price, it needs to take into consideration the rate of return it had from the customer and the demand of customer for the refurbished cell phone. The reduction in price can also be achieved if the firm is well aware of the trends in the marketplace and response of the customer to the same. The issue of electronic squander is a major concern for environment. Hence, if the government issues the laws in the favor of the process of refurbishment, it would make it easier for both the firm and customer to establish a smooth channel of interaction. Also at times, if a nation lacks laws promoting green supply chain becomes difficult for the firm to erect the resources required for refurbishment. The advent of technology can speed up the process of refurbishment, reduce non-value adding the task, and hence reduce the price charged by the firm. We have also considered the factor of occurrence of accidents or calamities, which if occurred in the present period will affect the pricing of product in the coming period. In this study, few internal and external factors are considered static for a period of time.

The following Fig. 1 represents the fault tree for the desired customer requirement. Where

I11 = Minimum total cost	I14 = Brand name
I12 = Firm's objective and policies	I15 = Product life cycle
I13 = Less degree of customization	I16 = Availability of refurbishing facilities and allocation of resources
I17 = Service support	E13 = Government legislation
E11 = Change in market trends	E14 = Consumer
E12 = Innovation and technology advancement	E15 = Occurrence of an accident and natural calamity
C1 = Refurbishing cost	C2 = Recovery/collecting cost
C3 = Storage cost	C4 = Transportation cost
C5 = Procurement cost/outsourcing cost	C6 = Marketing/promotion cost
C7 = Inspection and testing cost	C8 = Maintenance cost
C9 = Labor cost	P1 = Minimum depreciation cost
P2 = Time from recovery to reselling	S1 = Warranty
S2 = Consultant post go live service	S3 = Training service
S4 = Service rate	D1 = Demand
D2 = Return rate	

As per the steps described in the literature review, fault tree, Fig. 1, was evaluated with the help of fuzzy logic. Following are the results obtained by the analysis (Tables 6, 7 and 8).

Linguistic variables used are given below:

VH—Very high	FH—Fairly high
H—High	M—Medium
L—Low	FL—Fairly low
VL—Very low	

Therefore, the failure of the top event was obtained as 0.147375662. So the complement of this is the success probability. Therefore, the success probability of the reduced price of refurbished electronics than original was found as 0.852624338. This value suggests that the desirable customer requirement with the considered factors will be very successful (Tables 9, 10 and 11).

Another fault tree's result mentioned in this paper is the case of failure, where desirable customer requirement resulted in being unsuccessful.

Therefore, the failure of the top event was obtained as 0.57298. So the complement of this is the success probability. Therefore, the success probability is found as 0.42702. This value suggests that in this case firm will be unable to fulfill the customer need.

Table 6 Table showing linguistic variables of basic events and fuzzy failure probability

Name of event	Linguistic variable	FFR
Refurbishing cost	FH	0.00604703
Recovery/collecting cost	H	0.00500035
Storage cost	M	0.00366505
Transportation cost	H	0.00500035
Procurement cost/outsourcing cost	L	0.00500035
Marketing/promotion cost	VH	0.00500035
Inspection/sorting cost, disassembly cost	L	0.00500035
Maintenance of facilities cost	H	0.00500035
Labor cost	H	0.00500035
Min. depreciation cost	FH	0.00604703
Time from recovery to reselling	H	0.00500035
Firm's objective and policies	FL	0.01435737
Brand name	FL	0.01435737
Warranty	VL	0.00500035
Consultant post go live service	M	0.00366505
Training service	VL	0.00500035
Service rate	FL	0.01143428
Less degree of customization	FH	0.00500035
Availability of facilities and re-allocating resource	L	0.00500035
Return rate	VL	0.00500035
Demand	L	0.00500035
change in market trends	L	0.00500035
Government legislation and rules	M	0.00279664
Innovation and technology advancement	FL	0.00500035
Occurrence of accidents and natural calamities	M	0.00500035

3.4 Sensitivity Analysis

The two fault trees described in the earlier section were aimed to show the interaction of basic events with their preceding events and to determine the probability of success of the main event to occur, i.e., "reduced price of the refurbished cell phone than the original price of a brand new phone" in the respective trees. The objective of sensitivity analysis is to show how the changes in the probability of the basic events affect the total failure probability of the main event and hence the probability of success of the event to occur. Here, in the sensitivity analysis, the linguistic variables assigned to the basic event are kept the same for all the nine cases. The range assigned to each linguistic variable for the determination of the fuzzy number is divided into nine divisions. In short, each fluctuating critical factor has been assigned one

Table 7 Table showing the failure probability of events

Min. total cost	0.044714492
Product life cycle	0.011047375
Firm's objective	0.014357365
Brand name	0.014357365
Service support	0.025100014
Less degree of customization	0.005000345
Availability of facilities and re-allocating resource	0.005000345
Consumer	0.010000691
Change in market trends	0.005000345
Government legislation and rules	0.002796635
Innovation and technology advancement	0.005000345
Occurrence of accidents and natural calamities	0.005000345

Table 8 Table with the failure probability and success probability of the topmost event and the sub-events

Internal factors	0.119577301
External factors	0.027798361
Reduced price than original failure	**0.147375662**
Reduced price than original success	**0.852624338**

linguistic variable for the entire nine success trees but has a different fuzzy number (lying within the range of the assigned linguistic variable) for each of the nine fault trees analysis.

CASE: Changing the fuzzy numbers related to fluctuating critical factors.

Here, FT means fault tree, and numbering refers to different cases considered in the analysis. Table 12 consists of various combinations of fuzzy numbers used related to critical factor.

Here, the fuzzy number was set. For example, if a basic event has the linguistic variable, say FL (fairly low), then the fuzzy number assigned to it will be 0.225. As a part of this sensitivity analysis, the fuzzy number cannot be randomly from the range of linguistic variable but has a predetermined fuzzy number assigned to it.

Here in the study, we have considered few factors that are fluctuating critical factors and few that are static critical factors. As a matter of fact, we have carried the same fuzzy number of the static factors over all the nine fault trees. This was like stimulating a particular situation with some factors considered as constraints (i.e., constant over a period of time) and some as variables.

Table 9 Case of failure: desirable customer requirement resulted in being unsuccessful

Name of event	Linguistic term used	FFR
Refurbishing cost	M	0.005876
Recovery/collecting cost	FH	0.006014
Storage cost	M	0.122386
Transportation cost	FH	0.005
Procurement cost/outsourcing cost	FL	0.009284
Marketing/promotion cost	L	0.004498
Inspection/sorting cost, disassembly cost	H	0.005
Maintenance of facilities cost	L	0.005
Labor cost	M	0.003663
Min. depreciation cost	FH	0.005
Time from recovery to reselling	H	0.005
Firm's objective and policies	FL	0.014357
Brand name	FL	0.014357
Warranty	FL	0.009295
Consultant post go live service	VL	0.005
Training service	L	0.022945
Service rate	M	0.122386
Less degree of customization	FH	0.005
Availability of facilities and allocating resource	L	0.022945
Return rate	FL	0.00524
Demand	M	0.003665
Change in market trends	L	0.022945
Government legislation and rules	M	0.002797
Innovation and technology advancement	FL	0.022945
Occurrence of accidents and natural calamities	M	0.122386

The calculations were carried out as earlier, and results were presented in the table for this particular case.

Similarly, eight more fault trees were developed in accordance with the procedure mentioned above. The only difference lied in the changing of the fuzzy number for each linguistic variable in an ascending order. The results are presented in Table 13 (Tables 14, 15, 16, 17, 18, 19, 20, 21 and 22).

Table 10 Table showing the failure probability of events

Min. total cost	0.166722
Product life cycle	0.010001
Firm's objective	0.014357
Brand name	0.014357
Service support	0.159626
Less degree of customization	0.005
Availability of facilities and re-allocating resource	0.022945
Consumer	0.008905
Change in market trends	0.022945
Government legislation and rules	0.002797
Innovation and technology advancement	0.02294
Occurrence of accidents and natural calamities	0.122386

Table 11 Table with the failure probability and success probability of the topmost event and the sub-events

Internal factors	0.39301
External factors	0.17997
Reduced price than original failure	**0.57298**
Reduced price than original success	**0.42702**

Table 12 The division of the range of the linguistic variables. The fuzzy numbers here are assigned in ascending order for the sake of ease of calculations

	FT1	FT2	FT3	FT4	FT5	FT6	FT7	FT8	FT9
VL	0	0.025	0.05	0.075	0.1	0.125	0.15	0.175	0.2
L	0.1	0.125	0.15	0.175	0.2	0.225	0.25	0.275	0.3
FL	0.225	0.25	0.275	0.325	0.35	0.375	0.425	0.45	0.475
M	0.4	0.425	0.45	0.475	0.5	0.525	0.55	0.575	0.6
FH	0.525	0.55	0.575	0.625	0.65	0.675	0.725	0.75	0.775
H	0.7	0.725	0.75	0.775	0.8	0.825	0.85	0.875	0.9
VH	0.8	0.825	0.85	0.875	0.9	0.925	0.95	0.975	1.0

Table 13 Case 1: for fault tree 1 (FT1)

	FT1
VL	0
L	0.1
FL	0.225
M	0.4
FH	0.525
H	0.7
VH	0.8

Table 14 For FT1, table with the failure probability and success probability of the topmost event and the sub-event

Internal factors	0.146424256
External factors	0.063686993
Reduced price than original failure	**0.210111249**
Reduced price than original success	**0.789888751**

Table 15 For FT2, table with the failure probability and success probability of the topmost events and the sub-events

Internal factors	0.138652809
External factors	0.063686627
Reduced price than original failure	**0.202339435**
Reduced price than original success	**0.797660565**

Table 16 For FT3, table with the failure probability and success probability of the topmost events and the sub-events

Internal factors	0.173944612
External factors	0.087090844
Reduced price than original failure	**0.261035456**
Reduced price than original success	**0.738964544**

Table 17 For FT4, table with the failure probability and success probability of the topmost events and the sub-events

Internal factors	0.169859072
External factors	0.061699511
Reduced price than original failure	**0.231558583**
Reduced price than original success	**0.768441417**

Table 18 For FT5, table with the failure probability and success probability of the topmost events and the sub-events

Internal factors	0.159464604
External factors	0.057000657
Reduced price than original failure	**0.216465261**
Reduced price than original success	**0.783534739**

Table 19 For FT6, table with the failure probability and success probability of the topmost events and the sub-events

Internal factors	0.150069889
External factors	0.053112399
Reduced price than original failure	**0.203182288**
Reduced price than original success	**0.796817712**

Table 20 For FT7, table with the failure probability and success probability of the topmost events and the sub-events

Internal factors	0.174283496
External factors	0.068133444
Reduced price than original failure	**0.242416941**
Reduced price than original success	**0.757583059**

Table 21 For FT8, table with the failure probability and success probability of the topmost events and the sub-events

Internal factors	0.17751563
External factors	0.06598314
Reduced price than original failure	**0.24349877**
Reduced price than original success	**0.75650123**

Table 22 Summary of sensitivity analysis

Fault tree	Probability of failure	Probability of success
Fault tree 1	0.210111249	0.789888751
Fault tree 2	0.202339435	0.797660565
Fault tree 3	**0.147375662**	**0.852624338**
Fault tree 4	**0.261035456**	**0.738964544**
Fault tree 5	0.231558583	0.76441417
Fault tree 6	0.216465261	0.783534739
Fault tree 7	0.203182288	0.796817712
Fault tree 8	0.242416941	0.757583059
Fault tree 9	0.24349877	0.75650123

Bold values indicate the minimum limit and maximum limit

4 Conclusion

The study initially began with the aim to answer the two main important questions which included the most important customer requirement and the best combination of factors affecting the customer requirement responsible for creating customer value. The answers to these were satisfactorily achieved here. 40 basic customer requirements were determined from the literature and brainstorming. Now to test these requirements in the real-time scenario, a survey of 70 customers and personal interview of experts and academicians was conducted. The responses were then recorded and computed. Analytical hierarchy process was used for the purpose of weighing each customer requirement to each other and then prioritizing them. This led us to the determination of our most important customer requirement which was "reduced price of the refurbished cell phone than original cell phones." This prompted us to determine the factors which can help the firm to reduce price and improve sales. An exhaustive list of all internal and external factors was generated, which was then subjected to the fault tree analysis using the fuzzy number. Fault tree analysis is an effective way to determine the degree to which each factor can contribute to the failure of a particular event. Various combinations of these factors led us to different probabilities of failure and success of the event. Sensitivity analysis was also carried on a stipulated case, which provided us with nine fault trees. The method given in this paper in sensitivity analysis was slightly modified. The basic events are classified as constant critical factors (for a period of time) and fluctuating critical factors. A range of the linguistic variables were divided into nine subdivision, which was assigned to the factors in an incremental manner, i.e., each fluctuating critical factor has been assigned one linguistic variable for all analysis, but has a different fuzzy number (lying within the range of the assigned linguistic variable) for each of the nine fault tree analysis. This helped us in the determination of the range within which the probabilities of success and failure lied. On the basis of fault tree analysis, the probability of success with our contractedly assumed case lies in the range of 75–85%. And the best combination was taken to be the one with 85% of probability of success. The study also led us to an insight that "product life cycle" consistently has the minimum probability of failure throughout the nine fault trees. Hence, it will have the maximum probability of success, which means to ensure the success of the main event this factor should be specifically taken care of. The probability of failure is highly fluctuating for the factor concerning customer's behavior.

References

1. Anderson JC, Jain D, Chintagunta PK (1993) Understanding customer value in business markets: methods of customer value assessment. J Bus Bus Mark 1(1):3–30
2. Breivik K, Armitage JM, Wania F, Jones KC (2014) Tracking the global generation and exports of e-waste. Do existing estimates add up? Environ Sci Technol 48(15):8735–8743
3. Cepeda-Carrion I, Martelo-Landroguez S, Leal-Rodríguez AL, Leal-Millán A (2017) Critical processes of knowledge management: an approach toward the creation of customer value. Eur Res Manag Bus Econ 23(1):1–7
4. Chen A, Dietrich KN, Huo X, Ho SM (2011) Developmental neurotoxicants in e-waste: an emerging health concern. Environ Health Perspect 119(4):431
5. Jayant A, Gupta P, Garg SK, Khan M (2014) TOPSIS-AHP based approach for selection of reverse logistics service provider: a case study of mobile phone industry. Procedia Eng 97:2147–2156
6. Nagurney A, Toyasaki F (2003) Reverse supply chain management and electronic waste recycling: a multitiered network equilibrium framework for e-cycling. Appears in Transportation Research E (2005) 41:1–28
7. Ovchinnikov A (2011) Revenue and cost management for remanufactured products. Product Oper Manag 20(6):824–840
8. Rebentisch E, Schuh G, Riesener M, Gerlach M, Zeller P (2016) Determination of a customer value-oriented product portfolio. Procedia CIRP 50:82–87
9. Sabbaghi M, Behdad S, Zhuang J (2016) Managing consumer behavior toward the on-time return of the waste electrical and electronic equipment: a game theoretic approach. Int J Prod Econ 182:545–563
10. Schenkel M, Caniëls MC, Krikke H, van der Laan E (2015) Understanding value creation in closed loop supply chains–Past findings and future directions. J Manuf Syst 37(3):729–745

11. Schluep M, Hagelüken C, Meskers CE, Magalini F, Wang F, Müller E, Kuehr R, Maurer C, Sonnemann G (2009) Market potential of innovative e-waste recycling technologies in developing countries. Proc R 9

12. Van Weelden E, Mugge R, Bakker C (2016) Paving the way towards circular consumption: exploring consumer acceptance of refurbished mobile phones in the Dutch market. J Clean Prod 113:743–754

13. Wang G, Gunasekaran A (2017) Operations scheduling in reverse supply chains: identical demand and delivery deadlines. Int J Prod Econ 183:375–381

14. Yoo SH, Kim BC (2016) Joint pricing of new and refurbished items: a comparison of closed-loop supply chain models. Int J Prod Econ 182:132–143

Part III
Intelligent Communication
and Networking

Enhancing Performance of Proxy Cache Servers Using Daemon Process

Sachin Chavan, Naresh Lodha, Anukriti Rautela and Kunal Gupta

Abstract The significant increase in Internet users has led to an enormous rise in the network traffic which had a great impact on the characteristics of the network like reduced bandwidth, increased latency and higher response time for users. The most common method used is proxy cache server which is deployed to reduce the web traffic and workload on server. Proxy cache server stores the data locally but there is a problem of stale data. There are many instances when the proxy cache returns the data stored in its cache rather than returning the latest version from the original server. This paper is concerned with a methodology that addresses this issue of stale data by introducing Daemon process. Daemon is a background process that updates the data present in the cache periodically. It does not interfere with the processes of the user and updates the data when minimum number of users is online. Understanding and identifying the main characteristics is crucial part to plan techniques that help us save the bandwidth, improve access latency and response time. This paper focuses on the factors affecting the proxy cache server as well as on Daemon process.

Keywords Access latency · Stale data · Proxy cache · Mirroring · Duplicate suppression · Daemon process

S. Chavan (✉) · N. Lodha · A. Rautela · K. Gupta
Mukesh Patel School of Technology Management and Engineering, SVKMS NMIMS,
Shirpur Campus, Dhule, Maharashtra, India
e-mail: sachin.chavan@nmims.edu

N. Lodha
e-mail: nareshlodha.nmims@gmail.com

A. Rautela
e-mail: anukritirautela.nmims@gmail.com

K. Gupta
e-mail: kunalgupta.nmims@gmail.com

© Springer Nature Singapore Pte Ltd. 2019
M. L. Kolhe et al. (eds.), *Advances in Data and Information Sciences*, Lecture Notes
in Networks and Systems 39, https://doi.org/10.1007/978-981-13-0277-0_20

1 Introduction

With the fast changing technology and people around the globe having access to Internet, there has been significant growth in World Wide Web. There has been rise in number of users, and consumption of data has been increased. This has notably contributed to increase in web traffic. Web traffic leads to reduction in bandwidth and increased access latency [1]. So lately, there has been a lot of significant research going on regarding the methods to reduce web traffic and improve performance of web so as to decrease the response time and improve access latency [2]. The most common method is the use of web proxy cache. Web proxy caches have been established to reduce network traffic and workload on server and make the online communication as fast as possible [3] (Fig. 1).

A web proxy cache server is a system or software that runs on a computer which is mediator between client and server [1, 4]. The client requests certain pages and the web proxy forwards it to the original server if the requested pages are not present in proxy's cache (local storage). The main advantage of using a proxy server is they can be established anywhere in the Internet. Web caches divide the load on server thus reducing access latency and saving network bandwidth [5]. The basic function of web cache (both browser and proxy cache) is to store the most visited web pages. If the user requests the same pages next time, then it will be provided by the local server and not by the main server thus, reducing the response time [6]. Many times the content of the web pages stored in the server may be outdated or do not match with the new and updated content of the web server leading to the problem of stale data. It can occur when users try to extract the content locally for a span of time.

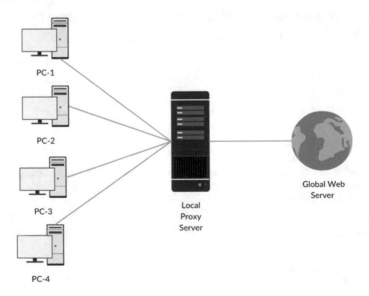

Fig. 1 Architecture of proxy cache server (*Source* [12])

In order to eradicate the problem of stale data, we propose Daemon process. Daemon process is a background process which is not in direct control of user. Daemon process runs periodically and checks the consistency of the data cached and the data at the web server. It updates information, which is done during slack time, i.e. time when web traffic is minimum [7]. Any new updates or changes made in a web page can be detected by MD5 hashing where data of every element is stored [8, 9]. This also helps in resolving the problem of content aliasing. This paper is organized in following sections: In Sect. 2, we represent some related works. Section 3 contains proposed methodology. Section 4 gives the conclusion of the work. In Sect. 5, we discuss about the future scope.

2 Related Work

2.1 Web Traffic

Web traffic is the amount of data that exchanged between the users and the website. Web traffic is used to measure the most popular websites or web pages within the websites. The increase in web traffic leads to increase in the response time and hence, the data is loaded at a slower rate [10].

2.2 Mirroring

Mirroring is used to keep several copies of the contents available on website or web pages on multiple servers in order to increase the accessibility of the content. This causes redundancy and, hence, ensures high availability of the web documents [11]. A mirrored site is the exact replica of the original site which is regularly updated and reflects the updates done on the actual site.

2.3 Caching

Caching is used to improve the response time as well as the performance of WWW Internet applications. Caching causes reduction in system load. Most techniques cache static content such as graphic and text files [10, 12].
Static caching: Static caching checks the log of yesterdays user requests to predict the requests of the user for the present day. Static caching improves cache performance by using compression techniques and also frees up the cache space. The cache servers performance is decided by two factors: byte hit ratio, which is the hit rate with respect

to the total number of bytes in the cache and hit ratio which shows the percentage of all accesses that are fulfilled by data in cache [3, 4].

Dynamic caching: Dynamic caching regularly updates the content on the local server in order to keep up with the updated content of the dynamic websites which helps in avoiding the problem of stale data [1, 3].

2.4 Duplicate Suppression

There are multiple instances when the occurrence of same content with different URLs is available on the Internet, i.e. these web pages are duplicates of each other but have different names. They are unwanted and use extra bytes over the network which increases the network cost, bandwidth and creates storage problem. If a cache finds a duplicate of a requested page, then the network cost can be reduced, and the storage problem can be avoided [13]. Prefetching is a technique in which the most likely data that will be requested in future is prefetched, i.e. the prediction of users future requests is done on the basis of past requests made by user. So, the cached data is used to satisfy the request, and hence, duplicate suppression is used [13].

2.5 Proxy Server

A proxy server acts as an intermediary between a server and an endpoint device such as the computer, from which a client is sending the request for a service. A user connects to the proxy server while requesting for a file or any other Internet resource available on a different server [5]. The proxy server on receiving a request performs a search in its local cache of previously visited pages. If the page is found, then it is returned to the user or else the proxy server acts as a client on behalf of the user and requests the page from the web server by using its own addresses. If the page is found by the web server, the proxy server relates it to the original request and sends it to the user or else the error message is displayed on the users device [7].

Few common types of proxies are mentioned below:

- Shared proxies: It serves multiple users.
- Dedicated proxies: It just serves one server at a time.
- Transparent proxies: It simply forwards the request without providing any anonymity to the user's information, i.e. the IP address of the user is revealed. It is found in a corporate office.
- Anonymous proxies: It provides reasonable anonymity for most users, i.e. the web pages receive the IP address of the proxy server rather than that of the client and allow the user to access the content that is blocked by firewall.
- High anonymity proxies: It does not identify itself as a proxy server, i.e. hides that it is being used by any clients.

- Forward proxies: It sends the requests of a client onward to a web server. Users access forward proxies by directly surfing to a web proxy address or by configuring their Internet settings.
- Reverse proxies: It passes requests from the Internet, to an isolated, private network through a firewall. It handles all requests for resources on destination servers without requiring any action on the part of the requester.

3 Proposed Methodology

3.1 Cache Hit

As the data requested by the user is available in the local storage/cache and should be updated as same as the origin web server which helps the user to get fresh and correct data or information which is been request by the user. Hence, it is known as cache hit [3, 13].

3.2 Cache Miss

As the data request by the user is not available in the local storage/cache or expired and the data which is available is not update to date as in the origin web server. Hence, it is known as the cache miss [9, 13].

3.3 Stale Data

Many sites on web server are updated frequently within a day, hour or week. So the data stored with the proxy server should be updated accordingly but this affects the performance of the server, so the server avoids to update the data frequently. Hence, it gives the data which is not updated as on the origin web server. This data is known as stale data [1, 12].

For all above factors, we purpose the methodology of using Daemon process in the proxy cache server which will help us to avoid the factors discussed above.

3.4 Algorithm

1. Start
2. Initialize Count = 0; It shows no process is updated.
3. Check Server the Time with local system,

   (a) If time is in between 2 to 5; Proceed to Step 3.
   (b) Else;
        Stop.

4. Check for Active Users,

   (a) If Number of Users are less than N (for example n = 50); Proceed to Step 5.
   (b) Else;
        Wait for m minutes (for example m = 15 min) and then go to Step 2.

5. Check for Update

   (a) If Yes,
        i. Download New Content.
        ii. Replace new content with old content.
        iii. Generate Hash and Store it in Table [9].
        iv. Count = Count + 1; Proceed to Step 6.
   (b) If No,
        Count = Count + 1;
        Proceed to Step 6.

6. Check for Count is less than number of process

   (a) If yes,
        i. Fetch next process.
        ii. Repeat Step 2, 3, 4 and 5 again
   (b) Else; Terminate the Daemon Process.

7. Stop

   (See Fig. 2).

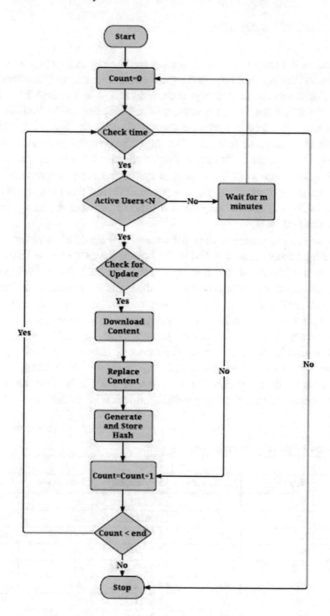

Fig. 2 Flow chart of Daemon process

4 Results and Discussion

The table shows the response time of search engine for some keywords. When the request is passed through the browser, it goes to cache memory when that particular page is not present in cache memory; then, the request is forwarded to main server and page is fetched from the main server. When that page is loaded on browser, a copy of that page is stored in cache memory and next time when it is requested by user reply given from cache memory. The table shows the figures for response time of search engine for some of the keywords of simple web search. The first column shows the response time when the page is retrieved from main server, and the second column shows the figures when the pages are retrieved from cache server present locally. The table also shows the response time of browser for image search with same keywords (Table 1).

Figure 3 shows the response time for several keywords of browser from clients which are using proxy cache hit. From the figure, the observation is that for some keywords, we are getting access latency reduced by half. For some keywords, the effect on access latency is very less or negligible. The results shown in the figure are for simple web search. Among these keywords, for some keywords, we are getting 50% or more than 50% reduction in access latency but in some cases, we are getting no effect on access latency. The reason for no effect on access latency is that the IMS requests send by the browser for freshness of data and if more updated copy is present on web server, then that updated copy fetched from main server (cache miss). The results shown in Fig. 4 are for image search. Among these keywords, for some keywords, we are getting 30–40% reduction in access latency but in some

Table 1 Response time of browser for some keywords

Web			Image		
Keywords	Web server	Cache server	Keywords	Web server	Cache server
SVKM	250	140	SVKM	230	200
NMIMS	140	130	NMIMS	300	100
RCPIT	250	120	RCPIT	350	150
CANNON	240	130	CANNON	250	100
SAMSUNG	210	140	SAMSUNG	640	200
NOKIA	250	190	NOKIA	240	120
MATLAB	240	160	MATLAB	280	160
OPERA	250	150	OPERA	120	120
SIEMENS	230	160	SIEMENS	310	100
MICROMAX	160	140	MICROMAX	190	110
MPSC	170	140	MPSC	180	100
UPSC	210	150	UPSC	150	140
IRCTC	160	140	IRCTC	330	90
RRB	260	120	RRB	310	70

cases, we are getting only 10–20% on access latency. The reason for no effect on access latency is that the IMS requests send by the browser for freshness of data and if more updated copy is present on web server then that updated copy fetched from main server.

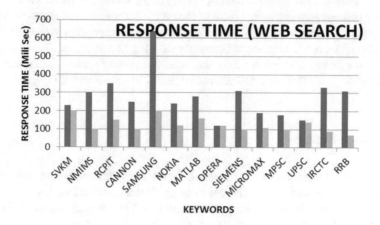

Fig. 3 Response time of browser for several keywords (web search)

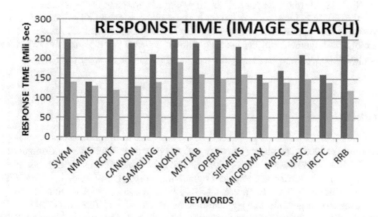

Fig. 4 Response time of browser for several keywords (web search)

5   Conclusion

Our paper proposes a novel methodology for implementing proxy cache server by using the suggested algorithm. The proposed methodology states that the Daemon process will improve the performance of proxy cache server in terms of cache hit. The Daemon process guarantees that most of the time whenever user accesses web pages, the user gets the fresh content of web page. As the Daemon process is scheduled in the downtime of server, it does not impose additional load on to server. Daemon process runs in background and updates the data available in proxy cache without interfering the user processes. The users are not interrupted due to schedule of Daemon process.

6   Future Scope

Currently, this Daemon process in system is working effectively with the websites which are static in nature, i.e. where the content is not updated frequently like www.w3schools.com while the system is not effective with the dynamic websites in which content is updated so often, such as www.msn.com. In cases of dynamic websites, the user will receive the stale data. Hence, an alternative method is needed to be proposed along with the Daemon process for overcoming this problem.

Acknowledgements We express our special thanks of gratitude to our Director Dr. Ram Gaud (SVKMS NMIMS, Shirpur Campus). Also we would like to thank our Associate Dean Dr. Nitin S. Choubey (MPSTME, Shirpur Campus) for his continuous support and guidance. We would like to acknowledge the support of the 'Computer Engineering Department' of Mukesh Patel School of Technology Management and Engineering without which the completion of project would not have been possible.

References

1. Sieminski A (2005) The impact of proxy caches on browser latency. Int J Comput Sci Appl
2. Chen W, Martin P, Hassanein HS (2003) Caching dynamic content on the web. In: Canadian conference on electrical and computer engineering, vol 2, pp 947–950, 4–7 May 2003
3. Patil SB, Chavan S, Patil P, Patil SR (2012) High quality design to enhance and improve performance of large scale web applications. Int J Comput Eng Technol (IJCET)
4. Chavan S, Chavan N (2013) Improving access latency of web browser by using content aliasing in proxy cache server. Int J Comput Eng Technol
5. E-Services Team (2006) Changing proxy server by the Robert Gordon University, School hill, Aberdeen, Scotland
6. USENIX Association (1997) In: Proceedings of the USENIX symposium on internet technologies and systems on USENIX symposium on internet technologies and systems, Berkeley, CA, USA
7. Bommepally K, Glisa TK, Prakash JJ, Singh SR, Murthy HA (2010) Internet activity analysis through proxy log. IEEE

8. Jarvinen K, Tommiska M, Skytta J (2005) Hardware implementation analysis of the MD5 hash algorithm. In: Proceedings of the 38th annual Hawaii international conference on system sciences, pp 298a–298a
9. Asaka T, Miwa H, Tanaka Y (1999) Hash-based query caching methods for distributed web caching in wide area networks. IEICE Trans Commun E82-B(6)
10. Breslau L, Cao P, Fan L, Phillips G, Shenker S (1999) Web caching and Zipf-like distributions: evidence and implications. In: INFOCOM'99, eighteenth annual joint conference of the IEEE computer and communications societies. Proceedings, vol 1. IEEE, New York, NY, pp 126–134
11. Bharat K, Broder A, Dean J, Henzinger MR (1999) A comparison of techniques to find mirrored hosts on the WWW. In: 4th ACM conference on digital libraries
12. Patil SB, Chavan S, Patil P (2012) High quality design and methodology aspects to enhance large scale web services. Int J Adv Eng Technol
13. Kelly T, Mogul J (2002) Aliasing on the world wide web: prevalence and performance implications. In: Proceedings of the 11th international conference on World Wide Web. ACM, New York, NY, USA, pp 281–292

Some Studies on Factors Influencing Maximum Association Number of Wi-Fi Access Point

Haribansh Mishra and Ratneshwer Gupta

Abstract Wi-Fi, which is standardized by IEEE as 802.11, is most popular wireless network protocol of the present time for WLAN deployment. In a professionally deployed WLAN, there are many Access Points (APs) placed at well-determined locations to provide good wireless connectivity. The decision of placement of APs is dependent on many factors. For successful WLAN deployment and maintenance, it is pertinent to know that what factors influence the maximum count of stations that can be connected to one single AP. This study tries to explore all the aspects that are responsible for deciding the maximum number of devices that can be connected to single AP without degrading the performance of network significantly. Such information will play pertinent role in placement of APs in different geographical locations in a cost-effective manner.

Keywords Wi-Fi · Access point association · Infrastructure mode deployment
WLAN optimization

1 Introduction

Wi-Fi, i.e., wireless fidelity is most popular wireless network protocol of the present time. More than other wireless protocols, this is a technology that is used prominently in home and business local area networks and as public hotspot networks. It is officially standardized as IEEE 802.11. Wi-Fi has become synonym of high-speed wireless Internet access on portable devices like laptop, tablets, and smartphones. One single device that connects to Wi-Fi network is also referred to as station in technical fraternity. In this paper, we will be using the term "device" and "station" interchangeably. If a person with Wi-Fi enabled device is within some Wi-Fi coverage

H. Mishra (✉)
DST-CIMS, Banaras Hindu University, Varanasi, India
e-mail: haribansh@bhu.ac.in

R. Gupta
SC&SS, Jawaharlal Nehru University, New Delhi, India
e-mail: ratnesh@mail.jnu.ac.in

© Springer Nature Singapore Pte Ltd. 2019
M. L. Kolhe et al. (eds.), *Advances in Data and Information Sciences*, Lecture Notes
in Networks and Systems 39, https://doi.org/10.1007/978-981-13-0277-0_21

area and want to access Internet, all one needs to do is put ON the Wi-Fi of his device. Device will normally scan the presence of Wi-Fi radio signals and may list one or more Wi-Fi network present in that area. These Wi-Fi radio signals one receives on his device are emitted by a device called Access Point (AP). APs are responsible to connect Wi-Fi devices to the backbone or distribution channel of the Wi-Fi network. The distribution channel of Wi-Fi networks is usually wired LAN. In a professionally deployed WLAN, there are many APs placed at well-determined locations to provide good wireless connectivity. Many stations can connect to single AP. This mode of configuring Wi-Fi network where stations connect to APs is called Infrastructure mode configuration. This is in contrast with ad hoc mode configuration where devices connect directly to each other without any access point. As the name suggests ad hoc mode is only for temporary uses and is not scalable to cater needs of bigger network.

Infrastructure mode is most used mode to connect devices where all devices within range of a central wireless access point connect through it. However, few very pertinent questions are worthy of contemplation regarding the count of stations that can be connected to one single AP, like what is the maximum count of devices that can be connected to a single AP? Is the count of devices that can connect to one AP a fixed number or may vary depending on some factors? What factors influences the count of devices that can be connected to one single AP. Increasing count of devices connecting to one single AP will obviously lead to performance degradation. So the viable count without degrading performance is important aspect to be considered by engineers deploying and administering WLAN. Hence during deployment of infrastructure mode Wi-Fi network, it is important to consider the optimum no of devices that can connect to one single access point retaining the Quality of Service experienced by the users.

The motivation behind this work is the fact that in spite of partial discussions on the issue being available in the literature, no study, to best of our knowledge, covers all the aspects comprehensively to give a single point of reference which addresses the issue of factors influencing maximum association of access point. This study tries to explore *all the concerned aspects that are responsible in deciding the optimum number of devices that can be connected to single AP* without degrading the performance of network significantly. The study will largely benefit the network engineers and administrators who are responsible for deploying WLAN or administering such an infrastructure. By having the knowledge of number of devices that an AP can be associated to, engineers will come up realistic budget allocation for provisioning infrastructure. Further, it will help to suggest a suitable number of APs in a given geographic area which consequently will result in better network access experience by the end users benefiting all the stakeholders. Our contribution in this work is as follows:

(1) Point out concern factors influencing maximum association to single AP.
(2) Discuss the reasons and extent of impact of each factor.
(3) Provide single point of reference to stakeholders in context of this issue.

2 Background Details and Related Work

In literature, the problem of factors influencing the optimum count of stations that can be connected to a single AP has been studied rather indirectly in terms of association control, channel selection, and management. Athanasiou et al. [1, 2] argue that the association process specified by the IEEE 802.11 standard does not cerebrate the circumstances of channel and the AP load in the process of association, which results into low user transmission rates and low throughput. They propose an association mechanism that is conscious of the uplink and downlink channel conditions. Further, they introduced a metric, to be used by station to associate with available APs. They also extended the functionality of association mechanism in a cross-layer manner where they take into account information from the routing layer. This makes QoS information of the backhaul, available to the end users. Broustis et al. came up with a framework called MDG (Measurement-Driven Guidelines). This simultaneously addresses three concerns: frequency allocation across APs, user affiliations across APs and adaptive power control for each AP [3]. They found that when these are used incoherently, it may reduce the performance by substantial amount. Athanasiou et al., in another study, propose an LAC (Load-Aware Channel Allocation) in which they point out that when the number of the clients associated with an AP is high, the influence of interference in the neighboring cells increases [4]. Xu et al. found that if access point (AP) is selected imprudently, it degrades client's throughput in such a way that it becomes difficult to mitigate it by other methods, like efficient rate adaptations [5]. They proposed an online AP association strategy which achieves a minimum throughput that is probably close to the optimum. Additionally, the strategy works efficiently in practice with rational computation and transmission overhead. Proposed association protocol was applied on the commercial hardware and is compatible with legacy APs without any alteration. Hong et al. pointed that the Received Signal Strength Indicator (RSSI) based AP selection algorithm, which is applied in most commercial IEEE 802.11 clients, results in an uneven dissemination of clients among the APs in the network [6]. This becomes the cause of serious performance degradation. They propose a new AP selection algorithm based on the estimation of available throughput calculated with a model based on the IEEE 802.11 Distributed Coordination Function (DCF). Proposed AP selection algorithm also considers hidden terminal problems and frame aggregation. Using this algorithm, the network can achieve up to 55.84% higher throughput compared to the traditional RSSI based approach. APs have to share a limited number of channels. Hence, densification of APs generates interference, contention and decreases the global throughput [7]. Optimizing the association process between APs and stations, alleviate this problem and increase the overall throughput and fairness between stations. Amer et al. gave a mathematical model for resolving this association optimization problem which can be tuned according to the CPU and time restraints of the WLAN controller [7]. Wang et al. pointed out that doing coverage adjustment to control association of stations may result in AP service cheating problem and AP service loophole problem. They proposed solution to this problem by providing a novel coverage adjustment algorithm [8].

Sajjadi et al. worked out meta-heuristic technique with backward compatibility to all existing Wi-Fi protocols, e.g., 802.11a/b/g/n/ac to provide proportional fairness. Here client gets associated to AP after extensive numerical analysis in terms of network throughput and fairness [9].

From the efforts done by researchers discussed above, it can be observed that RSSI alone cannot be the only indicator for a station to get associated with the AP, albeit there are several other factors. The channel load conditions, frequency allocation across APs, user affiliation across APs, consideration of hidden terminals, CPU and time constraints of WLAN controller are found to be major factors which will influence association problem. Hence, the optimum count of devices that will get associated to AP will be naturally dependent on these factors. In the present work, an attempt has been made to extend the above contribution further by discussing the factors influencing optimal count of devices that will get associated to an AP without degrading the network performance.

3   Association Process in Context of Wi-Fi Networks

Wi-Fi is a wireless local area networking protocol that allows Wi-Fi enabled devices to connect each other for data exchange without use of wires. Wi-Fi is a trademark of the Wi-Fi Alliance, an international association of companies. It is a type of wireless local area network (WLAN) protocol based on the IEEE 802.11 network standards. From the user's perspective, Wi-Fi is just a way for accessing Internet from a wireless capable device like a phone, tablet, or laptop. Almost every modern device supports Wi-Fi for Internet access and using shared network resources.

Wireless access points (APs or WAPs) or simply Access Points are networking hardware devices on wireless local area networks (WLANs). It usually connects to wired backbone through router. Nowadays it is mostly integral part of router. Access points are the central transmitter and receiver of wireless radio signals in Infrastructure mode configuration of Wi-Fi WLAN. Client stations like laptops, tablets, etc., connect to Internet through AP. It is also called base station owing to term which more commonly used in cellular networking.

Wi-Fi client devices use SSID (Service Set Identifier) to identify and get associated to (i.e., join) wireless networks. SSID uniquely names an 802.11 Wireless Local Area Network (WLAN) including home/office networks and public hotspots. SSID is a case-sensitive text string that can be as long as 32 characters consisting of letters and/or numbers. User trying to connect to a wireless network scans from their device like phones and laptops for available networks within the range. Then their device lists SSIDs of the network present around. Now user can initiate a new network connection by picking a name from the list. The subclasses of SSID are Broadcast SSID, Basic SSID, and Extended SSID. Broadcast SSID is also called ZERO LENGTH SSID and is used by stations to send probe requests. Basic SSID is generally MAC Address of an AP, and Extended SSID is also said only SSID and is the name of collection of APs in a single WLAN, i.e., NAME of WLAN.

Before considering the issue we are focusing on, i.e., "factors influencing maximum count of devices that can connect a single AP", it is imperative to know the exact methodology in which a device connects to AP. This understanding will be required to grasp clearly the concept of optimum count of devices that can be connected to one AP. A device can be associated to AP broadly in two ways: Actively or Passively. Although traditionally the devices connect to an AP passively, the detail of steps followed by the devices in both of the above ways of connection is described, respectively, as under.

A. *Connecting Actively*

In this method, station sends a probe frame with desired SSID. AP responds with its SSID again. Next, station sends associate request frame and finally AP responds with associate response frame.

B. *Connecting Passively*

Each AP advertises its presence by broadcasting beacon frames that carry SSID. Stations locate the AP by passively listening beacon frames with desired SSID. After this, station sends associated request with SSID on which AP responds with associate response frame.

4 Factors Influencing Maximum Number of Stations that Can Be Connected to a Single

The AP has the physical capacity to handle 2048 MAC addresses, but, because the AP is a shared medium and acts as a wireless hub, the performance of each user decreases as the number of users increases on an individual AP. In general, not more than 24 clients use to be associated with the AP because the throughput of the AP is reduced with each client that associates to the AP [10]. Through our study, the count of stations that can be connected to a single AP has been found to be dependent on many factors. Below is the list of major influencing factors that play a vital role in deciding optimal count of stations that can be connected to a single AP.

(1) Hardware of AP (CPU, Memory, Chipset),
(2) Load balancing across the spectrum,
(3) Applications used by stations,
(4) Type of device acting as station,
(5) Radio model,
(6) Coverage/throughput,
(7) Interference and hidden terminals,
(8) Protocol version and channels used by them, and
(9) Value of max association parameter of AP configuration.

The detailed discussion of above issues individually is in the following subsections.

4.1 Dependence on Hardware of AP (CPU, Memory)

Every AP has limited amount of memory and processing power of CPU. This limits the count of stations that can be associated with it. Increasing memory space and processing power of an AP will increase the count of associated stations provided other dependence factors are unaltered. Obviously, larger amount of memory can facilitate storing larger number of active connections which will directly affect the max number of stations connected to AP. Apart from this, having faster memory might be useful if underlying chipset is supportive to that. Fast Memory Access Technology from Intel significantly optimizes the use of available memory bandwidth and reduces the latency of the memory accesses. However, faster memory will not necessarily be useful if underlying chipset operates at a slower speed. This implies that the underlying circuit board design will also have an impact on the max association of stations to the AP. The circuit board should be appropriately compatible with the memory and CPU of the AP. Further, an AP with a faster CPU (with greater CPU cycles per unit time) will process data quicker than one with a slower CPU which will result in its ability to handle more simultaneous clients leading to larger client associations [11].

4.2 Dependence on Load Balancing Across the Spectrum

AP is generally manufactured to cater different bands in radio spectrum simultaneously. This helps them to utilize different radio spectrum simultaneously enabling them to shed the load across the available spectrum. In spectrum load balancing, also called band steering, the AP dictates whether the client uses 2.4 or 5 GHz spectrum. The technique is primarily used by a dual-radio AP to steer clients from 2.4 GHz band to 5 GHz band because the latter offers a cleaner radio environment. It can also be used to steer non-mission critical clients from 5 to 2.4 GHz. Band steering is also useful to steer clients between the two 5 GHz radios of a dual-radio AP that operates two different 5 GHz channels. Nowadays, the steering is done by the AP through some combination of sending disconnect (disassoc or deauth) messages to the client with vague reason codes (for example, Aruba AP sends error code 17) and ignoring the client's connection requests in the band where the AP does not want it to connect. Hence an AP can get connected to few stations on (say) 2.4 GHz band while to some other on 5 GHz band depending upon the support available on individual stations [12] (Fig. 1).

Spectrum Load Balancing (SLB) is featured as Adaptive Radio Management in commercial Aruba APs. SLB in APs is controlled by three configuration parameters: SLB Calculating Interval is mentioned in seconds. It determines how often spectrum load balancing must be calculated and have a default value of 30 s. SLB Neighbor Matching has value in percentage (%) for comparing client density of AP neighbors to determine the client load on a specific AP channel. Its default is 75%. SLB Threshold

Fig. 1 Load balancing
across the spectrum [12]

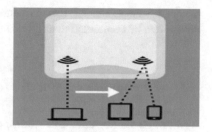

is number that indicates the number of clients on a channel. This value is used to calculate client count threshold. When the load for an AP reaches or exceeds the client count threshold, load balancing is enabled on the AP. Its default value is 2 [13]. By manipulating these configuration parameters of load balancing functionality of AP, increment of the count of station that can get connected to it can be done.

4.3 Dependence on Applications Used by Stations

Since the wireless medium is a shared medium, the data traffic generated by applications of stations also becomes a crucial factor of consideration. With stations running low data traffic application like email, APs can have high count of connected stations. However, with stations running high data traffic applications like streaming media, APs will have as low as count of three stations per AP for meaning full data transmission. Thus, the spread of applications running across a wireless AP affects the maximum number of clients that can be supported simultaneously since all clients running high throughput applications such as HD video streaming will hog the shared channel and hence a lesser number of clients will be supported as compared to when all clients are surfing the web. There will be much congestion within a channel if much traffic is created by applications and since access points are HUBs (i.e., single collision domain), only one device can trigger a frame at any given time. An optimum trade-off should be decided in each of cases for wireless QoS versus max count of stations [14].

4.4 Dependence on Type of Device Acting as Station

This factor is also actually guided by the amount of data traffic generated. We know different device obviously will generate different amount of data depending upon there processing power, data needs, display quality, etc. For example, PC may generate much more traffic than a tablet which in turn may generate much more traffic than a bar code reader or a simple WAP device. Hence, a greater number of tablets can be connected to AP then PC and higher number of bar code scanner or WAP device can be connected than tablets.

4.5 Dependence on Radio Model

APs can be single-radio, dual-radio, or triple-radio devices. The models which are single-radio devices will support a lesser number of clients whereas dual-radio access points will be able to support larger numbers of clients. Basically, it is the bands of radios on which simultaneous transmissions can be made from client nodes to the APs. The triple-radio models do not use their third radio for client traffic, so their theoretical client limit is similar to the dual-radio APs.

4.6 Dependence on Interference and Hidden Terminals

It has been found that the densification of APs generates interference, contention and decreases the global throughput, as APs have to share a limited number of channels. Optimizing the association step between APs and stations can alleviate this problem and increase the overall throughput and fairness between stations. It would be good if the association optimization problem is well modeled through a suitable neighborhood structure and be tuned according to the CPU and time constraints of the WLAN controller. Further, hidden terminal problems frequently occur in wireless networks with unlicensed frequencies, like IEEE 802.11 in the 2.4 GHz band. Considering frame aggregation, a major improvement in transmission efficiency for deployment of IEEE 802.11n standard WLAN can help eliminating hidden terminal problem [7]. With interference and hidden terminal issues taken care of more stations can get associated to an AP.

4.7 Dependence on Coverage/Throughput

Commercially available stations may select access point based on manufacturer, driver, type of card, etc., or a combination of these. Nevertheless, the most common AP affiliation mechanism remains Received Signal Strength Indicator (RSSI) where the station associates with the AP with strongest signal. But this type of association can lead to poor performance due to its lack of knowledge of the load on different APs. If stations are far away from an AP the client association to that AP will be low. On the other hand, if client is placed nearer to AP, the perceived RSSI will increase; hence, there will be an increase in max count of associated stations to that AP. However, it is to be noted that the wireless Quality of Service will depend on other factors like the volume of throughput required by stations, the density and

Table 1 No. of client per AP against protocol version [15]

Protocol	Type of web use	Amount of bandwidth allocated per device (kbps)	# of clients per AP
802.11b	A little over Web-Casual	600	13
802.11g	A little over Web-Casual	600	43
802.11n (1 spatial stream)	A little over Web-Casual	600	43
802.11n (2 spatial streams)	A little over Web-Casual	600	273
802.11b	Web-Casual	500	15
802.11g	Web-Casual	500	51
802.11n (1 spatial stream)	Web-Casual	500	51
802.11n (2 spatial streams)	Web-Casual	500	327

layout of stations around the access point, and the backhaul bandwidth available at AP. These factors will be constraints deciding the count of stations associated to an AP. Additionally, it should be noted that bigger devices like PCs will have powerful signal transmission than that of smaller devices like smartphones. So PCs although away from AP will be well connected in comparison to smartphone. Multiple site surveys should be performed to assess the RF environment for best approximation of max association count of stations [14, 15].

4.8 Dependence on Protocol Version and Channels Used by Them

Wi-Fi technology is based on 802.11 standards. This standard defines different versions of protocols like 802.11a, 802.11b, 802.11g, 802.11n, etc. Although they evolved as improvement, each of these has its pros and cons. The max no. of clients associated to an AP has been found to depend on which among these protocols is supported or is being used among particular station and client. Table 1 gives an idea of dependability and relations among protocol an idea of dependability and relations among protocol version, traffic type, bandwidth and number of clients that can be associated to AP [15].

With advent of Multiple-Input Multiple-Output (MIMO), we have even better client associations. MIMO is a wireless technology that uses multiple transmitters and receivers to transfer more data at the same time. All wireless products with

802.11n support MIMO. Hence the protocols used by AP model become an important deciding factor for maximum number of clients that can be associated to an AP.

4.9 Dependence on Value of "Max Association" Parameter of AP

This factor is actually used to mitigate rest of other factors discussed so far. Every device has a preconfigured max association limit which is generally 30 (however, recommend is of 25). When this limit is reached no other devices are allowed to join AP. This limit is configurable and decided based on rest of other factors discussed so far. If we know users will be using heavy data, then 'max association' should have a lower value and vice versa.

More APs with same SSID will be required in the WLAN for making the network available to increasing number of users. Once above factors are used to predict the optimum count of users that can get connected to a single AP, it will lead to deduction of number of APs that should be procured and installed in a WLAN for a given set of network users.

Association point of MN in case of two neighborhood APs

We wish to point out one more related aspect of WLAN APs here. Often there are more than one APs of same WLAN (or same SSID). Which AP should station connect from among two or more neighboring APs? One answer to this question is the one which is nearest. But what if the nearest AP has reached its max association limit or is catering stations with heavy data load? There is also a possibility of interference due to obstacles like wall or other environmental factors. Commercial Wi-Fi APs use RSSI as a measure to decide this. However, studies points that apart from this there are several factors that should be considered in making the decision of which AP should a station should associate with. Although the strength of signal still remains prominent criteria for selecting AP, distance of the station to AP is a related consideration factor. Next is the value of "max association" parameter of AP. If the station could be communicated that this value has reached its limit, then it can wisely try to get associated with other suitable AP. Similarly, if the station could be informed that the AP is busy serving heavy load, it will deter getting associated to it.

5 Conclusions

This work tries to explore the concern factors influencing Optimal Count of Stations that can be connected to a single AP. These considerations will help the network engineers in figuring out the number of APs that will be suffice to cater the users of Wi-Fi network without degrading the performance. It will also assist to plan the geographical layout of AP in the network. Future course ahead of this study would be to establish and draw concrete statistical relation among the pointed factors so that definitive approximation can be given for max association number of an AP. In future, we are planning to further investigate on congestion aspects of APs in a wireless network system.

References

1. Athanasiou G, Korakis T, Ercetin O, Tassiulas L (2007) Dynamic cross-layer association in 802.11-based mesh networks. In: IEEE INFOCOM 2007—26th IEEE international conference on computer communications, Anchorage, AK, pp 2090–2098
2. Athanasiou G, Korakis T, Ercetin O, Tassiulas L (2009) A cross-layer framework for association control in wireless mesh networks. IEEE Trans Mob Comput 8(1)
3. Broustis I, Papagiannaki K, Krishnamurthy SV, Faloutsos M, Mhatre VP (2010) Measurement-driven guidelines for 802.11 wlan design. IEEE/ACM Trans Netw 18(3):722–735
4. George A, Broustis I, Tassiulas L (2011) Efficient load-aware channel allocation in wireless access networks. J Comput Netw Commun 2011:1–13, Article ID 97205, x
5. Xu F, Zhu X, Tan C, Li Q, Yan G, Wu J (2013) SmartAssoc: decentralized access point selection algorithm to improve throughput. IEEE Trans Parallel Distrib Syst 24(12):2482–2491
6. Hong K, Kim J, Kim M, Lee S (2016) Channel measurement-based access point selection in IEEE 802.11 WLANs. Pervasive Mob Comput 30:58–70
7. AMER M, Busson A, Guérin Lassous I (2016) Association optimization in Wi-Fi networks: use of an access-based fairness. In: Proceedings of the 19th ACM international conference on modeling, analysis and simulation of wireless and mobile systems, pp 119–126
8. Wang S, Huang J, Cheng X, Chen B (2014) Coverage adjustment for load balancing with an AP service availability guarantee in WLANs. Springer Wireless Netw 475–491
9. Sajjadi D, Tanha M, Pan J (2016) Meta-heuristic solution for dynamic association control in virtualized multi-rate WLANs. In: Proceedings of IEEE 41st conference on local computer networks, pp 253–261
10. https://www.cisco.com/c/en/us/support/docs/wireless/aironet-1200-series/8103-ap-faq.html. Accessed 19 Jan 2010
11. Moore N (2015) how many devices can my access point support. https://boundless.aerohive.com/experts/How-many-devices-can-my-access-point-support.html. Accessed 03 Sept 2015
12. Chaskar H (2017) http://www.networkcomputing.com/wireless-infrastructure/improving-wifi-client-connectivity-management/585533370. Accessed 24 Aug 2017
13. Kumar A (2014) https://community.arubanetworks.com/t5/Controller-less-WLANs/What-is-spectrum-load-balancing-and-what-is-its-basic-function/ta-p/175862. Accessed 28 Jan 2014
14. Approximating maximum clients per access point. https://documentation.meraki.com/MR/WiFi_Basics_and_Best_Practices/Approximating_Maximum_Clients_per_Access_Point
15. Marek C (2013) Maximum recommended clients per access point. https://supportforums.cisco.com/t5/getting-started-with-wireless/maximum-recommended-clients-per-access-point/td-p/2075572. Accessed 25 Jan 2013

Evaluation of Replication Method in Shuffle-Exchange Network Reliability Performance

Nur Arzilawati Md Yunus, Mohamed Othman, Z. M. Hanapi and Y. L. Kweh

Abstract Nowadays, the Multistage Interconnection Network (MIN) is highly demand in switching process environment. MINs are well known for their cost efficiency which enables economic solution in communication and interconnection. Important features in interconnection topologies are improvement toward performance and increment of reliability. In general, the importance of the reliability will be proportionally increased with the increment of size as well as system complexity. In this paper, we proposed replicated shuffle-exchange network and compared with Benes and Shuffle-Exchange Network (SEN) topology to investigate the behavior analysis of MINs reliability. Three types of behaviors, namely, as basic stage, extra stage, and replicated stage, have been compared in this paper. The results show that replicated network provides highest reliability performance among all topologies measured in this paper.

Keywords Multistage interconnection network · Shuffle-exchange network Benes · Reliability performance · Terminal reliability

1 Introduction

A decade before MINs have been applied in field such as parallel computing architecture, MINs are known for its cost efficiency which enables economic solution in communication and interconnection [1]. MINs were operated by multiple stages with the input and output switches. The Switching Element (SE) in MINs has the ability to transmit input either straight or cross connection. These types of MINs can be differentiated by their topological representation, switching element connected, and the number of stages used in network configuration [2]. A reliable interconnec-

N. A. Md Yunus (✉) · M. Othman (✉) · Z. M. Hanapi · Y. L. Kweh
Faculty of Computer Science and Information Technology, Department of Communication
Technology and Network, Universiti Putra Malaysia, Selangor, Malaysia
e-mail: arzilawati@gmail.com

M. Othman
e-mail: mothman@upm.edu.my

© Springer Nature Singapore Pte Ltd. 2019
M. L. Kolhe et al. (eds.), *Advances in Data and Information Sciences*, Lecture Notes
in Networks and Systems 39, https://doi.org/10.1007/978-981-13-0277-0_22

tion network is a crucial factor to ensure overall system performance. The reliability performance is applied to measure the system's ability to convert information from input to output. Reliability itself can be known as the possibility of the network to work effectively according to specific period and environment [3]. The different types of behaviors in MINs could affect the reliability performance of the network. Several of multipath MINs have been developed that offer an alternate path in the network; however, the challenge in accepting the new design to enhance performance and reliability becomes a crucial task. Several performance measures have been implementing such as by adding the number of stages in the network and replicating the network topology [4]. The increment number of stage in the network, it is reliable to offer additional paths for source and destination pairs. It can increase the reliability to some extent. Alternatively, the replicating approach has been analyzed to provide a better reliability compared to increasing stages approach. This approach offers the opportunity to decrease the number of stages in the network while maintaining the reliability performance [5]. Therefore, this paper will investigate the behavior of non-blocking MINs known as SEN and Benes network toward the reliability performance. The topology measure in this paper consists of Shuffle-Exchange Network (SEN), Shuffle-Exchange Network with Additional Stage (SEN+), Replicated Shuffle-Exchange Network (RSEN), Benes Network, Extra-Stage Benes Network (EBN), and Replicated Benes Network (RBN). The performance measurement focused on the reliability performance called terminal reliability.

2 Background

The reliability of interconnection networks is the important measurement in system performance. Evaluation of the reliability were believe that each component were state either completely worked or completely failed. Reliable operation in multistage interconnection networks depends on their topology, network configuration, and number of stages in the system. Performance improvement and reliability increasing are two major attributes for multistage interconnection network topology. As the number of stages and system complexity increase, the reliability performance becomes an important issue. Therefore, the reliability performance has become a concern for researchers in the field of MINs. Mostly, the basic network provides lower reliability performance as compared to modification networks. Therefore, the researchers initiate the new approach by adding the stage into the network to increase the reliability performance [6]. In this approach, a fixed number of stages were added to the network. With advancement in the multiprocessing field, an increasing focus needs to be placed on multipath network. Replicating is a method to create the redundant path by replicating the network by L layer with the advantages of auxiliary links. The increasing number of layers in replicated MINs can improve the reliability compared to basic and additional stage network. Table 1 shows the comparison of MINs topology using the increasing stage and replicating network approach.

Table 1 Multistage interconnection network comparison

Topology	Additional stage	Replicated network	No. of stage	No. of layer
SEN+ [9, 12]	✓	–	4	–
SEN+ 2 [9, 12]	✓	–	5	–
Extra-stage gamma [13]	✓	–	5	–
Extra-stage cube [14]	✓	–	4	–
Extra Benes network [10]	✓	–	6	–
Replicated MIN [8, 10]	–	✓	–	2
Replicated Benes [10]	–	✓	–	2

2.1 Basic Network Topology

Shuffle-Exchange Network (SEN) consists of a unique path for each input and output pairs [3]. In this network, all SEs are assumed as a series connection. In this connection, the input for SE can be transmitted either straight or cross connection. The connections between stages are shuffle exchange. SEN consists of N input and output switches with n stage, where $n = \log_2 N$. SEN provides $(N/2) \log_2 N$ switching element in series [7]. Figure 1 represents the topological view of SEN.

Benes topology is built through an extension of inverse baseline. The first stage and the last stage of the inverse Baseline are merged. This leads to $2n - 1$ stages for Benes networks. When switching is performed in the network, the Benes networks can be classified as non-blocking MINs [8]. Compared to others non-blocking MINs, Benes has the smallest complexity packet switched in the network. Figure 2 shows the topology for Benes network.

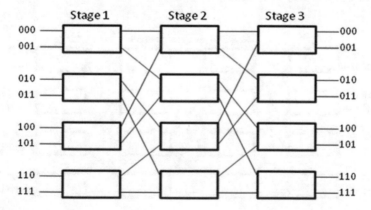

Fig. 1 Shuffle exchange network [3]

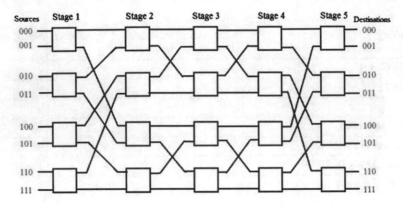

Fig. 2 Benes network [8]

2.2 Additional Stage Topology

Shuffle-Exchange Network with Additional Stage (SEN+) implementing by adding one extra stage to the network. SEN+ has four stages and N inputs-outputs pairs. Referring to SEN+ topology, the increment of stages leads to the increment of the number of switchs and links. It consists of $n = \log_2 N + 1$ stages with $N/2$ switching element for each stage. The additional stage of SEN+ has enabled extra path connection to provide double paths for communication process for each source and destination pairs [9]. SEN+ topology is illustrated in Fig. 3.

Extra-stage Benes Network (EBN) was designed to construct redundancy number of paths between each source and destination pairs as shown in Fig. 4. This network improves the path length for each source and destination. However, the additional stage method is not efficient to be used in Benes network because this approach decreases the reliability of this network [10].

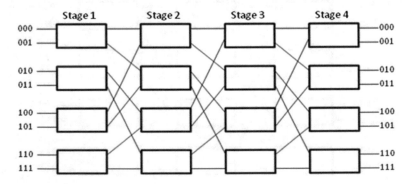

Fig. 3 Shuffle-exchange network with additional stage [9]

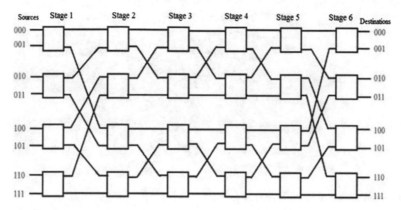

Fig. 4 Extra-stage Benes network [10]

2.3 Replicated Network Topology

The design pattern of the Replicated Benes Network (RBN) is a modification from Benes network by replicating the network to create the redundant path. The replicated Benes are arranged in L layer. Normally, for RBN topology, all switching elements are of size 2×2. It has $[2(\log_2 N) - 1]$ number of stages. Replicated Benes network is shown in Fig. 5.

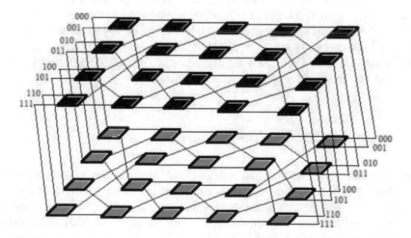

Fig. 5 Replicated Benes network [10]

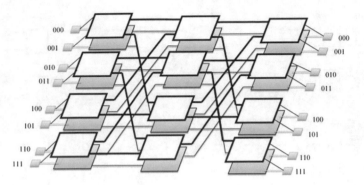

Fig. 6 Replicated shuffle-exchange network

3 Proposed Topology

In this section, we proposed a new topology named the replicated shuffle-exchange network by replicating the network on SENs topology. Replicated SEN enlarges regular SEN by replicating the network equal to L times. The corresponding input and output were synchronously connected [4]. Packets received by the inputs of the network distributed throughout the layers. Different to the MINs, the replicated MINs lead to the out of order packet sequence due to the availability of multipath for each source and destination pair. The rationale to send packets belongs to the same source is to avoid packet order destruction. For the reliability purposes, the increment of the number of layers will lead to a reliability improvement in the network. In a view of terminal reliability, the system can be said as a parallel system, where the system consists of two series systems in parallel [10]. Each series system comprises $\log_2 N$ of SE. Figures 6 and 7 show the three-dimensional view of the replicated SEN with two layers and the methodology for reliability analysis.

4 Terminal Reliability Measurement

Interconnection network topologies provide the ability to improved performance and the reliability in the network. The choice of interconnection network is dependent on a number of factors, such as the topology, routing algorithm, and communication properties of the network [11]. The reliability of an interconnection network is defined as measurement method to identify the capability of a system to convert information from the input to the output. Table 2 shows the details of each topology for SEN and Benes topology.

Terminal reliability can be described for the possibility of existing of at least one fault free path connecting a selected of inputs and outputs pairs [6]. The basic switching element for time-dependent reliability can be referred as r equal to (0.90–0.99).

Fig. 7 Methodology for reliability analysis

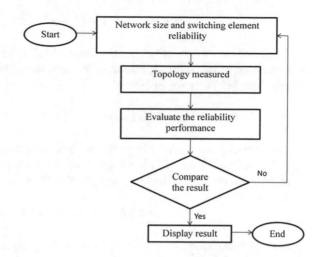

SEN consists of N input/output switches and $n = \log_2 N$ stage, where each stage has $N/2$ switching elements. The variable r can be defined as the possibility of a switch being operational. The terminal reliability for SEN topology is derived as follows:

$$SEN = r^{\log_2 N} \tag{1}$$

According to Benes topology as shown in Fig. 2, the terminal reliability for 8×8 can be calculated by the following equation:

$$Benes = r^2[1 - (1 - (r^2(1 - (1 - r)^2)))^2] \tag{2}$$

SEN+ has two paths MIN. It consists of $\log_2 N + 1$ stages, providing two paths for communication process for each source and destination pairs. The terminal reliability for SEN+ with $N \times N$ size is derived as follows:

Table 2 Topology comparison

Topology	Stage	Switching complexity	Layer	Network types
SEN	$\log_2 N$	$\left(\frac{N}{2}\right)\log_2 N$	–	Non-blocking
SEN+	$\log_2 N + 1$	$\left(\frac{N}{2}\right)\log_2 N + 1$	–	Non-blocking
R. SEN	$\log_2 N$	$\left(\frac{N}{2}\right)\log_2 N$	2	Non-blocking
Benes	$2(\log_2 N - 1)$	$\left(\frac{N}{2}\right)2(\log_2 N - 1)$	–	Non-blocking
E. Benes	$(2\log_2 N)$	$\left(\frac{N}{2}\right)2(\log_2 N)$	–	Non-blocking
R. Benes	$(2\log_2 N - 1)$	$\left(\frac{N}{2}\right)2(\log_2 N - 1)$	2	Non-blocking

$$SEN+ = r^2[1 - (1 - r^{(\log_2 N)-1})^2] \tag{3}$$

Similar to Benes network, for EBN, the terminal reliability for this topology can be formulated based on Fig. 4. Therefore, we have the following equation to calculate the terminal reliability for EBN:

$$Extra\ Stage\ Benes = r^2[1 - (1 - (r^2(1 - (1 - r)^2)^2))^2] \tag{4}$$

The time-dependent reliability of the basic switching element can be defined by r. By replicating the SEN topology, it will improve the reliability performance. Therefore, the terminal reliability for replicated SEN is derived as follows:

$$Rep.\ SEN = 1 - (1 - (r^{\log_2 N}))^2 \tag{5}$$

Furthermore, the replicated Benes arranged L layer as shown in Fig. 6. Therefore, the terminal reliability for 8×8 replicated Benes can be calculated as follows:

$$Rep.\ Benes = 1 - (1 - (r^2[1 - (1 - (r^2(1 - (1 - r)^2)))^2]))^2 \tag{6}$$

5 Results and Discussion

This paper focuses on the reliability measurement known as terminal reliability for the SEN and Benes network topologies. Table 3 shows the comparative analysis for MINs including the SEN, SEN+, replicated SEN, Benes, extra-stage Benes, and replicated Benes. From Table 3, the overall reliability performance shows that the replicated behavior provides better performance compared to extra stage and basic network. The basic stage for both networks provides the lowest reliability performance in the network. However, for basic stage comparison in Fig. 8, the results indicate that Benes network provides the highest reliability, since it provides more redundant path as compared to SEN topologies.

Table 3 Reliability performance comparison

TR	SEN	SEN+	Rep. SEN	Benes	Extra Benes	Rep. Benes
0.90	0.7289	0.7807	0.9265	0.7782	0.7757	0.9508
0.92	0.7786	0.8246	0.9510	0.8249	0.8235	0.9693
0.94	0.8305	0.8716	0.9712	0.8706	0.8702	0.9833
0.95	0.8573	0.8939	0.9796	0.8935	0.8931	0.9886
0.96	0.8847	0.9159	0.9867	0.9157	0.9155	0.9928
0.98	0.9411	0.9588	0.9965	0.9588	0.9588	0.9983
0.99	0.9702	0.9797	0.9991	0.9797	0.9797	0.9995

Fig. 8 Basic network comparison

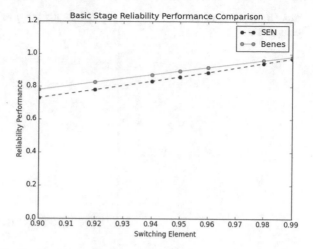

Fig. 9 Additional stage network comparison

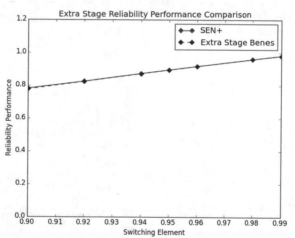

For extra-stage network comparison, Fig. 9 shows that SEN+ and extra-stage Benes network provide an equal reliability performance. In contrast to SEN, the Benes network reliability does not essentially affect by adding the extra stage in the network. The extra-stage network can increase the path length in the network; however, the approach is inefficient to be implement in certain network and does not always lead to increment of reliability performance. Additional extra stage in Benes network provides more complexity due to its configuration and redundant paths availability.

For replicated topology, the results indicate that terminal reliability for replicated Benes is superior compared to replicated SEN as shown in Fig. 10. The advantages of redundant paths in the replicated Benes enhance the reliability in the network. The result indicates that the reliability of the replicated Benes and replicated SEN is

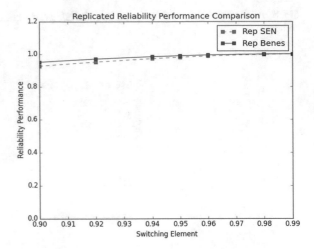

Fig. 10 Replicated network comparison

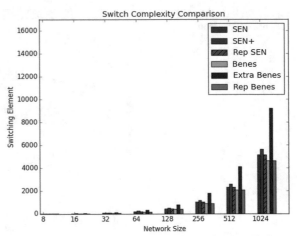

Fig. 11 Switch complexity comparison

superior to extra stage and basic stage topology because the network offers redundant paths for the network connectivity.

From Fig. 11, it shows that the additional stage has the highest switching element, and the replicated network has the lowest switching element used in both networks. In general, the switch complexity for basic stage network and replicated network is less than additional stage network. Based on Fig. 11, it is noted that for each size comparison, replicated Benes and replicated SEN have a lowest SEs compared to additional stage SEN and Benes topology. It can be seen that the switch complexity for basic stage network is equal to replicated network. Despite the equal switch complexity in the network, the replicated network for both networks leads to highest reliability performance for SEN and Benes topology. The advantages of lesser SE in the network will reduce the implementation cost of the network.

6 Conclusion

The network topology, the number of stages, and the types of network configuration are different for each interconnection network. The existed additional paths in interconnection network result in the increment of reliability performance. For basic stages, additional stage, and replicated network, the result comparison indicates highest terminal reliability for Benes as compared to SEN topology. The SEN topology itself has lower reliability performance due to the disadvantages of lesser stage and lesser path to route the message in the network compared to Benes topology. However, for overall performance, the behavior of replicated network provides a higher reliability for SEN and Benes topology. The additional stage for both networks enhances the reliability performance to small extent. It can be summarized that the replicated network provides the highest reliability compared to basic stage and additional stage network. We can see that the replicated behavior provides the availability of more redundant path that can lead to the network with highest reliability performance.

Acknowledgements This work was supported by the Universiti Putra Malaysia under the Geran Putra Berimpak: UPM/700-2/1/GPB/2017/9557900.

References

1. Tutsch D, Hommel G (2003) Multilayer multistage interconnection networks. In: Proceedings of 2003 design analysis and simulation of distributed system, Orlando, pp 155–162
2. Soni S, Dhaliwal AS, Jalota A (2014) Behavior analysis of omega network using multi-layer multi-stage interconnection network. Int J Eng Res Appl 4(4):127–130
3. Yunus NAM, Othman M, Hanapi ZM, Kweh YL (2016) Reliability review of interconnection networks. IETE Tech Rev 33(6):596–606
4. Aulah NS (2016) Reliability analysis of MUX-DEMUX replicated multistage interconnection networks. Exp Tech 19–22
5. Bistouni F, Jahanshahi M (2015) Pars network: a multistage interconnection network with fault tolerance capability. J Parallel Distrib Comput 75:168–183
6. Rajkumar S, Goyal NK (2015) Reliability analysis of multistage interconnection networks. Quality Reliab Eng Int 32(8):3051–3065
7. Bistouni F, Jahanshahi M (2014) Analyzing the reliability of shuffle-exchange networks using reliability block diagrams. Reliab Eng Syst Safety 132:97–106
8. Tutsch D (2006) Performance analysis of network architecture. Springer, Berlin, Heidelberg
9. Yunus NAM, Othman M (2015) Reliability evaluation for shuffle exchange interconnection network. Procedia Comput Sci 59:162–170
10. Jahanshahi M, Bistouni F (2015) Improving the reliability of Benes network for use in large-scale systems. Microelectr Reliab 55:679–695
11. Moudi M, Othman M, Kweh YL, Rahiman ARA (2016) x-folded TM: an efficient topology for interconnection networks. J Netw Comput Appl 73:27–34
12. Gunawan I (2008) Reliability analysis of shuffle-exchange systems. Reliab Eng Syst Safety 93(2):271–276
13. Gunawan I, Fard NS (2012) Terminal reliability assessment of gamma and extra-stage gamma networks. Int J Quality Reliab Manag 29:820–831
14. Gunawan I, Sellapan P, Lim CS (2006) Extra stage cube network reliability estimation using stratified sampling Monte Carlo method. Eng e-Trans 1(1):13–18

Rateless Distributed Source-Channel Coding of Correlated Binary Sources Using Raptor Codes

Saikat Majumder, Bibhudendra Acharya and Toshanlal Meenpal

Abstract The problem of distributed source-channel coding of a binary source with correlated side information at the decoder is addressed in this paper. We implement and evaluate the performance of rateless Raptor codes for distributed source-channel coding in binary symmetric (BSC) channel and additive white Gaussian noise (AWGN) channel. We adapt the conventional distributed source codec for error-free channel using Raptor codes to noisy channels. The proposed implementation is rateless, that is, the rate of the code can be adapted to varying channel conditions with incremental redundancy. This is achieved with built in rateless property of Raptor codes.

Keywords Raptor code · Source-channel coding · Rateless · Slepian–Wolf

1 Introduction

The theory of distributed coding of correlated sources has applications in many modern networks. It is specially relevant to energy- and bandwidth-sensitive technologies like Internet of Things (IoT) [1] and wireless sensor networks. In wireless sensor network, the objective is to compress and transmit large quantity of observations collected by many energy-constrained sensors. These wireless sensors may be deployed in hazardous and inaccessible regions where it may be difficult, if not impossible, to replace the depleted out batteries. Sensing and communication in such applications require algorithm which can utilize every bit of correlation in data for minimizing the transmission energy, while maintaining the reliability of communication.

S. Majumder · B. Acharya (✉) · T. Meenpal
Department of Electronics and Telecommunication, National Institute of Technology,
Raipur, Chhattisgarh, India
e-mail: bacharya.etc@nitrr.ac.in

S. Majumder
e-mail: smajumder.etc@nitrr.ac.in

T. Meenpal
e-mail: tmeenpal.etc@nitrr.ac.in

© Springer Nature Singapore Pte Ltd. 2019
M. L. Kolhe et al. (eds.), *Advances in Data and Information Sciences*, Lecture Notes
in Networks and Systems 39, https://doi.org/10.1007/978-981-13-0277-0_23

Consider the case of two correlated information sources that do not communicate with each other. Lossless compression of the output of these two sources, without exchanging information between them, is explained by Slepian–Wolf theorem [2]. According to Shannon's source coding theorem, the theoretical limit on total compression rate of two correlated sources A and B is given by their joint entropy $H(X^A, X^B)$. Results of Slepian and Wolf assert that separate compression suffers no loss of rate compared to joint compression. The admissible rate region for separate coding of two correlated sources is [2]

$$R_A \geq H(X^A | X^B) \tag{1}$$

$$R_B \geq H(X^B | X^B) \tag{2}$$

$$R_A + R_B \geq H(X^A, X^B), \tag{3}$$

where R_A and R_B are compression rates of source A and B, respectively. If the two encoders do not utilize the correlation between the sources, the rate region in such case is given as $R_A + R_B \geq H(X^A) + H(X^B)$ which is greater than the optimal value of $H(X^A, X^B)$.

A special scenario of correlated information sources is one in which a source B is fully known at the decoder. The source B is available at the decoder at rate $R_B = H(X^B)$. On the other hand, the source A is compressed using the estimate of the correlation between A and B, with rate $R_A \geq H(X^A | X^B)$. Such a scheme is commonly referred to as the asymmetric Slepian–Wolf coding or coding with decoder side information. First practical scheme for distributed source coding using conventional channel codes was proposed by Pradhan and Ramchandran in [3]. Authors in [4–7] proposed the method for distributed compression of binary sources using modern error correction codes and sum-product algorithm. Other significant contribution in the area of distributed source coding using channel codes can be found in [8–10] and the references therein.

Besides efficient compression, reliable transmission of the compressed information is required in case of wireless and other practical channels. In traditional communication systems, source and channel coding are performed separately. The redundancy at the sources is removed by the source encoder or data compression algorithm and the compressed data is channel encoded and transmitted through the channel. The source encoders and channel encoders can be combined into one joint source-channel encoder. This allows a single channel code for both source and channel encoding and decoding. Such a joint scheme can utilize the residual redundancy left out after source coding for improving the bit error rate of information transmitted through error-prone channel. Joint source-channel coding is also a requirement for distributed coding as source/channel separation principle does not necessarily hold for general multiterminal networks [11]. Source-channel coding of distributed correlated sources using irregular repeat accumulate (IRA) code were proposed in [7]. A rate adaptive method for distributed source-channel coding with decoder side information using extending and puncturing was given in [12]. In [13], source-channel decoder for a Markov source encoded with systematic IRA code was presented. In [14], authors combined entropy coding and channel coding into a single linear encoding stage. The use of systematic Raptor encoders was proposed, in order to obtain a continuum of coding rates with a single basic encoding algorithm.

Rateless codes are a class of forward error correction (FEC) codes in which an infinite stream of code can be generated from a block of data. Decoding can be done from any of the sections of the codeword, only if a certain minimum number of bits are received. Luby transform (LT) code [15] is one of the first rateless codes. In addition, LT codes are universal in sense that encoding and decoding times are asymptotically near optimal for all sorts of erasure channels. Raptor codes [16] were later introduced as rateless codes with linear time encoding and decoding complexity and they have capacity achieving performance over continuous channels. Subsequently, rateless codes were applied to many current interest topics. In [17], rateless space time block code was applied for massive MIMO systems. Application of rateless code in content delivery network was proposed by Zarlis et al. [18]. Results show that rateless codes are able to reduce packet loss up to 50%. In [19], a three-stage rateless coded protocol for a half-duplex time-division two-way relay system was proposed, where two terminals send messages to each other through a relay between them. In the proposed protocol, each terminal takes one of the first two stages, respectively, to encode its message using rateless code and broadcast the result until the relay acknowledges successful decoding.

Rateless coding allows communication systems to adapt their transmission rates to the channel conditions, without channel state information at the transmitter or encoder. Rateless codes like Raptor codes generate an infinite number of bits at the transmitter and stop only after receiving an ACK from the decoder that the information has been successfully decoded. Conventional codes require the knowledge of the capacity of the channel for selection of code rate. For rapidly varying and unknown channel, this information may not be available and message needs to be transmitted at very low rate, resulting in wastage of energy and bandwidth.

In this paper, we apply rateless codes for distributed source-channel coding. Out of the two correlated sources, one source X^B is assumed to be available as side information at the receiver. The other source X^A is encoded with systematic Raptor code and is transmitted with rate R_A. Since transmission rate is less than actual entropy $H(X^A)$ of the source, compression is achieved and hence, Raptor code, in this context, acts as both source and channel code. The proposed scheme incrementally transmits encoded bits, which depends both on source redundancy, i.e., amount of correlation between the two sources and quality of channel. We verify the scheme with simulation on binary symmetric channel and AWGN channel. Simulation results show that data compression is achieved by encoding with Raptor code for different values of source correlation and compression ratio increases with decrease correlation between sources increase,

The rest of the paper is organized as follows. Section 2 briefly discusses systematic and non-systematic encoding operation by Raptor code. In Sect. 3 we explain the proposed rateless source-channel coding scheme and its decoder. In Sect. 4, we give the performance results obtained through simulation on BSC and AWGN channel. Finally, Sect. 5 is conclusion.

2   Raptor Codes

Raptor codes have been initially designed for rateless communication over erasure channel with unknown capacity [16]. Its analysis has been extended to Gaussian channels in [20, 21]. Rateless codes can produce an endless stream of bits till the information is successfully decoded at the receiver. Raptor codes are concatenation of outer linear code C (LDPC code here) and an inner Luby transform (LT) code. It is LT part of the code, which makes the code rateless. According to the convention of Raptor codes, *input symbols* are intermediate symbols between LDPC and LT code, whereas *output symbols* are output of LT code, as shown in Fig. 1. The output node degrees of LT code is given by generator polynomial $\Omega(x) = \sum_{m=1}^{n} \Omega_m x^m$. Here Ω_m is the probability that an output symbol is connected to m input symbols. A Raptor code is specified by the triplet $(K, C, \Omega(x))$, where K is the length of the information vector.

Raptor code encoding is done in two steps [16]. Length K information vector is first encoded with a weak (high rate) LDPC code to produce n number of output symbols. The LDPC encoded symbols (or input symbols) are then LT coded to form output $C = SG_{ldpc}X^A$, where G_{ldpc} is LDPC generator matrix, and S is the LT code matrix for first N output symbols. Similar to [6], we use systematic Raptor code [16] for the proposed source-channel coding scheme. Systematic Raptor code is obtained by applying Gaussian elimination to SG_{ldpc} to obtain an invertible submatrix R of SG_{ldpc}. At the encoder, input to LDPC code is obtained as $R^{-1}X^A$. This produces $C = [X^A|P] = [c_1, ..., c_N] = SG_{ldpc}X^A$ as output symbols, where X^A is the systematic part, and P is the parity part of the Raptor code. For source-channel coding two different transmission schemes are possible [7]. When only the parity part is transmitted, $C^{tx} = P$, we have a non-systematic source-channel code (NSSCC) with compression ratio of $R_A = (N - K)/K$. In case of systematic source-channel code (SSCC), the usual code vector $C^{tx} = [X^A|P]$ is transmitted. The compression in this case is $R_A = N/K$.

3   Rateless Source-Channel Coding and Decoding

We consider the system of Fig. 2 for distributed source-channel coding with decoder side information. There are two correlated sources A and B, generating correlated binary vectors $X^A = [x^A(1), ..., x^A(K)]$ and $X^B = [x^B(1), ..., x^B(K)]$, respectively, of length K. Both the sources produce independent and identically distributed (i.i.d.) binary random variable, with correlation defined by $P(X_i^A|X_i^B) = P(X_i^A \neq X_i^B) = p \leq 0.5$. Source B of rate $R_B = H(X^B)$ is available at the receiver without any compression, where H indicates binary entropy function. Source A is Raptor encoded to vector $C = [c(1), c(2), ...]$ of rate $R_A \geq H(p)/C_{ch}$, where C_{ch} is the capacity of the channel. The channel capacity for BSC with crossover probability of q is $C_{bsc} = 1 - H(q)$. Since, in many practical situations, where the capacity of the channel or amount of correlation between sources is not known, the encoder transmits incre-

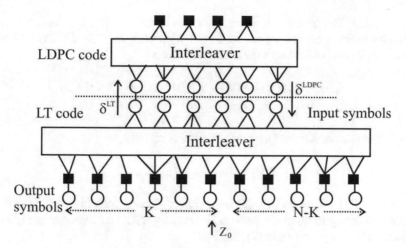

Fig. 1 Factor graph of Raptor code. Black squares represent parity-check nodes and circles are variable nodes associated with input and output symbols

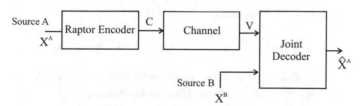

Fig. 2 Block diagram of distributed source-channel coding using systematic Raptor code. For NSSCC, C is purely parity bits obtained by removing systematic bits. Source B is available as side information

mentally transmits bits till number of received bits are sufficient for successful decoding.

Raptor codes can be decoded iteratively on graph using sum-product algorithm. Figure 1 shows tanner graph for systematic Raptor code, with K output bits being systematic bits and $N - K$ parity bits. Let $Z_o(i)$, $i = 1, ..., N$ be the LLR received from equivalent channel at the ith bit of LT code. $\delta_{lt}(i)$ is the total LLR generated at the ith input node of LT code and passed to corresponding variable node of LDPC code. Similarly, $\delta_{ldpc}(i)$ is the message from variable nodes of LDPC code to input node of LT code. Let the received vector at decoder be $V = [v(1), v(2), ...]$.

In case of NSSCC, decoder estimates systematic bits from side information X^B and parity bits from the signal V received through the channel. LLR contribution due to correlation between sources is

$$Z_o(i) = \log(P(c(i) = 0|x^B(i))/P(c(i) = 1|x^B(i))) \qquad (4)$$
$$= (1 - 2x^A(i))\log(1 - p/p), \qquad (5)$$

for $i = 1, ..., K$. For transmission through binary symmetric channel, let $v(i) \in \{\pm 1\}$ be the transmitted bit corresponding to $(i + K)$th codeword bit. LLR contribution due to signal received through the actual channel is

$$Z_o(i) = \log(P(c(i) = 0|v(i - K))/P(c(i) = 1|v(i - K))) \tag{6}$$

$$= (v(i - K))\log(1 - q/q), \tag{7}$$

for $i = K + 1, ..., N$. Similarly, in case of AWGN channel,

$$Z_o(i) = 4v(i - K)\sqrt{E_s}/N_0, \tag{8}$$

where $E_s = R_c E_b$, $R_c = K/(N - K)$ is the rate of channel code in NSSCC mode.

Though not optimum for Raptor code setup, for SSCC we calculate the LLR for systematic bits as proposed in [7]. For BSC,

$$Z_o(i) = (1 - 2x^A(i))\log(1 - p/p) + (v(i))\log(1 - q/q), \tag{9}$$

for $i = 1, ..., K$. For AWGN channel,

$$Z_o(i) = (1 - 2x^A(i))\log(1 - p/p) + 4v(i)\sqrt{E_s}/N_0, \tag{10}$$

for $i = 1, ..., K$ and $R_c = K/N$ is the rate of channel code in systematic transmission mode. The LLR equation for parity bits remains unchanged.

4   Simulation Results

In order to illustrate the behavior of the proposed scheme, we simulate the NSSCC and SSCC schemes for BSC and AWGN channels. Similar to [6], we use Raptor code with LT code of degree distribution $\Omega(X) = 0.008x + 0.494x^2 + 0.166x^3 + 0.073x^4 + 0.083x^5 + 0.056x^8 + 0.037x^9 + 0.056x^{19} + 0.025x^{65} + 0.003x^{66}$. The precode is regular $(2,100)$ LDPC code. We use information of block length $K = 9801$ bits and block length of Raptor code varied from 12000 to 21000 bits.

Simulation results are provided for different compression ratios (for different rates of the code) to find the corresponding bit error rate. In Fig. 3, we report the results for binary symmetric channel of channel crossover probability $q = 0.05$. For different value of source correlation parameter p, the non-systematic part is transmitted through physical channel. The side information with appropriate LLR acts as the systematic part of the code. Non-systematic bits, in incremental blocks of 250 bits, transmitted till bit error rate (BER) is lower than 1×10^{-5}. The figure shows different values of compression ratio achievable by varying p and constant channel parameter. For example, compression ratios for $p = 0.05, 0.10, 0.15$ is approximately 0.55, 0.80, and 0.95, respectively.

Fig. 3 BER versus compression ratio for binary symmetric channel (channel crossover probability of 0.05) for various values of $p = P(X^A|X^B)$ in NSSCC mode

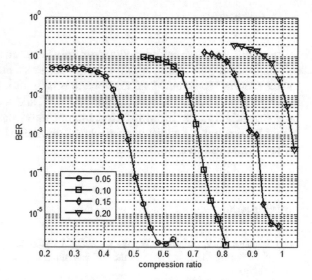

Fig. 4 BER versus compression ratio for various values of $p = P(X^A|X^B)$ and AWGN channel of $E_b/N_0 = 1$ dB. Solid curves are simulation results for NSSCC and dashed curves are for SSCC

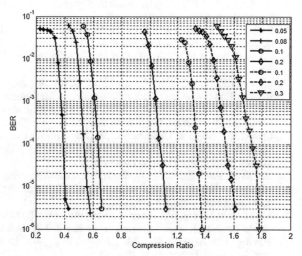

Figure 4 shows the results for AWGN channel with E_b/N_0 of 1 db. The scheme was simulated for both NSSCC and SSCC. It can be seen that better compression is achievable through NSSCC scheme compared to SSCC scheme. The performance gap is large for small value of p, but reduces with increase in p. Similar phenomena was also reported in [7] for IRA code. It is to be noted that we did not optimize Raptor code degree distribution for particular channel and better results can be obtained.

5   Conclusion

In this paper, we have introduced a distributed asymmetric source-channel coding scheme with Raptor codes, which extends earlier scheme with fixed rate codes to rateless codes. We obtained values of compression ratios for different source correlation parameters. Results were obtained for BSC and AWGN channel by incrementally transmitting bits till error-free communication was achieved. Such a scheme with ARQ can be applied for cases in which correlation between the sources or channel state information is not known.

References

1. Centenaro M, Vangelista L, Zanella A, Zorzi M (2016) Long-range communications in unlicensed bands: the rising stars in the IoT and smart city scenarios. IEEE Wireless Commun 23(5):60–67
2. Slepian D, Wolf JK (1973) Noiseless coding of correlated information sources. IEEE Trans Inf Theory 19(4):471–480
3. Pradhan SS, Ramchandran K (2003) Distributed source coding using syndromes (DISCUS): design and construction. IEEE Trans Inf Theory 49(3):626–643
4. Garcia-Frias J, Zhao Y (2001) Compression of correlated binary sources using turbo codes. IEEE Commun Lett 5(10):417–419
5. Liveris AD, Xiong Z, Georghiades CN (2002) Compression of binary sources with side information using low-density parity-check codes. IEEE Commun Lett 6(10):440–442
6. Fresia M, Vandendorpe L, Poor HV (2009) Distributed source coding using Raptor codes for hidden Markov sources. IEEE Trans Signal Process 57(7):2868–2875
7. Liveris AD, Xiong X, Georghiades CN (2002) Joint source-channel coding of binary sources with side information at the decoder using IRA codes. In: Proceedings of IEEE workshop on multimedia signal processing, pp 53–56
8. Sartipi M, Fekri F (2005) Distributed source coding using LDPC codes: lossy and lossless cases with unknown correlation parameter. In: Proceedings of Allerton conference on communication, control and computing
9. Stankovic V, Liveris AD, Xiong X, Georghiades CN (2006) On code design for Slepian-Wolf problem and lossless multiterminal network. IEEE Trans Inf Theory 52(4):1495–1507
10. Varodayan D, Aaron A, Girod B (2006) Rate-adaptive Codes for distributed source coding. Elsevier Signal Process 86:3123–3130
11. Al Gamal A, Kim Y-H (2011) Network information theory. Cambridge University Press, New York
12. Majumder S, Verma S (2011) Rate adaptive distributed source-channel coding using IRA code for wireless sensor networks. In: Proceedings of AIM 2011, CCIS, vol 147, pp 207–213
13. Majumder S, Verma S (2012) Joint source-channel decoding of IRA code for hidden Markov source. In: 1st IEEE international conference on recent advances in information technology (RAIT), Dhanbad, pp 220–223
14. Bursalioglu OY, Fresia M, Caire G, Poor HV (2008) Lossy joint source-channel coding using Raptor codes. Int J Digital Multimed Broadcast, vol 2008, Article ID 124685
15. Luby M (2002) LT codes. In: Proceedings of the 43rd annual IEEE symposium on foundations of computer science. IEEE, pp 271–280
16. Shokrollahi A (2006) Raptor Codes. IEEE Trans Inf Theory 52(6):2551–2567
17. Alqahtani AH, Sulyman AI, Alsanie A (2014) Rateless space time block code for massive MIMO systems. Int J Antennas Propag 2014

18. Zarlis M, Sitorus SP, Al-Akaidi M (2017) Applying a rateless code in content delivery networks. In: IOP conference series: materials science and engineering, vol 237, no 1. IOP Publishing, p 012016
19. Zhang Y, Zhang Z, Yin R, Yu G, Wang W (2013) Joint network-channel coding with rateless code in two-way relay systems. IEEE Trans Wireless Commun 12(7):3158–3169
20. Etesami O, Shokrollahi A (2006) Raptor codes on binary memoryless symmetric channels. IEEE Trans Inf Theory 52(5):2033–2051
21. Venkiah A, Poulliat C, Declercq D (2009) Jointly decoded Raptor codes: analysis and design for the BIAWGN channel. EURASIP J Wireless Commun Netw 2009(1):657970

Home Automation Using Internet of Things

Vibha Lohan and Rishi Pal Singh

Abstract Internet of Things is the internetworking of devices such as heart monitoring, automobiles with built-in sensors, environmental monitoring, home automation and lightning that enable these devices to gather, communicate, share, receive, and transfer data over a network. The Internet of Things is about making the surrounding environment smart including the buildings used for living beings like homes and hospitals. In this paper, various sensors are deployed in home such as motion sensors, luminance sensors and temperature sensors, is discussed. Algorithm to save the energy of lights and air conditioners of the entire home is also implemented in this paper. Energy consumption before and after implementation of system model has been calculated which showed that a total of 20.78% energy has been saved.

Keywords Internet of things (IOT) · Internet · Future internet · Smart home
Radio frequency identification (RFID)

1 Introduction

Internet of Things (IOT) represents general concept of the internetworking or interconnecting of devices that contains sensors, softwares and actuators. It enables all these devices to gather and transfer data over a network. The term IOT provided new life to radio frequency identification (RFID) and was introduced to help radio frequency identification technology to particularly identify interoperable united objects. IOT is a combination of two words, Internet and things. Internet is a network of networks for sharing and communicating information worldwide. Things in the Internet of Things have the capability to send and receive information to a network barring the intervention of humans and it contains physical objects that have a particular identi-

V. Lohan (✉) · R. P. Singh
Department of Computer Science and Engineering, Guru Jambheshwar University of Science and Technology, Hisar, Haryana, India
e-mail: vibhalohan22@gmail.com

R. P. Singh
e-mail: pal_rishi@yahoo.com

© Springer Nature Singapore Pte Ltd. 2019
M. L. Kolhe et al. (eds.), *Advances in Data and Information Sciences*, Lecture Notes in Networks and Systems 39, https://doi.org/10.1007/978-981-13-0277-0_24

fier with an embedded system. IOT is a "global infrastructure of networks that have self configuring abilities and standard protocols for communication where virtual as well as physical. Thing has virtual personalities, identities, intelligent interfaces, physical attributes which are combined into a network of information" [1].

Today, IOT is an evolving technology that contains devices, sensors, and objects. They all connected to the Internet. The IOT is about making the surroundings smart including the buildings used for living beings like home, hospital, etc. Today, the energy dependency for homes is very high. Therefore, the objective of this paper is automation of homes using sensors like temperature sensors, motion sensors, and luminance sensors to save energy. In this paper, a good architecture for energy efficient automation, which can save energy, is proposed. For this a home model with a total area of 874 ft^2 is taken. In this home model, four persons are considered. Then three types of sensors deploy in the home. First type of sensors is motion sensors. Motion sensors are used in the system model to track the position of the persons living in home. Second type of sensors is luminance sensors. Luminance sensors are used to check the luminance level of the areas in the home. Third type of sensors is temperature sensors. Temperature sensors are used in the system model to check the temperature of the home. Then, the deployment of the sensors distance between the neighboring sensors has to be calculated. The algorithm is implemented and collection of energy consumption of lights and air conditioners of the entire home has to be done.

The remaining paper is organized as follows. Section 2 describes existing related work for energy efficient smart homes. Section 3 presents the proposed home model. Design of the system is described in Sect. 4. Section 5 presents the implementation of system. Section 6 explains the results. Section 7 discusses the conclusion and future work.

2   Related Work

Existing work associated with energy-efficient smart homes can be classified into three categories. The classifications of smart homes are as follows. Energy of homes can be saved by scheduling tasks, by tracking the position of the persons and by controlling devices using various types of sensors. The energy dependency for homes is very high, so one can save energy by using one of these methods.

The first method to conserve energy in homes is by scheduling tasks smartly. Baraka et al. [2] presented an energy-efficient system using task scheduling. They showed that the system is scalable and flexible as well as it is an energy-saving system. Hence it can be used in big houses to save energy. Pedrasa et al. [3] introduced more efficient and upgraded distributed energy resource operation schedule. They select and upgrade the particle swarm optimization for solving optimization problem. The enhanced schedules of distributed energy resources are then investigated by comparing the enhanced algorithm against each distributed energy resource schedule. Another task scheduling method is introduced by considering dynamic electricity price. An energy-efficient algorithm proposed for scheduling house appliances to save energy. This algorithm provides result in 10 s, which is necessary for house appliances [4]. Energy consumption scheduling devices are used in smart meters. An algorithm is proposed to balance the load of the whole home [5].

The second method for conserving energy in homes is by tracking the position of the persons. Wang et al. [6] presented and implemented an energy- saving method for smart homes based on IOT. Their system initially collects information of the position of person and energy consumption of house appliances. The system finds out the waste electricity based on person's location. It also enhanced the lifestyle of the people. Mainetti et al. [7] proposed a location-based architecture of heterogeneous systems for home automation. The environment of home can be automatically controlled by rules of the person. It also supports development of new services for developing a home smart. Roy et al. [8] took an information theory to design a management scheme for location based smart home. Information theory contains Asymptotic Equipartition Property that helps to find out the future location of object. Another location based smart home using information theory is also proposed. It also reduced the operating cost of homes [9]. Subbu and Thomas [10] described a smart home control system that needs a smartphone to find out the location of person to control appliances of home according to his location. A location-based distributed access control is also described for homes [11]. A system based on user location and user-defined rules are described for managing smart home [12].

The third method to save energy in homes is by using various types of sensors. Sensors can collect information easily and they can also manage home devices. An enhanced routing protocol Link Quality Indicator Based Routing (LQIR) is developed to enhance the performance of sensors [13]. Lu et al. [14] demonstrated low-cost sensing technology and sleeping patterns of sensors to conserve energy. They deployed sensors in eight homes and compared results with previous approaches. Song et al. [15] presented a home appliances control system using hybrid sensor networks.

There are other methods for home automation system that contains networked components which are coordinated and have requirements such as moderate cost, future-prof, installation overhead, security, user interaction, and connectivity. 6LoW-PAN and IPv6, a single network serves the increasing aspects of home automation while concepts from the web give benefits for both users as well as developers [16]. Home automation is done in IOT for elderly people by using the combination of Keep-In-Touch technology and closed-loop healthcare services [17]. Much of work has been done in home automation using Wi-Fi-based system [18], cloud network and mobile devices [19], ZigBee [20], Android-based smart phone [21], IOT with web services and cloud computing [22]. In addition, many other typical smart home architecture solutions are also proposed based on IOT [23, 24].

3 Proposed Model of the System

3.1 Home Model

The home model is displayed in Fig. 1. The model of the home consists a total area of 874 ft^2. The home model comprises three rooms, one living room, one kitchen, and two washrooms. The area of first room, second room, and third room are 120, 140, and 120 ft^2. The area of the living room and kitchen are 319 and 50 ft^2. The area of washrooms is 55 and 50 ft^2. Each room, i.e., first room, second room, third room, living room, and kitchen contains 11 W LED bulb and washrooms contains five watts LED bulb. Five-star air conditioner of 1.7 KW is placed in each room. In this home model, four persons are considered. According to the presence of persons, lights and air conditioners of the rooms are switched on or off. The system model consists of three types of sensors. First type of sensors is motion sensors. Motion sensors are used in the system model to track the motion of the persons living in home. Second type of sensors is luminance sensors. Luminance sensors are used to check the luminance level of the areas in the home. Third type of sensors is temperature sensors. Temperature sensors are used in the system model to check the temperature of the home. Each room (three rooms and one living room) contains one motion sensor, one luminance sensor, and temperature sensor. Kitchen and washrooms contain motion sensor and luminance sensor. If a person in home leaves living room and enters into the first room, the motion sensor detects the position of the person and the light and air conditioner of the living room are switched off. As the person enters into the first room, the motion sensor of the room detects the entry of the person. After tracing the person's entry, the luminance sensor of the room checks the luminance level and if the luminance level is below a threshold level then it switch on the light of the room, otherwise the room light is in off state. Further, the temperature sensor checks the temperature of the room and if it is above a threshold level then it switches on the air conditioner of the room, otherwise the room air conditioner is in off state.

3.2 Problem Definition

The IOT is about making the surroundings smart including the buildings used for living beings like home, hospital etc. Today, the energy dependency for homes is very high. Therefore, there is urgency for automation of homes using sensors like temperature sensors, motion sensors, and luminance sensors. The challenge is to give good architecture for energy-efficient automation, which can save energy. The following problems are to be solved from this home model. First, the data of energy consumption (lights and air conditioner) of model home is collected. Second, placements of sensors like motion sensor, luminance sensor, and temperature sensor in the home model. Third, distance between the sensors according to the coverage of the sensors so that the sensors can cover the entire home and sensors can easily exchange infor-

Fig. 1 Model of home with sensors

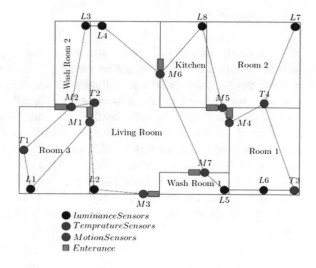

luminanceSensors
TempratureSensors
MotionSensors
Enterance

mation from one sensor to another. Fourth, collect data of energy consumption after placements of sensors in home. At Last, calculate the energy consumption change before and after the placement of sensors.

4 Design of the System

4.1 System Design

In this part of the paper, the system model is explained. First, collection of energy consumption of lights and air conditioners has been done for 30 days. The energy dependency and energy consumption for homes are very high. Hence, designing of a system model that is based on three types of sensors which are motion sensors, luminance sensors, and temperature sensors is necessary. Our home model consists of seven motion sensors, eight luminance sensors, and four temperature sensors. All the 19 sensors are placed in the home model in such a way that it can cover the entire area of the home. Then distance between the sensors is calculated. Finally, according to the motion of the persons, luminance level and temperature the lights and air conditioners of the home are switched on and off. After that calculation of energy consumption of lights and air conditioners has to be done again to see whether energy is saved or not.

4.2 Algorithm

The algorithm controls lights and air conditioners of the home. The data of the motion sensors is collected in motion variable. The motion is a Boolean variable, which contains two values 0 and 1. If value of motion is 1, it means that a person or persons enter. The data of the luminance sensors is collected in L_2 variable. L_2 is the present luminance level. L_1 is the sufficient luminance level at which a person can see easily. $(L_1 - \delta(delta))$ is the threshold level below which lights are necessary. The data of the temperature sensors is collected in Temp variable. Temp is the present temperature level. Initially, motion sensor checks the motion of the person, if it detects the motion variable changes to 1. Then it checks the value of L_2. If the value of L_2 is less than or equal to $(L_1 - \delta(delta))$, it switches on the light. After that it checks the temperature level. If Temp is greater than 25 °C, it switches on the air conditioner.

Algorithm:-
For each room motion sensor check the position of the persons
If motion = 1 (a person or persons enters into the room)
  If $L_2 <= L_1-\delta$ (delta)
    Switch on the light of the room.
  If Temp>= 25 degrees Celsius
    Switch on the air conditioner of the room.
End if

For kitchen and washrooms motion sensors check the position of the person
If motion = 1 (a person enters)
  If $L_2 <= L_1-\delta$ (delta)
    Switch on the light of the room.
End if

5 Implementation of the System

Implementation of the system is done using MATLAB. All the sensors are deployed in the home as displays in Fig. 2. The motion sensors are placed in the following coordinates: 10, 9, 7; 5, 12, 7; 18, 0, 7; 21, 15, 7; 28, 5, 7; 32, 9, 7; 32, 12, 7. The luminance sensors are placed in the following coordinates: 0, 0, 7; 10, 0, 7; 10, 22, 7; 11, 22, 7; 32, 0, 7; 37, 0, 7; 42, 22, 7; 28, 22, 7. The temperature sensors are placed in the following coordinates: 0, 6, 7; 10, 12, 7; 42, 0, 7; 39, 12, 7. After the placement of sensors distance between the neighbor sensors are calculated by using Euclidean distance method. The Euclidean distance between the neighboring sensors must be small than the coverage of the sensors, so that it can cover the entire area of the home and neighboring sensors can exchange data from one sensor to another. Then the

Fig. 2 Deployment of sensors

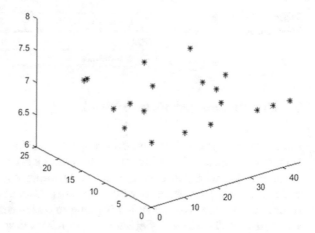

Table 1 Energy consumption by lights and air conditioners

Energy consumption	Before implementation of system (KWh/month)	After implementation of system (KWh/month)
Lights	16.17	3.30
Air conditioner	2040.00	1632.00

algorithm is implemented and energy consumption of lights and air conditioners of the entire home has calculated.

6 Results

The data of energy consumption by lights and air conditioners of model home is collected separately. Table 1 displays the energy consumption of lights and air conditioners before and after implementation of the system in the entire home separately for 30 days. The home consists of total eight LED bulbs and four air conditioners. The total energy consumption before implementation of system in one month is 2056.17 KWh/month. The total energy consumption after implementation of system in one month is 1635.30 KWh/month. Hence, 12.87 KWh/month of energy has been saved in lights and 408 KWh/month of energy has been saved in air conditioners as shown in Table 2. Finally, 79.59% of energy in lights and 20% of energy in air conditioners has been saved. 420.87 KWh/month of total energy or 20.78% total energy has been saved.

Table 2 Energy saved by lights and air conditioners

Energy consumption	Energy saved (KWh/month)
Lights	12.87
Air conditioner	408.00

7 Conclusion and Future Work

Internet of Things provides a general concept of the integration of Internet with sensors, radio frequency identification, and smart objects. In Internet of Things devices are embedded with networking, identifying, processing capabilities, and sensing which allows them to interact with each other. Overall it must be considered as a future Internet because everything is going to be connected in a network. Existing work associated with energy-efficient smart homes based on Internet of Things can be classified into three categories. The classifications of smart homes are as follows. Energy of homes can be saved by scheduling tasks, by tracking the position of the persons and by controlling devices using various types of sensors. In this paper, conservation of energy of home is done by deploying various sensors such as motion sensors, luminance sensors, and temperature sensors in home and by implementing an algorithm to save the energy of lights and air conditioners of the entire home. Then calculation of before and after implementation of system model showed that 79.59% of energy in lights and 20% of energy in air conditioners has been saved and a total of 20.78% energy has been saved. In future by same way one can save more energy by applying this algorithm on other appliances of home.

References

1. Van Kranenburg R (2008) The internet of things: a critique of ambient technology and the all-seeing network of RFID. Institute of Network Cultures, Amsterdam, The Netherlands
2. Baraka K, Ghobril M, Malek S, Kanj R, Kayssi A (2013) Low cost arduino/android-based energy-efficient home automation system with smart task scheduling. In: 2013 IEEE fifth international conference on computational intelligence, communication systems and networks, pp 296–301
3. Pedrasa MAA, Spooner TD, MacGill IF (2010) Coordinated scheduling of residential distributed energy resources to optimize smart home energy services. IEEE Trans Smart Grid 1(2):134–143
4. Chen X, Wei T, Hu S (2013) Uncertainty-aware household appliance scheduling considering dynamic electricity pricing in smart home. IEEE Trans Smart Grid 4(2):932–941
5. Mohsenian-Rad AH, Wong VW, Jatskevich J, Schober R (2010) Optimal and autonomous incentive-based energy consumption scheduling algorithm for smart grid. Innov Smart Grid Technol 1–6
6. Wang J, Cheng Z, Jing L, Ozawa Y, Zhou Y (2012) A location-aware lifestyle improvement system to save energy in smart home. In: 2012 4th international conference on awareness science and technology, pp 109–114

7. Mainetti L, Mighali V, Patrono L (2015) A location-aware architecture for heterogeneous building automation systems. In: 2015 IFIP/IEEE international symposium on integrated network management, pp 1065–1070
8. Roy A, Bhaumik SD, Bhattacharya A, Basu K, Cook DJ, Das SK (2003) Location aware resource management in smart homes. In: Proceedings of the first IEEE international conference on pervasive computing and communications, pp 481–488
9. Roy A, Das SK, Basu K (2007) A predictive framework for location-aware resource management in smart homes. IEEE Trans Mob Comput 6(11):1270–1283
10. Subbu KP, Thomas N (2014) A location aware personalized smart control system. In: International symposium on information processing in sensor networks, pp 309–310
11. Moreno MV, Hernandez JL, Skarmeta AF (2014) A new location-aware authorization mechanism for indoor environments. In: 2014 28th international conference on advanced information networking and applications workshops, pp 791–796
12. Mainetti L, Mighali V, Patrono L (2015) An IoT-based user-centric ecosystem for heterogeneous smart home environments. In: IEEE international conference on communications, pp 704–709
13. Suh C, Ko YB (2008) Design and implementation of intelligent home control systems based on active sensor networks. IEEE Trans Consum Electron 54(3):1065–1070
14. Lu J, Sookoor T, Srinivasan V, Gao G, Holben B, Stankovic J, Field E, Whitehouse K (2010) The smart thermostat: using occupancy sensors to save energy in homes. In: 8th ACM conference on embedded networked sensor systems, pp 211–224
15. Song G, Wei Z, Zhang W, Song A (2007) A hybrid sensor network system for home monitoring applications. IEEE Trans Consum Electron 53(4):1434–1439
16. Kovatsch M, Weiss M, Guinard D (2010) Embedding internet technology for home automation. In: IEEE conference on emerging technologies and factory automation, pp 1–8
17. Dohr A, Modre-Opsrian R, Drobics M, Hayn D, Schreier G (2010) The internet of things for ambient assisted living. In: IEEE seventh international conference on information technology: new generations, pp 804–809
18. ElShafee A, Hamed KA (2012) Design and implementation of a Wi-Fi based home automation system. World Acad Sci Eng Technol 6(8):2177–2180
19. Dickey N, Banks D, Sukittanon S (2012) Home automation using cloud network and mobile devices. In: 2012 IEEE proceedings of Southeastcon, pp 1–4
20. Han DM, Lim JH (2010) Design and implementation of smart home energy management systems based on zigbee. IEEE Trans Consum Electron 56(3):1417–1425
21. Piyare R (2013) Internet of things: ubiquitous home control and monitoring system using android based smart phone. Int J Internet Things 2(1):5–11
22. Soliman M, Abiodun T, Hamouda T, Zhou J, Lung CH (2013) Smart home: integrating internet of things with web services and cloud computing. In: 2013 IEEE 5th international conference on cloud computing technology and science, vol 2, pp 317–320
23. Song Y, Han B, Zhang X, Yang D (2012) Modeling and simulation of smart home scenarios based on Internet of Things. In: 2012 3rd IEEE international conference on network infrastructure and digital content, pp 596–600
24. Bing K, Fu L, Zhuo Y, Yanlei L (2011) Design of an Internet of Things-based smart home system. In: 2011 2nd IEEE international conference on intelligent control and information processing, vol 2, pp 921–924

Improved Prophet Routing Algorithm for Opportunistic Networks

Sanjay Kumar, Prernamayee Tripathy, Kartikey Dwivedi
and Sudhakar Pandey

Abstract The Delay Tolerant Network paradigm was proposed to address communication issues in challenged environments where no end-to-end connectivity exists. DTN is called Opportunistic Network as intermediate node always looks for the best opportunity to relay a message from source to destination. In Vehicular Delay Tolerant networks, some nodes follow a fixed path while others follow a random path. In this paper, after the study and analysis of various protocols, a new protocol has been proposed by making changes in PRoPHET Routing Algorithm to maximize and improve the delivery probability and minimize the number of messages being dropped and the overhead ratio. A comparison has been provided between the original and proposed PRoPHET Algorithm.

Keywords Opportunistic network · Routing algorithm · Predictable path
Delivery probability

1 Introduction

Delay Tolerant Networks (DTNs) [1] have been found suitable for the environment where end-to-end communication link between source to destination does not exist for most of the time. Since the network is highly inconsistent, communication in such an environment is not continuous but opportunistic [2] in nature, i.e., the messages are

S. Kumar (✉) · P. Tripathy · S. Pandey
Department of Information Technology, National Institute of Technology, Raipur, India
e-mail: skumar.it@nitrr.ac.in

P. Tripathy
e-mail: tripathyprernamayee95@gmail.com

S. Pandey
e-mail: spandey.it@nitrr.ac.in

K. Dwivedi
Department of Electronics and Communication,
Manipal Institute of Technology, Manipal, India
e-mail: kartikey18747@gmail.com

© Springer Nature Singapore Pte Ltd. 2019
M. L. Kolhe et al. (eds.), *Advances in Data and Information Sciences*, Lecture Notes
in Networks and Systems 39, https://doi.org/10.1007/978-981-13-0277-0_25

delivered only when a connection is established. In networks like these, instantaneous end-to-end paths are difficult to establish. The routing protocols use the "store carry forward" approach where the data is moved from node to node until it finally reaches the destination. Some protocols make multiple copies and share them with the other nodes and so on until it reaches the destination node while some protocols (e.g., PRoPHET) make use of a buffer to store the Delivery Probabilities of the nodes and forward the data only when the probabilities are higher.

Vehicular Delay Tolerant Networks (VDTNs) [3] are considered as a type of DTNs where vehicles are used to carry the messages when direct communication link between two communicating nodes is not present. There are various types of vehicles available on the roads like cars, trams, etc., including the pedestrians. The movement patterns of different vehicles are different. The pedestrians and cars tend to move randomly and hence follow an opportunistic path while trams or trains follow a predefined path or a predictable path. Since the probability of two nodes interacting while on a predictable path is greater than nodes following a random path, the Delivery Probability calculation for both the groups can be done in different ways that gives the best and optimal results. In this research paper, this property has been used to make modifications in the existing routing algorithms [4] like PRoPHET and can also be applied to the other variants of the PRoPHET Routing algorithm like the PRoPHET V2 Router or PRoPHET Router with Estimation, etc.

In this paper, we have presented modified PRoPHET routing algorithm and compared it with the original PRoPHET algorithm. Results have been compared and analysed through simulations on ONE simulator. The rest of the paper is organized as follows: Sect. 2 describes three most widely used routing algorithms by researchers, Epidemic, Spray and Wait (S&W) and PRoPHET. Section 3 describes the proposed algorithm. Section 4 describes the simulation setup in Opportunistic Network Environment (ONE) for our experiments. The comparison of results of the original and proposed algorithm has been discussed in Sect. 5. Section 6 concludes the paper and gives the direction towards the future work.

2 Related Work

2.1 Epidemic Routing

Epidemic routing is based on the concept of flooding in intermittently connected mobile networks. It works on eventual delivery of messages [5] where it has minimal knowledge of topology and connectivity of the network. The messages are delivered eventually through periodic pair-wise connectivity of the nodes.

When two devices or hosts come into the communication range of each other, they exchange their summary vectors and compare them to determine the missing messages. The host can then request the other host to provide copies of messages it does not have. This way all messages are eventually spread to all nodes which eventually reach its destination.

2.2 Spray and Wait Routing

The Spray and Wait routing [6] is an improvement over flooding-based techniques like Epidemic routing.
It consists of the following two phases:

a. Spray Phase
b. Wait Phase

It initially starts spreading the copies of messages in a manner similar to epidemic routing. When it is made sure that enough copies have been spread so that there is some probability of them reaching the destination, the source node stops spreading the messages. It enters Wait phase and lets other nodes having a copy transmit the message directly to the destination.

2.3 PRoPHET Routing

The PRoPHET Routing protocol [7] is a probabilistic routing protocol that uses the history of previous encounters with the other nodes. For each known destination node y, a delivery predictability $P(x, y)$ is maintained at every node x. When two nodes meet, they exchange the summary vector, probability value for the destination node and decision regarding transmission of message based on the fact that message is transmitted to the other node only when the Delivery Probability of the latter node is better than the former. The Delivery Probability (DP) calculation is done in three steps as proposed in [8].

2.4 Other Work

Various different routing algorithms have been proposed that work on single-copy- or multiple-copy-based schemes. Some make use of global positioning devices such as GeOpps [9].

The PRoPHET algorithm has been modified to produce other variants of the algorithm like the Advanced PRoPHET algorithm [10], PRoPHET V2 [11] algorithm, PRoPHET+ [12] etc.

3   Proposed Method

In the real world, some nodes move around in predictable path while some nodes tend to move randomly. This property has been used in this paper to introduce some changes in the PRoPHET algorithm and a modified version is presented.

In the ONE Simulator environment [13, 14], the nodes are

1. Pedestrians
2. Cars, buses and trucks
3. Trams and trains

Now these trams and trains use Map-based Mobility for moving in the simulator and hence have a predictable path while other groups like pedestrians and cars use Shortest Path Map-based movements. The movement pattern of the pedestrians and the cars is random since they can take any path on the road. The trams on the other hand have a predefined route laid out for them and have to follow that particular path. So, the probability of a random node meeting the tram node in the predefined path for the tram is more than the probability of two random nodes coming into the communication range of each other. Hence we propose changing the value of the initialization and the aging constant for the nodes with predictable path.

The number of groups is divided into two sets –

1. **Opportunistic**: The nodes that do not have a predictable path like pedestrians and cars.
2. **Predictable**: The nodes that do have a predictable path like the trams.

The calculation of the Delivery Predictabilities is done as follows. Whenever a node is encountered the metric is updated for the Opportunistic group as

$$P(x, y) = P(x, y) \text{ old} + (1 - P(x, y) \text{ old}) \times \text{Pinit} \tag{1}$$

And for the Predictable group as

$$P(x, y) = P(x, y) \text{ old} + (1 - P(x, y) \text{ old}) \times \text{Pinit}_{new} \tag{2}$$

The new value of the **Pinit** (*Pinit$_{new}$*) can be derived through a genetic algorithm [15] method involving various simulations, where we take values from 0.76 to 0.99. The values can be seeded at an interval of 0.1 which provides higher probability for a better performance. The PRoPHET [7] algorithm has defined the value for this initialization constant as 0.75. Our simulations proved that the delivery probabilities give the best results when the initialization constant for predictable group is set as 0.85.

If a pair of nodes does not encounter each other for a long time, the Predictability values age with the passage of time. The metric can then be updated for the opportunistic group as

$$P(x, y) = P(x, y) \text{ old} \times \alpha^k \tag{3}$$

And for the Predictable group as

$$P(x,\ y) = P(x, y)\ old\ \times \alpha_{new}^k \qquad (4)$$

The value of α_{new} can be obtained in a similar manner as the initialization constant. The original value is taken to be 0.98. We have increased this value for the predictable group to 0.9999 based on simulations.

Changing the value of the Scaling constant, however, had no effect on the Delivery Predictability values of the opportunistic and Predictable groups. Hence the Transitivity formula remains the same for both the groups.

$$P(x, z) = P(x, z)\ old + (1 - P(x, z)\ old) \times P(x, y) \times P(y, z) \times \beta \qquad (5)$$

This can be implemented by making changes in the code of the routing protocol. We select the tram groups only from the hosts using the group ID [16] and set the new values for the Predictable groups.

4 Simulation Setup

We have implemented the PRoPHET Routing protocol along with the modified version of it in the ONE (Opportunistic Network Environment) Simulator [13, 14] to compare the results. We have used the same simulation setup as described in [17]. We have used a part of the Helsinki downtown area (4500 × 3500 m) which comes by default in ONE simulator. Maps of other cities can also be used, e.g., Raipur. In addition to the roads and other paths, tracks for trams have been laid out in the map. We have used six groups of mobile nodes having 40 hosts each. Each group is identified with a group ID. Groups 1 and 3 are a group of pedestrians while Group 2 is cars that drive on the roads. The pedestrians move at random speeds of 0.5–1.5 m/s with pause times of 0–120 s. The cars move at 10–50 km/h with pause times of 0–120 s. Groups 4, 5, 6 are tram groups with speeds of 7–10 m/s and pausing time of 10–30 s. The common setting for all groups uses the shortest path map-based movement and the tram groups use the map route movement mobility model. The default buffer size is 5 M and TTL is 300 min, both of which can be changed according to the requirements as presented in the results below. We use a buffer size of 2, 3, 4 and 5 MB and message TTL of 60, 90, 120, 150 min respectively. A timescale of 1500 is assumed for the PRoPHET router with estimation routing protocol, the Javadoc of which can be found in [16]. The simulations are run for 5000 s instead of 43200 s with an Update Interval of 0.1 s.

5   Simulation Results

The performance of the PRoPHET and the proposed PRoPHET algorithm are shown in the results. The modified code can be used as a generalized idea and can be applied to the other variants of the PRoPHET algorithm. We applied the generalized idea in the PRoPHET Router with Estimation code and have presented the results comparing the original and the proposed methods.

Figure 1 shows the comparison of Delivery Probability as a function of Bundles Time to live for the original and the proposed PRoPHET algorithm. It can be seen that for the simulations performed, the PRoPHET algorithm gives the same value of 0.2024 for DP for all the TTL. But the proposed PRoPHET algorithm gives DP of 0.2143 for TTL 60 and 0.2202 for TTL 90, 120, and 150 min. Hence it can be seen that our proposed method clearly gives better results for Delivery Probabilities as a function of TTL.

Figure 2 below shows the comparison of Delivery Probabilities as a function of Buffer size for the original and proposed methods. It can be seen that the PRoPHET algorithm gives a minimum value of 0.1250 and a maximum value of 0.2024 for buffer sizes 2, 3, 4, and 5 MB. The proposed algorithm, on the other hand, gives a minimum value of 0.1429 and a maximum value of 0.2202 and hence gives better results.

Figures 3 and 4 compare the Overhead Ratios of the two algorithms with bundles TTL and Buffer size. The proposed algorithm visibly outperforms in both the cases. It gives a minimum value of 37.8056 and a maximum value of 40.5676 for Overhead Ratios against TTL 60, 90, 120, and 150 min. These values are lesser than the PRoPHET algorithm which gives a minimum value of 39.9118 and a maximum value of 44.2353 for the same. When compared against the Buffer size, the proposed algorithm provides a minimum value of 40.5676 and a maximum value of 48.7083

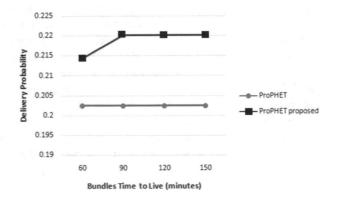

Fig. 1   Delivery probability as a function of bundles TTL

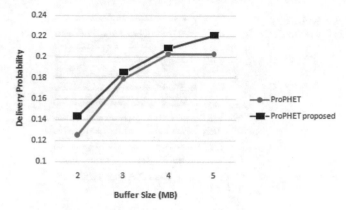

Fig. 2 Delivery probability as a function of buffer size

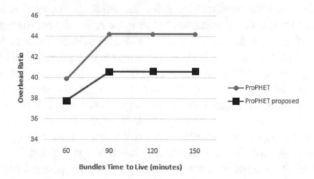

Fig. 3 Overhead ratio as a function of bundles TTL

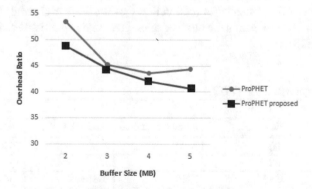

Fig. 4 Overhead ratio as a function of buffer size

for Overhead Ratio. The PRoPHET algorithm gives the minimum and maximum value of 44.2353 and 53.3333, respectively.

Fig. 5 Delivery probability as a function of bundles TTL (est.)

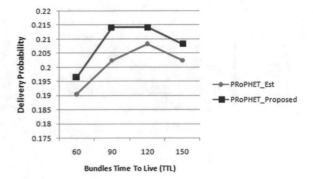

Figure 5 represents the comparison of Delivery Probabilities for PRoPHET Router with Estimation and the modified code for the same. It shows that the Delivery Probabilities as a function of bundles TTL give better results for the modified code as compared to the original code for PRoPHET Router with estimation. The modified code gives a minimum value of 0.1964 and a maximum value of 0.2143 while the original code gives the minimum value of 0.1905 and a maximum value of 0.2083.

Tables 1 and 2 below represent a summary of message stats report for the original and proposed PRoPHET algorithm for message TTL 60 and 150 min and buffer sizes 2 and 5 MB. Here, "Sim. Time" stands for the total simulation time which in our case was 5000 s. The variables "created" represent the number of messages created by the simulator, "started" represents the messages started for transmission, "relayed" indicates relayed by nodes, "aborted" indicates aborted between nodes, and "dropped" represents dropped by network respectively. The variable "delivered" means the number of messages successfully delivered to the final destination. "Delivery Probability" stands for total delivery of message probability whereas the "Overhead Ratio" represents the number of created copies per delivered message as described in [18].

Table 1 Message stats report (TTL)

Msg TTL	TTL 60		TTL 150	
	PRoPHET	Proposed method	PRoPHET	Proposed method
Sim. Time	5000	5000	5000	5000
Created	168	168	168	168
Started	3325	3346	3538	3564
Relayed	1391	1397	1538	1538
Aborted	1931	1946	1996	2022
Dropped	1111	1089	847	835
Delivered	34	36	34	37
Delivery prob.	0.2024	0.2143	0.2024	0.2202
Overhead ratio	39.9118	37.8056	44.2353	40.5676

Table 2 Message stats report (buffer size)

Buffer	2 MB		5 MB	
	PRoPHET	Proposed method	PRoPHET	Proposed method
Sim. Time	5000	5000	5000	5000
Created	168	168	168	168
Started	2863	2914	3538	3564
Relayed	1141	1193	1538	1538
Aborted	1719	1718	1996	2022
Dropped	966	967	847	835
Delivered	21	24	34	37
Delivery prob.	0.1250	0.1429	0.2024	0.2202
Overhead ratio	53.3333	48.7083	44.2353	40.5676

Our simulation results have shown that the modified version of both the algorithms give better results than the original algorithms.

6 Conclusion

In this paper, we have proposed a generalized method to improve the performance of routing algorithms in opportunistic networks and evaluated their performances through simulations. We have made use of the property that in a real-time scenario, some groups like the pedestrians, cars, etc., follow an opportunistic path while certain groups like trams and trains, etc., follow predictable paths. Since the probability of two nodes interacting while on a predictable path is greater than any random path, we made use of the Genetic Algorithm to derive new initialization and aging constants for the groups following predictable paths. We have observed that changing the initialization and aging constants for nodes with predictable paths like that of trams and trains can result in greater Delivery Predictabilities and lesser Overhead Ratios with respect to Buffer size and Bundles Time to Live. This idea can also be extended to other variants of the PRoPHET algorithm like "PRoPHET Router with Estimation" (as shown through simulation results) where the initialization constant, Pinit and aging constant, α can be modified with the help of a Genetic Algorithm and an optimum value for both can be obtained for which the algorithm provides best results. Similarly, the initialization constant, PEncMax and the aging constant, α can be modified for the "PRoPHET V2 Router" algorithm and better results can be obtained for the same.

References

1. Fall K (2003) A delay-tolerant network architecture for challenged internets. In: Proceedings of ACM SIGCOMM. ACM Press, USA, pp 24–27
2. Pelusi L, Passarella A, Conti M (2006) Opportunistic networking: data forwarding in disconnected mobile ad hoc networks. IEEE Commun Mag 44:134–141
3. Shao Y, Liu C, Wu J (2009) Delay-tolerant networks in VANETs. In: Olariu S, Weigle MC (eds) Vehicular networks: from theory to practice. Chapman & Hall/CRC Computer & Information Science Series, pp 1–36 (Chapter 10)
4. Jain S, Fall K, Patra R (2004) Routing in a delay tolerant network. In: Proceedings of ACM SIGCOMM
5. Vahdat A, Becker D (2000) Epidemic routing for partially connected ad hoc networks. Technical Report CS-200006, Duke University
6. Spyropoulos T, Konstantinos P, Raghavendra CS (2005) Spray and wait: an efficient routing scheme for intermittently connected mobile networks. In: Proceedings of the 2005 ACM SIGCOMM workshop on delay-tolerant networking. ACM
7. Lindgren A, Doria A, Schelen O (2003) Probabilistic routing in intermittently connected networks. ACM Mob Comput Commun Rev 7:19–20
8. Sok P, Tan S, Kim K (2013) PRoPHET routing protocol based on neighbor node distance using a community mobility model in delay tolerant networks. In: 2013 IEEE 10th international conference on high performance computing and communications & 2013 IEEE international conference on embedded and ubiquitous computing (HPCC_EUC). IEEE, pp 1233–1240
9. Leontiadis I, Mascolo C (2007) Geopps: geographical opportunistic routing for vehicular networks. In: 2007 IEEE international symposium on a world of wireless, mobile and multimedia networks. Ieee
10. Xue J, Li J, Cao Y, Fang J (2009) Advanced prophet routing in delay tolerant network. In: International conference on communication software and networks, 2009 (ICCSN'09). IEEE, pp 411–413
11. Grasic S et al (2011) The evolution of a DTN routing protocol-PRoPHETv2. In: Proceedings of the 6th ACM workshop on challenged networks. ACM
12. Huang TK, Lee CK, Chen LJ (2010) PRoPHET+: an adaptive prophet-based routing protocol for opportunistic network. In: 2010 24th IEEE international conference on advanced information networking and applications (AINA). IEEE, pp 112–119
13. Keränen A (2008) Opportunistic network environment simulator. Special assignment report, Helsinki University of Technology, Department of Communications and Networking
14. Keränen A, Ott J, Kärkkäinen T (2009) The ONE simulator for DTN protocol evaluation. In: Proceedings of the 2nd international conference on simulation tools and techniques. ICST (Institute for Computer Sciences, Social-Informatics and Telecommunications Engineering)
15. Genetic algorithm. https://en.wikipedia.org/wiki/Genetic_algorithm
16. Project page of the ONE simulator (2008). http://www.netlab.tkk.fi/tutkimus/dtn/theone
17. Keränen A, Ott J (2007) Increasing reality for DTN protocol simulations. Tech. rep., Helsinki University of Technology, Networking Laboratory
18. Metric description of the ONE simulator reports. http://agoes.web.id/metric-description-from-simulator/

SPMC-PRRP: A Predicted Region Based Cache Replacement Policy

Ajay K. Gupta and Udai Shanker

Abstract Earlier cache replacement policies are solely based on the temporal features of client's access pattern. As mobile client moves to different locations, their access pattern shows not only temporal locality rather it shows spatial locality as well. The previous policies have not evolved any accurate next location prediction policy that can be used in cost computation of data items. To overcome this limitation of previous policies, this paper proposes a Cache Replacement Policy (SPMC-PRRP) based on mobility prediction. To predict accurate future location of moving client, mobility rules have also been framed based on similarities between user's movement data. Here, removal of noise (random movements) in mobile users' profiles in trajectories has been done to be used in above policy. Simulation results show that proposed policy achieves up to 5% performance improvement compared to previous well-known replacement policies via improving the accuracy in next location prediction.

Keywords Mobile computing · Location-dependent data · Cache replacement
Predicted region · Root-mean-squared distance

1 Introduction

Earlier cache replacement policies in mobile environment [1, 2] solely based on the temporal features of client's access pattern. The conventional cache replacement policy such as LFU, LRU, LRU-K [3] have been widely used in various applications in past. The working principle of these policies are that the access pattern shows temporal locality, i.e., the future access pattern dependent on only past access pattern rather than spatial information. These conventional replacement policies results into poor

A. K. Gupta (✉) · U. Shanker
Department of Computer Science & Engineering, M.M.M. University of Technology,
Gorakhpur, India
e-mail: ajay25g@gmail.com

U. Shanker
e-mail: udaigkp@gmail.com

© Springer Nature Singapore Pte Ltd. 2019
M. L. Kolhe et al. (eds.), *Advances in Data and Information Sciences*, Lecture Notes
in Networks and Systems 39, https://doi.org/10.1007/978-981-13-0277-0_26

cache hit ratio when used in location-dependent information system (LDIS). Initially, Manhattan replacement policy [4] was given to support location-dependent services. The distance between mobile current location and cached data items origin location is defined as Manhattan distance. The data items are evicted and replaced from the cache which has largest Manhattan distance. In urban environments, the Manhattan Distance cache replacement policy is best suitable policy for location-dependent queries. This policy has some disadvantages that it does not consider mobile clients temporal access locality and the client movement directions while taking decisions of cache replacement. Another policy for cache replacement FAR [5] additionally dismisses the client's temporal access properties. In the event when mobile clients' direction updated frequently, it will make unpredicted effect on membership of objects as it will show frequent switching between the in-direction set (towards valid scope) and the out-direction sets (moving far from valid scope). The factors that are considered in LDIS are distance between mobile nodes current location, movement direction, and valid scope area.

The remainder of this paper is organized as follows. The next section provides preliminaries about predicted region based cache replacement policy. In Sect. 3, group mobility behavior based mobility prediction model is proposed to be used in estimating the anticipated distance between location of data item and that of client for cache replacement. This is followed by simulation model in Sect. 4. Finally, Sect. 5 has a discussion about possible future research scope and conclusion of the paper.

2 Predicted Region Definition

Predicted Region [6] radius length L_r is estimated using root-mean-squared distance between cached data item's valid scope C_i (for all cached data items) and mobile client current location (C_m) and make the current location at the time of estimating predicted region as the center of predicted region. The advantage of this policy is that there is no need to estimate predicted region every time the moving client changing its direction or speed. Rather, region is estimated only at end of prediction interval. To estimate the predicted region, all the cached data item's valid scope (C_i) and mobile client current location (C_m) at the time of query issue are considered. Here, the time of the issued query is different from the time of estimating predicted region. The time of the issued query is determined from query interval and the time of estimating predicted region is determined from prediction interval.

$$L_r = \sqrt{\frac{\sum_{i=1}^{K}(C_i - C_m)^2)}{K}} \tag{1}$$

L_r Predicted Region Radius
$(C_i - C_m)$ Valid scope (c_i) and client's current location (c_m) Distance
K Total number of cached data objects

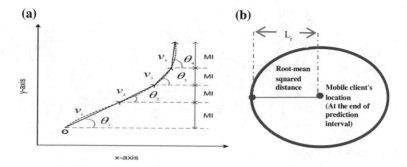

Fig. 1 **a** Client's path (Discrete model) **b** root-mean-square distance-based predicted region

The data distance w.r.t. region should be used in such a way that the cached data items nearest to the client current location are not evicted from it. So, SPMC-based PRRP has suggested a modification to previous algorithms in that, it has used SPMC model for predicting next location of the user and then find distance between the valid scope reference point of data item from the predicted client location at the time of query issue (Fig. 1).

3 Proposed Model: Cache Replacement in Predicted Region Using Mobility Rules

The accuracy of mobility prediction degrades in LDIS when it involves a lot of random movements [7, 8]. Thus, these random movements must be reduced to get accurate mobility prediction. Here, a sequential pattern mining method in moving client's movement histories for the coverage region is employed to find frequent mobility patterns. The paper investigates clustering technique to extract similar mobility behaviors in users moving histories. The detail of the SPMC for next location prediction in predicted region is described below.

3.1 SPMC Based Next Location Estimation

Given a Location's coverage region graph (G), client's mobility history log file (D), a set of timestamps (T), a minimum confidence threshold ($conf_{min}$.) and a minimum support threshold $supp_{min}$., The problem of spatiotemporal data mining based mobility prediction in predicted region is comprised of five steps.

1. In the first step, reduction of noise/outliers depicting random movements is performed. Discovery of all frequent mobility patterns in trajectory datasets D

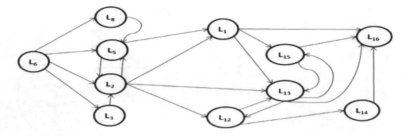

Fig. 2 Location's coverage region graph (G)

Table 1 Predefined timestamps

Timestamps	Time interval	Timestamps	Time interval
t_1	0:00–2:14	t_7	13:00–15:44
t_2	2:15–4:29	t_8	15:45–17:59
t_3	4:30–6:44	t_9	18:00–20:14
t_4	6:45–8:59	t_{10}	20:15–22:29
t_5	9:00–11:14	t_{11}	22:30–23:59
t_6	11:15–13:29		

satisfying $supp_{min}$ from the generated trajectory log file of a node mobility history is done in this step.

2. The second phase is to Partition frequent mobility patterns L_k into n similar groups or clusters, i.e., $C = \{C_1, C_2, \ldots, C_n\}$ [9].
3. Group (G) determination of current trajectory P is performed in the third phase.
4. The fourth step is to use the mined frequent mobility patterns to generate mobility rules R. The predefined threshold value $conf_{min}$ is used to filter the mobility rules [10, 11], and then all the filtered rules are used for next cell prediction in the predicted region.
5. Find the next cell_id in the predicted region using matched rule in previous phase.

3.1.1 First Phase: Discovering Spatio-temporal Frequent Mobility Patterns

In this paper, a dataset generator is used in the experiment that shows movement behaviors of client's around the coverage region as in Fig. 2. To reduce the understanding complexity of movement, a small set of possible locations is considered in example. The predicted regions are divided into cells. Each cell is treated as a location in our model. In the given predicted region, the current location of the Moving client at the start of prediction interval is assumed to be L_6. After L6, next Location can be on among L_2, L_3, L_5, and L_8. The possible movement trajectory sequences to reach L_{16}, can be given by depicted region graph (Table 1).

Table 2 Table log file of mobility history

Sequence ID	Mobility patterns
1.	$<(L_5, t_3), (L_1, t_5), (L_{15}, t_7)>$
2.	$<(L_5, t_3), (L_1, t_5), (L_{13}, t_9)>$
3.	$<(L_5, t_3), (L_2, t_4), (L_1, t_5), (L_{15}, t_7)>$
4.	$<(L_5, t_3), (L_2, t_4), (L_1, t_5), (L_{13}, t_9)>$
5.	$<(L_2, t_4), (L_1, t_5), (L_{15}, t_7)>$
6.	$<(L_2, t_4), (L_1, t_5), (L_{13}, t_9)>$
7.	$<(L_3, t_2), (L_2, t_4), (L_{12}, t_6), (L_{13}, t_9)>$
8.	$<(L_3, t_2), (L_2, t_4), (L_{12}, t_6), (L_{14}, t_8)>$
9.	$<(L_4, t_1), (L_3, t_2), (L_2, t_4), (L_{12}, t_6), (L_{13}, t_9)>$
10.	$<(L_4, t_1), (L_3, t_2), (L_2, t_4), (L_{12}, t_6), (L_{14}, t_8)>$

Table 3 Mobility 1-patterns and frequent mobility 1-patterns $(k = 1)$

C_1		L_1	
Candidates 1-patterns	Support	Frequent 1-patterns	Support
$<(L_5, t_3)>$	4	$<(L_5, t_3)>$	4
$<(L_2, t_4)>$	8	$<(L_2, t_4)>$	8
$<(L_3, t_2)>$	4	$<(L_3, t_2)>$	4
$<(L_4, t_1)>$	2	$<(L_1, t_5)>$	6
$<(L_1, t_5)>$	6	$<(L_{15}, t_7)>$	3
$<(L_{15}, t_7)>$	3	$<(L_{13}, t_9)>$	3
$<(L_{13}, t_9)>$	3	$<(L_{12}, t_6)>$	3
$<(L_{12}, t_6)>$	3		
$<(L_{14}, t_8)>$	2		

For understanding the working phenomenon of frequent pattern mining, a client's log file of mobility history having 11 location and corresponding timestamps for 1 day are taken as the dataset. Here $supp_{min} = 3$ (Tables 2 and 3).

In the above table, the candidate 1-patterns $<(L_4, t_2)>$, $<(L_{14}, t_8)>$ support value is less than minimum defined threshold i.e. $supp_{min} = 3$. Thus they are treated as noise or outliers and are removed from further computation (Table 4).

Table 4 Mobility patterns and frequent mobility patterns (k = 2)

C_2		L_2	
Candidates 2-patterns	Support	Frequent 2-patterns	Support
$<(L_5, t_3), (L_2, t_4)>$	2	$<(L_2, t_4), (L_{13}, t_9)>$	4
$<(L_5, t_3), (L_1, t_5)>$	2	$<(L_2, t_4), (L_{12}, t_6)>$	4
$<(L_5, t_3), (L_{13}, t_9)>$	2	$<(L_1, t_5), (L_{15}, t_7)>$	3
$<(L_2, t_4), (L_1, t_5)>$	2	$<(L_1, t_5), (L_{13}, t_9)>$	3
$<(L_2, t_4), (L_{13}, t_9)>$	4		
$<(L_2, t_4), (L_{12}, t_6)>$	4		
$<(L_3, t_2), (L_5, t_3)>$	0		
$<(L_3, t_2), (L_2, t_4)>$	2		
$<(L_1, t_5), (L_{15}, t_7)>$	3		
$<(L_1, t_5), (L_{13}, t_9)>$	3		
$<(L_{15}, t_7), (L_{13}, t_9)>$	0		
$<(L_{12}, t_6), (L_{13}, t_9)>$	2		

Frequent Mobility k-patterns Discovering (k ≥ 2)

For $k \geq 2$, candidate k-patterns, i.e., C_k are discovered as follows.

Let P_{k-1} be the frequent $(k-1)$-pattern in the form of $P_{k-1} = <(l_1, t_1), (l_2, t_2)\dots (l_{k-1}, t_{k-1})>$. Let $V(l_{k-1})$ be the set of all neighboring cells of cell l_{k-1} in coverage region.

For all the frequent $(k-1)$-patterns in P_{k-1} following steps are carried out.

For each $v \in V(l_{k-1})$ satisfying $t_k \geq t_{k-1}$, a candidate k-pattern C is generated by attaching $p = (v, t_k)$ to the end. Where l_{k-1} has timestamp t_{k-1} and v has timestamp t_k.

$$C_k = C_k \cup C \tag{2}$$

Then, *supp*$_{min}$ is used to filter the candidate k-patterns.

$$P_k = \{C \mid C \in C_k \text{ and } support\ (C) \geq supp_{min}\} \tag{3}$$

The same procedure is repeated till the $L_k \neq \phi$

Algorithm 1: Frequent Mobility Patterns (L) Discovery Algorithm

Input:	Candidate mobility 1- pattern and Log file of mobility history (D)
	*Supp*min: Minimum support threshold.
	G= A directed coverage region graph.
Output:	L = frequent mobility pattern set.

Begin

    // Let C_k is a set of candidate K- patterns

    // Let P_k is a set of frequent mobility *K-patterns*

    $P_1 \leftarrow$ frequent-mobility.1-pattern set.

      $k = 1$

    Repeat

        For k>=2 and each frequent mobility k- pattern \\ Candidate pattern (C_{k+1}) Generation for l_k

            $P=<(c_1, t_1), (c_2, t_2), ..., (c_k, t_k)> \in L_k$ do

            $V(c_k) \leftarrow \{v | v$ is the neighboring cells of $c_k\}$

            For Each $v \in V(c_k)$ do

                $P(c_k) \leftarrow \{p=(v, t_{k+1}) | <(v, t_{k+1})> \in L_1$ and $t_{k+1} \geq t_k\}$

                For all $p=(v, t_{k+1}) \in P(c_k)$ do

                $C=<(c_1, t_1), (c_2, t_2), ..., (c_k, t_k), (v, t_{k+1})>$

                $C_{k+1}= C_{k+1} \cup C$

                End For

            End For

        End For

        For Candidate-k+1 pattern $c \in C_{k+1}$ do

        $L_{k+1} \leftarrow \{c | c \in C_{k+1}$ and c.support$\geq$ supp$_{min}\}$

        $L = L \cup L_{k+1}$

        End For

        $k = k+1$

    Until $L_k = \emptyset$

    Return L

End

3.1.2 Mining Temporal Weighted Mobility Rules

A mobility rule(R) for two frequent mobility patterns namely A & B is in the form of R: $A \rightarrow B$ such that $A \cap B = \emptyset$.

For frequent mobility pattern set S all the possible mobility rules are given as follows:

$$\mathbf{Z} \rightarrow (\mathbf{S} - \mathbf{Z}) \quad \text{for all} (Z \subset S) \& (Z \neq \emptyset)$$

For example Let $P = <(c1, t1), (c2, t2), ..., (ck, tk) >$ be the frequent mobility pattern, where k>1.

Then it will have following rules:

R1: $<(c1, t1)> \rightarrow <(c2, t2), ..., (ck, tk)>$

R2: $<(c1, t1), (c2, t2)> \rightarrow <(c3, t3), ..., (ck, tk)>$

...

R2: $<(c1, t1), (c2, t2), ..., (ck - 1, tk - 1)> \rightarrow <(ck, tk)>$

All the rules are associated with a confidence value. The confidence value is computed by using following equation. A confidence threshold conf$_{min}$, are used to filter the frequent mobility rules.

$$\mathbf{Confidence(R)} = \frac{support(A \cup B)}{support(A)} \times 100 \tag{4}$$

Algorithm 2: Mobility Rules Formulation Algorithm

Input:	*L:* Frequent mobility patterns set.
	conf$_{min}$: Minimum confidence threshold.
Output:	*Rules* : frequent mobility rules set.

Begin

 Rules ← ϕ.

 For each frequent mobility *k*-pattern. P_k=<$(c_1, t_1), (c_2, t_2), ..., (c_k, t_k)$> ∈ L_k, k>=2

  P_l ← P_k.

  Repeat.

   //A ← (*l*-1) sub pattern of P_l.

   A ← <$(c_1, t_1), (c_2, t_2), ..., (c_{l-1}, t_{l-1})$>.

   conf = support.(P_k)/support.(A).

   If conf ≥ *conf$_{min}$* then.

    R ← {<$(c_1, t_1), (c_2, t_2), ..., (c_{l-1}, t_{l-1})$>→<$(c_l, t_l), ..., (c_k, t_k)$>} //R ← (A → P_k-A).

    R.w = (RuleTime-MinTime) / (MaxTime-MinTime)×100.

    Rules = Rules ∪ R.

   Else break.

   End if.

   l = l-1.

  Until l>1.

 End For.

 Return *Rules.*

End

The goal of frequent mobility patterns mining is to formulate the mobility rules for near places of user's current moving position. Here, the rules generated from the most recent mobility patterns are more important than that of generated from the older one. To do so, each rule r_i is assigned a temporal weighted value w_i. Let Min$_{Time}$ and Max$_{Time}$ is the minimum and maximum time recorded in the mobile client's mobility log file, respectively. Rule$_{Time}$ is the time of the last point of the rule's tail. The weighted value is calculated by (5).

$$weight(R) = \frac{Rule_{Time} - Min_{Time}}{Max_{Time} - Min_{Time}} * 100 \tag{5}$$

Algorithm 2: Location_Prediction(R, P)

Input:	moving client's current trajectory.
	$P = <(c_1,t_1),(c_2,t_2),(c_3,t_3),\ldots\ldots,(c_{i-1},t_{i-1})>$.
	The set of generated rules, R.
Output:	Cell_ID

Begin

Predicted cell, Cell_ID.

$i = 0$.

For each rule r in R:$<(a_1,t_1'),(a_2,t_2'),\ldots,(a_j,t_j')>$ → $<(a_{j+1},t_{j+1}'),(a_{j+2},t_{j+2}'),\ldots,(a_9,t_n')>$.

    If $<(a_1,t_1'),(a_2,t_2'),\ldots,(a_j,t_j')>$ is contained by $P = <(c_1,t_1),(c_2,t_2),(c_3,t_3),\ldots,(c_{i-1},t_{i-1})>$ and $a_j = c_{i-1}$.

        $T_{diff} = 1/(|t_{i-1} - t_j'|+1)$.

        r.matchingscore = r.w + T_{diff}

        r.score=r.confidence + r.support + r.matchingscore.

        MatchingRules = MatchingRules ∪ r .

        NextCells[l] = $(a_{j+1}$, r.score) //Add the $(a_{j+1}$, r.score) tuple to the array of possible next cells.

        $i = i +1$.

    End If.

End for.

NextCells = Sort(NextCells) // NextCells descending order sorting according to matching score.

Cell_ID = NextCells[0] // highest matching score prediction cell extraction.

Return Cell_ID // Cell ID one amongst $L_5, L_3, L_2, L_4, L_7, L_{10}, L_9, L_8$.

End

3.2 Data Item's Cost Estimation

If at any instant of time, the cache becomes full then the replacement policy will take following steps for cached data objects:

1. Estimate access probability, data distance based on predicted Region, data size, and valid scope area.
2. Estimate the predicted Region radius (r).
3. Compute the distance d_i between client location and corresponding cached ith data item's value$_i$ valid scope. Using this distance find whether $v_{i,j}$ (jth value of ith data item) valid scope lies inside or outside of the predicted region.
   If $d_i \leq r$ then the data value$_i$ lies within the predicted region.
   Otherwise, the data value$_i$ lies outside the predicted region.
4. Evaluate the cost function, Cost$_{i, j}$ for each data item. The lowest cost valued data item is evicted for replacement purposes.

The aim of the replacement policy is to find out proper order of objects that can be evicted from cache such that overall cost of evicted data can be minimized. The problem can be formally defined as below.

Find S, such that

$$\text{Min} \left(\sum_{di \in S} \text{cost(i)} \right) \quad \text{and} \quad \sum_{di \in S} \text{objSize} > d_{new} \tag{6}$$

Broader data item's valid scope region makes higher probability of being queried again rather than that is in smaller region. If the size of data items is less, then more number of data items can be placed in the cache. If the distance between data item's valid scope to the client current location is larger then, the probability that it can enter into valid scope area in near future will be low. Our replacement policy SPMC-PRRP chooses the victim as data item having small valid scope area, low access probability, large data size and who have a large distance from valid scope of data item from the client location as defined in PPRRP [12–14]. The following equation shows object d_i cost of client m's cache.

$$Cost_i = \begin{cases} \dfrac{1}{\min(L_r, D(vs(d_i)))} \cdot \dfrac{P_i.A(vs(d_i))}{S_i} \cdot \dfrac{\lambda_i}{\mu_i} \; if \; vs(d_i) \in \; pred\_Reg \\[4mm] \dfrac{1}{D'(vs(d_i))} \cdot \dfrac{P_i.A(vs(d_i))}{S_i} \cdot \dfrac{\lambda_i}{\mu_i} \qquad if \; vs(d_i) \notin \; pred\_Reg \end{cases} \qquad (7)$$

Here, P_i is the ith data item access probability, $A(vs(d_i)$ is the srea of the attached valid scope $(v_{i,j})$. Exponential aging method [6, 7] is employed to estimate/update the access probability for each data item d_i. The initial value of P_i is 0. $D(vs(d_i))$ is the distance between the client's predicted next locations $L_{am} = (Lx_{am}, Ly_{am})$ for future queries data object (d_i) and the reference point $L_i = (Lx_i, Ly_i)$ of valid scope for d_i at the time of query issued. It is estimated using following equation:

$$D(vs(d_i)) = |(L_{am} - L_i| = \sqrt{(Lx_{am} - Lx_i)^2 + (Ly_{am} - Ly_i)^2} \qquad (8)$$

$D'(vs(d_i))$ is the distance between predicted region center and reference point of valid scope for data object di's. This is estimated using following equation:

$$D'(vs(d_i)) = |L_p - L_i| = \sqrt{(Lx_p - Lx_i)^2 + (Ly_p - Ly_i)^2} \qquad (9)$$

S_i is the size for data item d_i, λ_i is the Average query rate of d_i, V_m is the velocity of client m, μ_i is the Average update rate of d_i. Lx_p is the center of predicted region. Third term $\frac{\lambda_i}{\mu_i}$ in this equation is the query rate to update rate ratio of data object d_i. Objects that were updated rarely and queried frequently will have high query rate to update rate ratio.

4   Simulation Model

In this paper, a dataset generator is used in the experiment that shows movement behaviors of client's around the coverage region. Outlier ratio O is used to define the weights for a number of random trajectories in the total number of trajectories N. The execution model for client/server and system are similar to that discussed in [15]. The simulator is implemented in c++ using CSIM. The model considers a rectangle of fixed size as a service area. The Zipf distribution is followed for access pattern of different items. The query inter-arrival time is exponentially distributed. The client-server system works on FCFS principle. The initial Voronoi Diagrams

Fig. 3 Scope distribution

[16] with 110 randomly distributed points as shown in Fig. 3 is used for data item's scope distribution. The range of client's motion speed/velocity v_m is from v_{min} to v_{max}. The client's motion speed follows randomly distributed quadratic equation in terms of time t, which is defined the Eq. (11).

$$\mathbf{v_m} = 2 * \mathbf{rand}(1) * \mathbf{t} + \mathbf{rand}(0) * 10, \tag{10}$$

where rand(0), rand(1) are uniformly distributed random function having a range of 0–1. The range of size for data value is from S_{min} to S_{max}. The equation for variation in size is given by the following equation:

$$\mathbf{S_i} = \mathbf{S_{min}} + \lfloor rand(0) * (S_{max} - S_{min}) \rfloor \, \mathbf{i} = 1, 2, \ldots, \mathbf{itemNum}; \tag{11}$$

The input parameters are taken as defined in default simulation parameter table. The input parameters are taken as defined in default simulation parameter Table 1. Recall [17, 18] measure is used to evaluate the accuracy of mobility prediction. The recall is computed using a ratio of number of correct locations to the number of issued request (Fig. 4, Table 5).

5 Conclusions

A cache replacement policy SPMC-PRRP is proposed in the paper that consists of Sequential Pattern Mining and Clustering in Predicted Region based Cache Replacement Policy. The mobility rules result into an accurate next location prediction which can be used in estimating the distance between data item's valid scope reference point to the anticipated next location of the client. The cache replacement cost function for eviction of data item uses the next location prediction for effective cost computation of valued data items.

Table 5 The default parameters settings for simulation model

Parameter	Details	Parameter values	Parameter	Details	Parameter values
Rect_Size.	Rectangular service area size	9000 m * 9000 m	Downlink_Band	Downlink channel bandwidth	144 kbps
Item_Num.	Data items count in the database	110	Prediction_Interval	Time interval after which predicted region is again evaluated	100.0 s
Scope_Num.	Count of unique values for each item at various locations	220	Query_Interval	Queries average time interval	50.0 s
S_{min}	Data value minimum size	64 bytes	CacheSize_Ratio	Cache size to database size ratio	10%
S_{max}	Data value maximum size	1024 bytes	Min_Speed	Client's minimum moving speed	10 mps
K	Number of cluster for clustering algorithm	5	Max_Speed	Client's maximum moving speed	20 mps
Θ	Zipf access distribution, skewness parameter	0.5	A	A constant factor to weight the importance of the most recent access to the probability estimate	0.70
N	Number of trajectories (dataset size)	5000	$conf_{min}$	Minimum confidence threshold	50%
sup_{min}	Minimum support threshold	30%	B	Combination weight for dissimilarity measure	0.5
Uplink_Band	Uplink channel bandwidth	19.2 kbps	O	Outlier ratio	Vary

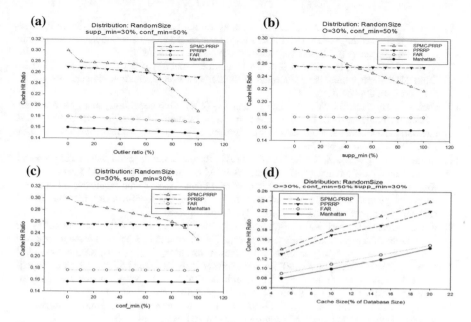

Fig. 4 **a** Cache hit ratio versus outlier ratio (o), **b** cache hit ratio versus minimum support threshold (supp$_{min}$), **c** cache hit ratio versus minimum confidence threshold (conf$_{min}$), **d** cache hit ratio versus cache size

References

1. Swaroop V, Shanker U (2010) Mobile distributed real-time database systems: research challenges. In: Proceedings of international conference on computer and communication technology (ICCCT 2010). Motilal Nehru National Institute of Technology, Allahabad, India, pp 421–424
2. Swaroop V, Gupta GK, Shanker U (2011) Issues in mobile distributed real-time databases: performance & review. Int J Eng Sci Technol (IJEST) 3:3504–3517
3. O'Neil EJ, O'Neil PE, Weikum G (1993) The LRU-K page replacement algorithm for database disk buffering. In: Proceedings ACM SIGMOD conference, vol 1. pp 297–306
4. Dar S, Franklin MJ, Tan M (1996) Semantic data caching and replacement. In: Proceedings of 22th international conference on very large data bases, vol 22. Morgan Kaufmann Publications Inc., San Francisco, CA, USA, pp 333–341
5. Ren Q, Dunham MH (2000) Using semantic caching to manage location dependent data in mobile computing. In: 6th ACM/IEEE mobile computing and networking (MobiCom), vol 3. Boston, MA, USA, pp 210–221
6. Gupta AK, Shanker U (2017) Modified predicted region based cache replacement policy for location dependent data in mobile environment. In: Proceedings of the 6th International Conference on Smart Computing and Communications (ICSCC 2017), 7–8 December 2017. NIT, Kurukshetra, pp 917–924. https://doi.org/10.1016/j.procs.2017.12.117
7. Gupta AK, Shanker U (2017) SPMC-CRP: a cache replacement policy for location dependent data in mobile environment. In: Proceedings of the 6th International Conference on Smart Computing and Communications (ICSCC 2017), 7–8 December 2017. NIT, Kurukshetra, pp 632–639. https://doi.org/10.1016/j.procs.2017.12.081

8. Xiao X, Zheng Y, Luo Q, Xie X (2010) Finding similar users using category-based location history. In: Proceedings of 18th SIGSPATIAL international conference on advanced geographic information systems (GIS), vol 10. San Jose, CA, USA, pp 34–42

9. Jeong J, Leconte M, Proutiere A (2016) Cluster-aided mobility predictions. In: IEEE INFO-COM 2016—35th annual IEEE international conference on computer communications, vol 35. San Franciso, CA, USA, pp 1–9

10. Niehoefer, Burda R, Wietfeld C, Bauer F, Lueert O (2009) GPS community map generation for enhanced routing methods based on trace-collection by mobile phones. In: Proceedings—1st international conference on advances in satellite and space communications SPACOMM 2009, vol 1. Colmar, France, pp 156–161

11. Zheng Y, Zhang L, Ma Z, Xie X, Ma W-Y (2011) Recommending friends and locations based on individual location history. ACM Trans Web (ACM, New York, NY, USA) 5:1–44

12. Kumar A, Misra M, Sarje AK (2008) A predicted region based cache replacement policy for location dependent data in mobile environment. In: 10th international-research-institute student seminar in computer science, vol 7. IIIT Hyderabad, pp 1–8

13. Kumar A, Misra M, Sarje AK (2006) A new cost function based cache replacement policy for location-dependent data in mobile environment. In: 5th annual international research institute student seminar in computer science, vol 5. Indian Institute of Technology, Kanpur, pp 1–8

14. Kumar A, Misra M, Sarje AK (2007) A weighted cache replacement policy for location dependent data in mobile environments. In: Proceedings of 2007 ACM symposium on applied computing SAC'07, vol 7. Seoul, Republic of Korea, pp 920–924

15. Zheng B, Xu J, Member S, Lee DL (2002) Cache invalidation and replacement strategies for location-dependent data in mobile environments, vol 51. pp 1141–1153

16. Rourke JO' (1998G) Computational geometry in C. Cambridge University Press, New York, NY

17. Yavas G, Katsaros D, Ulusoy O, Manolopoulos Y (2005) A data mining approach for location prediction in mobile environments. J Data Knowl Eng 54:121–146

18. Duong TV, Tran DQ (2012) An effective approach for mobility prediction in wireless network based on temporal weighted mobility rule. Int J Comput Sci Telecommun 3:29–36

5G Cellular: Concept, Research Work and Enabling Technologies

Vinay Kumar, Sadanand Yadav, D. N. Sandeep, S. B. Dhok,
Rabindra Kumar Barik and Harishchandra Dubey

Abstract The enormous growth in communication technology is resulting in an excessively connected network where billions of connected devices produce massive data flow. The upcoming 5th generation mobile technology needs a major paradigm shift so as to fulfil the growing demand for reliable, ubiquitous connectivity, augmented bandwidth, lower latency and improved energy efficiency. All new mobile technologies for 5G are expected to be operational by 2020. This paper presents the basic concepts, working research groups for 5G, standards and the various enabling technologies. Besides, the paper presents a comparative analysis of 5G over the contemporary cellular technologies (LTE, GSM) based on various relevant parameters. This review is helpful for the new researchers, aspiring to work in the field of 5G technology.

Keywords 5G · Cellular communication · Research group in 5G · Enabling technology · Layered architecture

V. Kumar (✉) · S. Yadav · D. N. Sandeep
Department of ECE, Visvesvaraya National Institute of Technology,
Nagpur 440010, India
e-mail: vk@ece.vnit.ac.in

S. Yadav
e-mail: sadanand.0501@gmail.com

D. N. Sandeep
e-mail: sandysmilent@gmail.com

S. B. Dhok
Center for VLSI and Nanotechnology, Visvesvaraya National Institute of Technology,
Nagpur 440010, India
e-mail: sanjaydhok@gmail.com

R. K. Barik
School of Computer Application, KIIT University, Bhubaneswar 751024, India
e-mail: rabindra.mnnit@gmail.com

H. Dubey
Electrical Engineering, The University of Texas at Dallas, Richardson, USA
e-mail: harishchandra.dubey@utdallas.edu

© Springer Nature Singapore Pte Ltd. 2019
M. L. Kolhe et al. (eds.), *Advances in Data and Information Sciences*, Lecture Notes in Networks and Systems 39, https://doi.org/10.1007/978-981-13-0277-0_27

1   Introduction

In the last two decades, the cellular communication technologies have witnessed a rapid evolution from 2G (global system of mobile communications (GSM)) to 4G (large term evolution (LTE)). This evolution is necessitated due to an increasing demand of the bandwidth and low latency by consumers [1–3]. The parameters which become points of interest in this evolution are throughput, jitter, interchannel interference, connectivity, scalability, energy efficiency and compatibility. It is assumed that around 50 billion devices will be connected to the future global IP network. The society will be fully equipped with a network which will try to get the control of machines and equipment by a single click with almost zero time delay, such a challenging demand has fulfilled by 5G technology. Real-time controlling of machines should be possible with the 5G since Internet of Things (IoT) is also emerging rapidly [4, 5]. By 2020, the society will be connected by means of IoT and intelligent and integrated sensor systems. Smart living class of people require a ubiquitous and uninterrupted mobile connectivity for IoT control commands and to upload the data of daily activities. Thus, there will be a huge flow of uplink data. Also, it is assumed that vehicular ad hoc networks (VANETs) will be integrated with cellular network, which will lead to smarter and safer transportation system. As the device count is increasing, the data offloading in unlicensed band will also play an important role in network load balancing, reduction in overheads (control signalling) [6, 7]. Hence, 5G is expected to provide a smooth compatibility with dense heterogeneous networks to fulfil the real-time traffic so that end user can experience a smoother connectivity of the network [8–10].

The remaining part of the paper is compiled as follows. The basic concept of 5G network architecture is explained in Sect. 2. A comparison of 5G technology with the previous technologies is detailed in Sect. 3. Section 4 presents the key technologies and trends which help enable the 5G technology. In Sect. 5, a list of various research groups working in developing 5G is provided. Along with a future scope, the paper is concluded in Sect. 6.

2   5G Network Architecture and Protocol Stack

This section of the paper presents the protocol stack and network architecture of 5G. The 5G network architecture comprises RANs (radio access networks), aggregators, IP network, Nanocore, cloud computing, etc. Figure 1 shows 5G network architecture in which different RANs can use same Nanocore for communication using flat IP concept. GSM, GPRS/EDGE, UMTS, LTE, LTE-advanced, WiMax, Wi-Fi, etc., are supported by RANs [11, 12]. Flat IP architecture is able to identify devices using symbolic names which are not possible in hierarchical architecture that uses normal IP addresses. This architecture results in a reduction of the number of network elements in data path and results in reduction of cost and latency to a great extent. Aggregator

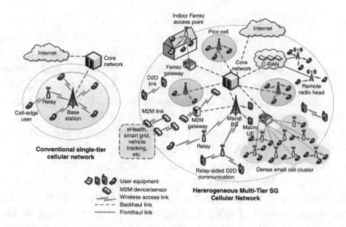

Fig. 1 5G network architecture (adapted from [13])

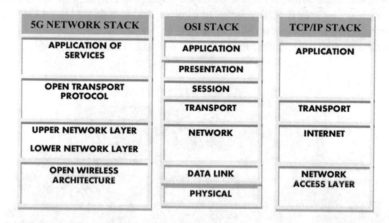

Fig. 2 Comparison of protocol stack for 5G technology with OSI and TCP model

collects the entire RAN traffic and routes or directs it to a gateway. Its location is at the place of Base Station Controller/Radio Network Controller (BSC/RNC). Nanocore comprises Nanotechnology, cloud computing, all IP architecture. Cloud computing makes use of web and remote central servers to upkeep data and user applications. It allows end user to use application without any kind of installation and with the use of Internet, the end user can access the files from any computer across the world [1, 4].

The 5G technology protocol stack consists of four layers. A comparison with OSI and TCP/IP stack is given in Fig. 2.

OWA Layer: Open wireless architecture layer is analogous to data link layer and physical layer of the OSI layer stack. The planned OWA is to present open interface baseband processing modules to sustain various current and futuristic wireless communication standards.

Network Layer: Network layer is split into two layers, viz. lower network layer and upper network layer. The aim of this is to direct data from source IP system/device to the intentional device. The network layer in OSI model is the most important layer as it does the advancement of packet, directing it via buffer routers, because it has a clear and complete idea of the address of neighbouring network. The network layer in addition deals with QoS, i.e. quality of service; it makes out and promotes the local server domain information towards the next level that is the Transport layer. The Network layer performs the function of connection modelling, i.e. to have connection-less communication, host addressing and forwarding the message or information. It also deals with the requests coming from Transport layer and generates the request to Data link layer within the network architecture. As the demand for data rate is increasing day by day and the communication should be robust. The disadvantages of 4G technology should be overcome, so introduction of 5G is there in which the terminating end part being able to use diverse wireless technologies; meanwhile, it should be capable to unite distinct flows from diverse technologies. The user mobility factor must be considered so that good QoS should be maintained. The advantage of 5G over 4G is that it will increase the bandwidth that is around 400 times faster than existing wireless technology. It will use ultra-

Fig. 3 Enabling 5G technologies and trends

wideband and smart antennas with software-defined radio, encryption and flexibility (Fig. 3).

OTP Layer: OTP layer combines the functionality of both session layer and transport layer of OSI stack. The transport layer in the wireless network or cellular network has different configuration than that of wired network. In the wireless networks, because of high bit error fraction in the wireless interface losses occur. So, to overcome this problem, TCP reforms and version are presented for the cellular or wireless networks that will again transmit the damaged or missing TCP packets via the wireless or cellular links. For the 5th generation mobile stations, it will be appropriate that the transport layer can be downloaded and installed in the networking system. These cell phones must be having the ability to download and instal (e.g. TCP, RTP, RTCP, etc. or other new protocols) versions that are battered to a precise wireless technology that has to be installed at the various base stations.

Application Layer: It converts the data as per the required appropriate format. It is used for encryptions and decryptions of the data. Also, it is capable of selecting the best wireless connection for given service. This layer deals with graphical interface of technology with end user. The application layer is the ending of data transfer, i.e. now data is shown to the user in graphical manner by use of graphical user interface (GUI). For ease of end user in 5G, it may be integrated by means of adding the

Table 1 Comparison of 5G with contemporary cellular technology

S.No	Parameters	2G	3G	4G	5G
1	Start of technology	1981	1990	2000	2014
2	Deployment of technology	1999	2002	2010	2020
3	Modulation scheme	GMSK	64QAM	QAM (OFDM)	FSK-QAM
4	Type of technology	Digital cellular	Broadband technology	IP technology	IP and broadband technology
5	Multiplexing	TDMA	CDMA	CDMA	CDMA
6	Bandwidth (MHz)	25	25	100	800
7	Operating frequency	900–1800 MHz	2100 MHz	850–1800 MHz	28 GHz
8	Type of data support	Low-speed voice data	High-speed voice data	High-speed voice data for a mobile node	Voice and high speed of data for a highly mobile node
9	Cost of technology	Very low	High	More than 3G	More than 4G
10	Data rate	Up to 64 Kbps	Up to 2 Mbps	Up to 200 Mbps	Up to 36 Gbps

voice with graphics, i.e. application will speak with end user. Also, the end user can respond to that by means of voice or GUI. For variety of available networks, intelligent management of QoS is the absolute request from 5G MT regarding the applications (Table 1).

In today scenario, the user selects the wireless interface for any specific Internet service in the mobile phones itself, without any QoS history to choose the optimal wireless connection. The 5G phones will facilitate quality testing of service and will store measured data in information database in the mobile terminal (MT). Intelligent algorithms that run in MT will provide the optimal wireless connectivity upon required QoS and personal cost constraints. The QoS parameters are jitter, delay, bandwidth, reliability, losses. All these parameters will be stored in the MT and based upon these networks are selected researchers are working on the optimization of application layer and they have proposed some technique which is as follows. For machine-type communications (MTC) or IoT, contention-based random access is the major challenge, and 5G includes MTC.

3 Comparison of 5G with Other Generations

This section is dedicated for comparison of 5G technology with its contemporary cellular communication techniques. Researchers are facing various challenges for achieving 5G performances from its contemporary cellular communication. The challenges are as follows: multimode user terminals, network infrastructure, QoS and QoE, security (jamming and spoofing), charging and billing, and data encryption [14].

4 Key Enabling Technologies

The major technological findings that can bring renaissance to wireless cellular communication networks comprise (1) a wireless software-defined network, (2) energy harvesting networks, (3) cloud RAN, (4) network function virtualization, (5) millimetre wave spectrum, (6) massive MIMO, (7) heterogeneous multi-tier networks, (8) network ultra-densification, (9) cloud computing and big data, (10) scalable-IoT, (11) device-to-device (D2D) connectivity with high mobility, (12) green communication and (13) novel radio access techniques [13, 15, 16].

D2D Communication: It is kind of technology which enables device-to-device communication without the need of infrastructures such as base stations or access points. The two most known D2D techniques are Bluetooth and Wi-Fi-Direct which both work in the unlicensed industrial, scientific and medical (ISM) bands [17] (Table 2).

Millimetre Wave Technology: It shows an enormous potential to enable a throughput of gigabit per second with its large available bandwidth of 5G technol-

Table 2 Requirements and enabling solutions for 5G cellular technology

S.No	5G expectation	Proposals
1	High data rate and throughput enhancement (approximately throughput is increased 1000 times as compared with 4G, data rate approximately 10 Gb/s)	Spectral reuse and using a different band (e.g. mm-wave communication), multi-tier network, ultra-densification, massive-MIMO, D2D communication and C-RAN
2	Low latency (between 2 and 5 ms)	Full-duplex communication, mobile cloud computing and Big data, C-RAN and D2D communication
3	Energy efficiency improvement (energy consumption is decreased 1000 times per bit)	Ultra-densification, wireless charging, energy harvesting D2D communication and green communication
4	Autonomous applications and network management	M2M communication, self-organizing and cognitive network
5	High scalability (to be able to accommodate up to 50 billion devices	Wireless software-defined networking, mobile cloud computing and massive MIMO
6	A very high connectivity	Enhancing the connectivity for cell edge users, D2D communications
7	Standardization of security on authentication, authorization and accounting	Big data, wireless software-defined networking and mobile cloud computing
8	Advanced applications (e.g. smart city)	Network virtualization, C-RAN and M2M communication

ogy. Millimetre wave refers to the band of frequencies at 30,300 GHz, the unlicensed millimetre wave frequency bands can provide a quantity of spectrum resource that can help meet the needs for low latency, high data rate, scalable connectivity for a large number of users.

Massive MIMO: Massive MIMO is an important enabling technology for 5G networks. Unlike multi-user MIMO (MU-MIMO) of 4G system which uses only a few number of antenna components built on user terminals and base stations, massive MIMO incorporates a large number of antennas at base stations in order to increase the system throughput and capacity.

Bigdata and Mobile Cloud Computing: Data storing methods also will undergo evolution in 5G because of facing higher network throughput. The traditional data storing technique will no longer be able to handle the exponential increase of data flow especially when a huge data is needed to be downloaded. Of late, the cloud storage is getting popular due to providing on-demand service. Users can use cloud storage

Table 3 Summary of working research group for 5G standard

S.No	Research group	Outcomes
1	Mobile and Wireless Communications Enablers for the Twenty-Twenty Information Society (METIS)	High-level architectures, channel modelling and direct device-to-device communication [19]
2	5G Infrastructure Public Private Partnership (5G-PPP)	Development of high capacity ubiquitous infrastructure for both mobile and fixed networks [20]
3	5th Generation Non-orthogonal Waveforms for Asynchronous Signalling (5GNOW)	Development of unified frame structure, latency at ultra-low level and ultra-high reliability [21]
4	Enhanced Multicarrier Technology for Professional Ad-Hoc and Cell-Based Communications (EMPhAtiC)	Development of highly flexible and efficient filter-bank processing structure, MIMO transceiver, MIMO transmission and channel estimation, equalization and synchronization functionalities [22]
5	Energy Efficient E-band Transceiver for Backhaul of the Future Networks (E3NETWORK)	Aims at high spectral and energy efficiency by using modern digital multilevel modulations and highly integrated circuits (advanced SiGe BiCMOS technology) in the RF analogue front-end [23]
6	PHYsical LAyer Wireless Security (PHYLAWS)	Privacy enhancement of radio interface in wireless networks using physical layer security and secrecy coding approaches [24]
7	Full-Duplex Radios for Local Access (DUPLO)	This project is built upon radio transceiver technology wherein the carrier frequency can be reused simultaneously for transmission and reception [25]
8	Connectivity Management for Energy Optimised Wireless Dense Networks (CROWD)	Project [26] The aim of this project is to integrate the dense heterogeneous networks with wireless/ wired return capabilities

(continued)

Table 3 (continued)

S.No	Research group	Outcomes
9	Dense Cooperative Wireless Cloud Network (DIWINE)	Project utilizes the paradigm of virtual relay centred self-contained wireless cloud, which has a simple and unambiguous interface to terminals, in solving the problem of wireless communications in densely interfering ad hoc networks [27]
10	Network of Excellence in Wireless Communications (NEWCOM)	Working on tightest upper limits of wireless networks multihop communication. Calculation of energy and efficiency in wireless networks [28]
11	New York University Wireless (NYU Wireless)	Working on mm wave [29]
12	5G Innovation Center (5GIC)	Recent outcome is wireless speed of 1 Tbps in point-to-point communication [30]
13	Software-Defined Access Using Low-Energy Subsystems (SODALES)	The project [31] is aimed at developing a novel wireless access interconnection service having a low-cost 10 Gbps fixed access target and to offer transparent transport services for the mobile and fixed subscribers

and thus save in the local storage of their devices. Also, mobile cloud computing will become a major method for computing data at a large scale [3].

5 Research Groups

This section of paper is devoted to research group working on 5G standards. A few working groups are Mobile and Wireless Communications Enablers for the Twenty-Twenty Information Society (METIS), Enhanced Multicarrier Technology for Professional Ad-Hoc and Cell-Based Communications (EMPhAtiC), 5G Innovation Center at the University of Surrey, NYU WIRELESS, and the Electronics and Telecommunications Research Institute (ETRI), Korea [14, 18]. These groups are researching different technical and probable standardization aspects of 5G which is mentioned in Table 3.

6 Conclusion and Future Directions

In this paper, the 5th generation wireless network layered architecture is explained. The 5th generation cellular network is explained as an open platform for designing devices on different layers of OSI model. With the help of enabling this technology, we can get best quality of service with lowest cost. A new revolution of 5G-based technologies is now beginning because the new technology is giving a tough completion to the ordinary desktops and laptops whose marketplace and values are now changing with this up gradation. There have been a lot of improvements and up gradation starting from 1G to 2G and from 2G to 3G, and 4G to 5th Generation which is shown in the above table in the networking world of telecommunications. The 5th generation cellular phones and networks will be having access to various different wireless and cellular technologies at the same time along with the terminal that will be having the ability to blend different flows from various technologies present. The 5G technology network will be giving very high resolution and without any delay for the crazy and demanding cell phone users. Users now will be able to watch TV channels at high-definition clarity in their respective cellular phones without having any buffering or interruption. The 5th Generation cellular phones can also be a type

Table 4 Summary of various futures scope for 5G

S.No	Outcomes	
1	Performance metric optimization	For a better assessment of 5G, more performance metrics should be considered like energy efficiency, spectral efficiency, latency, user fairness, implementation complexity, etc.
2	Realistic channel model	Some realistic channel models are required other than conventional channels
3	Reducing signal processing complexity of massive MIMO	Since the transmitted and received signals are lengthy in nature, the search algorithm must be efficient to reduce the signal processing complexity
4	Cognitive radio network interference management	The main challenge is to practically manage the mutual interference of cognitive radio and primary system

of tablet PC which will result in evolution of many mobile embedded technologies [6] (Table 4).

Acknowledgements The authors express thanks to Ajinkya Ramdas Puranik, Akshay Kulkarni, Ankit Waghmare, Amit Waghmare and Vivek Pathak for their valuable contributions in this work.

References

1. Govil J, Govil J (2008) 5G: functionalities development and an analysis of mobile wireless grid. In: First international conference on emerging trends in engineering and technology, ICETET'08. IEEE
2. Gohil A, Modi H, Patel SK (2013) 5G technology of mobile communication: a survey. In: International conference on intelligent systems and signal processing (ISSP). IEEE
3. Agiwal M, Roy A, Saxena N (2016) Next generation 5G wireless networks: a comprehensive survey. IEEE Commun Surv Tutor 18(3):1617 1655
4. Munoz R, Mayoral A, Vilalta R, Casellas R, Martinez R, Lopez V (2016) The need for a transport API in 5G networks: the control orchestration protocol. In: Optical fiber communications conference and exhibition (OFC). IEEE, pp 1–3
5. Petrov I, Janevski T (2016) Design of novel 5G transport protocol. In: 2016 international conference in wireless networks and mobile communications (WINCOM). IEEE, pp 29–33
6. Rao S, Kumar V, Kumar S, Yadav S, Ancha VK, Tripathi R (2017) Power efficient and coordinated eICIC-CPC-ABS method for downlink in LTE-advanced heterogeneous networks. Phys Commun (Elsevier)
7. Pedapolu PK, Kumar P, Harish V, Venturi S, Bharti SK, Kumar V, Kumar S (2017) Mobile phone users speed estimation using WiFi signal-to-noise ratio. In: Proceedings of the 18th ACM international symposium on mobile ad hoc networking and computing, p 32
8. Boviz D, El Mghazli Y (2016) Fronthaul for 5G: low bit-rate design enabling joint transmission and reception. In: Globecom workshops (GC Wkshps). IEEE, pp 1–6
9. Sharawi MS, Podilchak SK, Hussain MT, Antar YM (2017) Dielectric resonator based MIMO antenna system enabling millimetre-wave mobile devices. IET Microw Antennas Propag 11(2):287–293
10. Chao H, Chen Y, Wu J, Zhang H (2016) Distribution reshaping for massive access control in cellular networks. In: 84th in vehicular technology conference (VTC-Fall). IEEE, pp 1–5
11. Gupta A, Jha RK (2015) A survey of 5G network: architecture and emerging technologies. IEEE Access 3:1206–1232
12. Wang CX, Haider F, Gao X, You XH, Yang Y, Yuan D, Hepsaydir E (2016) Cellular architecture and key technologies for 5G wireless communication networks. IEEE Commun Mag 52(2):122–130
13. Hossain E, Hasan M (2015) 5G cellular: key enabling technologies and research challenges. IEEE Instrum Meas Mag 18(3):11–21
14. Mitra RN, Agrawal DP (2015) 5G mobile technology: a survey. ICT Express (Elsevier) 1(3):132–137
15. Akyildiz IF, Nie S, Lin SC, Chandrasekaran M (2016) 5G roadmap: 10 key enabling technologies. Comput Netw (Elsevier) 106:17–48
16. Wei L, Hu RQ, Qian Y, Wu G (2016) Key elements to enable millimeter wave communications for 5G wireless systems. IEEE Wirel Commun 21(6):136–143
17. Shen X (2015) Device-to-device communication in 5G cellular networks. IEEE Netw 29(2):2–3
18. Pirinen P (2014) A brief overview of 5G research activities. In: 1st international conference on 5G for ubiquitous connectivity (5GU), IEEE, pp 17–22
19. Project Coordinator: Afif Osseiran Ericsson AB, FP7 Integrating Project METIS (ICT 317669). https://www.metis2020.com/documents/deliverables/

20. The 5G Infrastructure Public Private Partnership. http://5g-ppp.eu/
21. Internet Resource, 5GNOW Deliverable 2.2. http://www.5gnow.eu/download/5GNOW_D2.2_v1.0.pdf
22. Internet Resource, EMPhAtiC Deliverable 4.1: http://www.ict-emphatic.eu/images/deliverables/deliverable_d4.1_final.pdf
23. FP7 STReP project E3NETWORK (lCT 317957). http://www.ict-e3network.eu/
24. FP7 STReP project PHYLAWS (ICT 317562). http://www.phylaws-ict.org
25. FP7 STReP project DUPLO (lCT 316369). http://www.fp7-duplo.eu/
26. FP7 STReP project CROWD (ICT 318115). http://www.ict-crowd.eu/
27. FP7 STReP project DIWlNE (ICT 318177). http://diwine-project.eu/
28. Internet Resource, NEWCOM Deliverables 23.3. http://www.newcom-project.eu/images/Delivarables/D23.3-Secondreportontoolsandtheirintegrationontheexperimentalsetups.pdf
29. Rappaport TS, Sun S, Mayzus R, Zhao H, Azar Y, Wang K, Gutierrez F (2013) Millimeter wave mobile communications for 5G cellular: it will work. IEEE Access 1:335–349
30. Internet Resource, 5GIC. http://www.surrey.ac.uk/5gic
31. FP7 STReP project SODALES (ICT 318600). http://www.fp7-sodales.eu/

Modeling and System-Level Computer Simulation Approach for Optimization of Single-Loop CT Sigma Delta ADC

Anil Kumar Sahu, Vivek Kumar Chandra and G. R. Sinha

Abstract The purpose of this paper is to modeling and computer simulation of optimized GUI-based system-level design of Sigma Delta Modulator (SDM) based analog-to-digital converter (ADC) in which a top-down approach of designing is mainly focused to improve the resolution of single-loop continuos-time sigma delta because system-level simulation is an important part of system design and can help to estimate power consumption and noise analysis accurately. The presented GUI for the noise calculation and transistor level circuit design and output simulation of Sigma delta modulator based ADC along with its various components like operational trans-conductance amplifier (OTA), Analog-to-Digital converter (ADC), Digital-to-Analog converter is well taken care here to achieve optimized performance. The simulation results show performance of modulator in transistor level and circuit design that are compatible with each other where output was generated successfully in transistor level circuit design along with its waveforms.GUI (Graphical User Interface) allows us to calculate the noise parameters and power consumption parameters. The resolution of modulator is mainly affected by the noise parameters as thermal noise, clock jitter noise, and slew rate of amplifiers, quantization noise, which leads for consumption of power. The modulators offers phase margin 92.84°, gain bandwidth product of 31.89 kHz, Unity gain bandwidth of 1.2295 kHz, That consuming power of 1.430 mW. Even with the high order of modulator, the complexity increases and the device get bulky. When the technology reduces there is possibility for improvement and getting good results. Thus, the CT $\sum\Delta$ modulator architecture is well designed and implemented using Tanner EDA tool and with the help of GUI the noise and power consumption parameters have been analyzed.

A. K. Sahu (✉)
SSGI (FET), SSTC, Bhilai, India
e-mail: anilsahu82@gmail.com

V. K. Chandra
C.S.I.T, Durg, India
e-mail: vivekchandra1@rediffmail.com

G. R. Sinha
IIIT Bangalore, CMR Technical Campus Hyderabad, Hyderabad, India
e-mail: drgrsinha@ieee.org

© Springer Nature Singapore Pte Ltd. 2019
M. L. Kolhe et al. (eds.), *Advances in Data and Information Sciences*, Lecture Notes in Networks and Systems 39, https://doi.org/10.1007/978-981-13-0277-0_28

Keywords CT ADC · Power estimation of ADC · Noise calculation of ADC
System-level design (GUI)

1 Introduction

In the present world, the continuous-time sigma delta modulator have been used because of their lower power requirements. Analog VLSI is in a great demand today because analog design in proved to be fundamental necessary in various complex and high performance systems [1]. All the natural occurring signals are in an analog form, for example the photo cell in video camera produces a current that is low as few electrons per microsecond and a seismograph sensor has an output voltage ranging from a few microvolt's for very small vibration of earth to hundreds of mill volt of heavy earthquakes [2]. All of these signals must eventually undergo extensive processing in digital domain and each of these signals consists of an analog-to-digital convertor and a digital signals processor [3].

Conventional analog Continuous-Time $\sum\Delta$ ADC involves a comparator circuit in the loop as the quantizer with high speed and low noise, thus it poses a design challenge in nanoscale technology [4]. In the nanometer-scale technology, the design of ADCs become more complex due to the low supply voltage that comes along the technology scaling and require complex analog buildings blocks [5].

In a practical, the electrical version of natural signals may be prohibitively small for direct digitization by ADC. The signals also some time phase a problem by unwanted, out of band interferes. Moreover when digitizing an analog signal there is always a need of ADC, anti-aliasing, and reconstruction filter. New application continues to appear in which speed and power consumption requirements often demands the use of high speed analog front end. Also an integrated circuit becomes larger due to system integration, it is much more likely that at least some portion of total modern integrated circuit will include analog circuitry required to interface to the real world [6].

1.1 Specialty of Continuous-Time $\sum\Delta$ Modulator Based ADC

We have chosen to employ a continuous-Time filter inside our modulator loop rather than a discrete time (DT) filter due to the possible reasons which are as follows:

- In a DT, $\sum\Delta$ modulator has a maximum clock rate limited both by OPAMP bandwidths and by the truth that circuit waveforms need several time constants to settle clock periods. For a $\sum\Delta$ modulator built in a process with maximum transistor speed, the maximum clock rate. While, waveform varies continuously in a CT $\sum\Delta$ modulator, and the restriction on OPAMP bandwidths are relaxed. And a CT modulator can be clocked up to an order of magnitude faster [7].

- In a DT modulator, large glitches exist on OPAMP virtual ground nodes due to switching transients. This is not the case in a CT modulator as it is possible to kept OPAMP virtual grounds very quiet.
- One problem with working in the DT domain is aliasing is that signals get separated by a multiple numbers of the sampling frequency. DT $\sum\Delta$ modulators usually require a separate filter at their inputs to attenuate aliases sufficiently whereas CT $\sum\Delta$ modulator has free anti-aliasing [8].
- The switched capacitor (discrete time) filter is used as a loop filter in the forward path in a discrete time $\sum\Delta$ ADC, while the loop filter is used a continuous-time (LC, gm-C or active RC) filter in a continuous-time Analog-to-Digital Converter [9].

1.2 Application's of Analog-to-Digital Converter

The architecture of ADC depends on voltage-domain creates difficulty because of low input supply voltage with upcoming technology. Sigma delta converters basically depend on a principle of oversampling. That means the sampling frequency (Fs) is much greater than message signal (Fm). As compared with nyquist rate ADCs, oversampling ADCs having high resolution in spite of it uses analog components in digital signal processing for performing analog-to-digital conversion, and due to the oversampling sigma delta ADCs; they do not need anti-alias filtering, which is the prime requirement of nyquist rate ADCs. Thus, higher order with better and higher linearity is no used and generally avoided [10].

2 Problem Statement

After reviewing the papers, we found that there are some limitations in the process taken. Some of them are

- **Linearity**: There is problem of linearity when you design a filter for low frequency application. Filter designed using OTA-C method required trans-conductance of OTA in nano-Siemens range, so problem of linearity is there.
- **Noise problem**: At low frequency there is problem of noise. At low frequency flicker noise become dominant as compared to other type of noise.
- **Change of trans-conductance during process variation**: Since the filter design using multiple OTA. There is problem of trans-conductance variation from one to another.
- **Power consumption and complex circuit**: Design circuitry is very complex to increase the chip area and power consumption.

- **Stability**: Due to some amount of injected quantization noise into the loop filter, the modulator's stability is degraded.
- **Noise shaping**: Noise shaping is limited with the lower number of quantization levels.

3 Method of GUI Implementation for Noise and Power Calculation of CT Sigma Delta ADC

3.1 Power Consumption of DAC

A current of $I_{DAC}(t)$ is entered into integrator of modulator ADC and P_{DAC} is estimated as

$$P_{DAC} = 2 \times \text{mean} \left\{ I_{DAC}(t)^2 \right\} \times R_{DAC}. \tag{1}$$

where $R_{DAC} = R$ (Equivalent DAC resistance) and Mean{.} = averaging function.

The total power consumption of DAC's in continuous-time sigma delta modulator ADC can be written as

$$P_{DACTotal} \approx \frac{V_{FS}^2}{R} \left(\frac{1}{4} + \frac{1}{\sqrt{3} X 2^{Nc}} \right) \left(1 + \frac{\prod^2}{0.3 g_1^2 OSR^3} \sum_{i=2}^{nDAC} \left(\frac{g_i}{g_1} \right)^2 \right), \tag{2}$$

where g_i is the gain corresponding to Ith feedback path and N_{DAC} = number of continuous-time DAC's.

3.2 Power Consumption of Clock Generation

Power consumption in clock generation stage depends on Supply voltage and clock frequency [11]. Power overhead is due to both on chip and off chip clock generation. Power of clock generation circuit is expressed as

$$P_{clk} = K_{clk} \frac{V dd}{\sigma_t^2} f_s = K_{clk} \frac{V dd}{\sigma_t^2} x 2 B W x OSR, \tag{3}$$

where K_{clk} = constant having a dimension of ampere multiplied by cubic power of second.

3.3 Power Consumption of Quantizer

Generally fast latches and pre-amplifier are required in the design of quantizer. When the GWB is high it further needs high value of trans-conductance [12]. Therefore, power consumption increases which is given as

$$P_{quant} \approx K_{quqnt} V_{dd} 2^{Nc} Fs + K_{DEM} V_{dd} 2^{Nc} f_s, \tag{4}$$

where Kquant and K_{DEM} = constants of quantizer and DEM power terms.

3.4 Power Consumption of Digital Circuits

The power consumption of digital circuit is given as

$$P_{digital} \approx L_c X n_{Go} X P_{inv} V_{fs} \tag{5}$$

The total power consumption of a continuous-time digital circuit is given as

$$P_{CT} \approx P_{Amp} + P_{DACTotal} + P_{clk} + P_{quant} + P_{digital} \tag{6}$$

There are various types of noise present in the circuit, which account for some amount of power. The errors which are mitigate by the system or circuit level techniques constitutes.

A low part of the total noise [5]. This is less than 25%. When we go for the calculation of thermal noise and the clock jitter noise, they account for the 75% of the total noise. The graphical user interface is prepared for the noise calculation of the modulator [13]. The delay can be reduced or compensated by using a fast path around the quantized which is used in the modulator. If the fully differential architecture is used, the mismatch in the rise and fall edge of the DAC output can be reduced (Fig. 1).

The results are obtained after executing the MATLAB code and prepared a graphical user interface for the noise calculation [4]. For different values of all the parameters there are various values of noise parameters. The different values of bandwidth and power are tabulated with values of Lc, Nc, and OSR (Fig. 2).

3.5 GUI of Power Consumption

For the calculation of power consumption, the different parameters are there on which the overall power consumption depends. The analog supply voltage plays a vital role in it (Figs. 3 and 4).

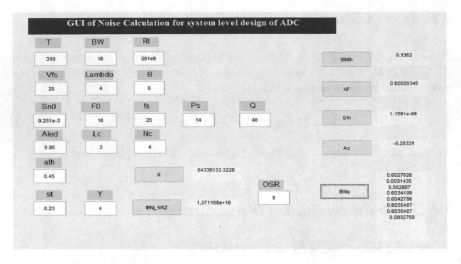

Fig. 1 Noise analysis based on modeling of CT ADC

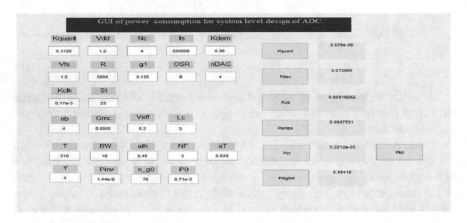

Fig. 2 Power consumption analysis based on modeling of CT ADC

4   Results and Discussion

As we have prepared a graphical user interface for the noise calculation presented in Table 1. For different values of all the parameters we are getting various values of noise parameters. Table 2 shows summary of modulator parameter (Fig. 5).

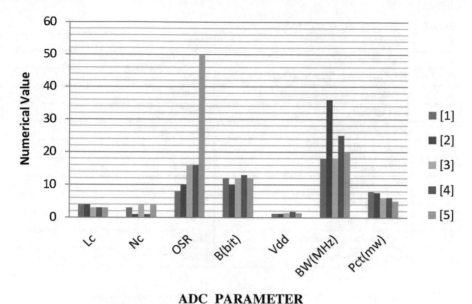

ADC PARAMETER

Fig. 3 CT ADC design parameter and numerical value

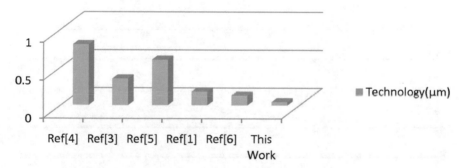

Fig. 4 Comparison of CMOS technology with previous work

Table 1 Noise calculation of continuous-time sigma delta modulator ADC

Refs.	BW (MHz)	B(bit)	Lc	Nc	IBN_th	IBN_NRZ	IBN_q
[1]	18	8	3	4	0.113	0.24	2.7
[2]	20	8	3	4	0.240	0.42	3.1
[3]	25	16	3	5	0.340	0.48	3.2
[4]	36	16	4	1	0.140	0.36	3.4

Table 2 Design summery of CT modulator parameter

No.	Parameter	Value
1.	Transfer function	12.1100 m
2.	Total harmonic distortion (THD)	1.52%
3.	Average power consumption	143 mW
4.	Bias voltage	2.5000e−001
5.	Total output noise	1.0222p sq V/Hz
6.	Power supply	1.2 v
7.	Technology	0.45 nm
8.	Offset voltage	0.9 v
9.	Sampling Frequency	40 MHz
10.	Phase margin	92.84
11.	Gain bandwidth product	31.89 k
12.	Unity gain bandwidth	1.2295 k

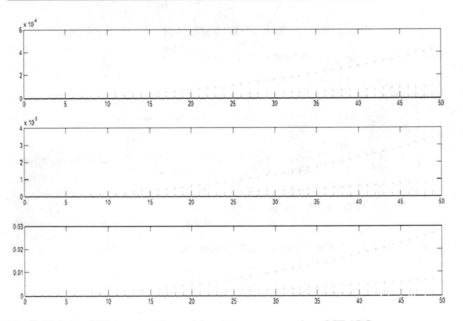

Fig. 5 Variation OSR (oversampling ratio) and power consumption of CT ADC

5 Conclusion

A low power and optimized system-level design of continuous-time sigma delta modulator based ADC is built in 0.45μ CMOS technology using T-SPICE simulator of Mentor Graphic Tanner EDA 16.3 electronic design automation tool. This design provides 1.43 mW power consumption and harmonic distortion on 1.52%. Total noise is 1.022 p sq V/Hz, which is better than previous work. Power consumption and noise calculation has done using MATLAB. In this paper, noise parameters and power consumption in single-stage sigma delta modulator based ADC are analytically extracted. The power consumption of different possible design of CT sigma delta modulator based ADC are compared to choose the best one. GUI allows us to get the power values without actually simulating the circuit design. Further, the transistor level design can be implemented and the author is free to use any tool for the design. We selected different values of OSR, Lc, Nc, and various parameter and got different values of power and in recent results for Lc $= 3$, Nc $= 4$, B $= 12$, we got the power as 5.05 mW.

References

1. Shettigar P, Pavan S (2012) Design techniques for wideband single-bit continuous-time $\Sigma\Delta$-modulators with FIR feedback DACs. IEEE J. Solid-State Circuits 47(12):2865–2879
2. Mitteregger G, Ebner C, Mechnig S, Blon T, Holuigue C, Romani E (2006) A 20-mW 640-MHz CMOS continuous-time $\Sigma\Delta$ ADC with 20-MHz signal bandwidth, 80-dB dynamic range and 12-bit ENOB. IEEE J Solid-State Circuits 41(12):2641–2649
3. Arifuddin SM, Chenna Kesava RK, Sattar SA (2012) Design of low power $\sum\Delta$ ADC. Int J VLSI Des Commun (VLSICS) 3(4):67–79
4. Pakniat H, Yavari M (2015) System level design and optimization of single-loop sigma delta modulators for high resolution wideband applications. Microelectron J (Elsevier) 46(11):1073–1081
5. Solis-Bustos S, Sliva-Martinez J, Sanchez-Sinencio E (2000) A 60-d B dynamic-range CMOS sixth order 2.4 Hz Low pass filter for medical applications. IEEE Trans Circuit Syst II Analog Dig Sig Process 47(12):1391–1398
6. Xu ZT, Zhang XL, Chen JZ, Hu SG, Yu Q, Liu Y (2013) VCO-based continuous-time sigma delta ADC based on a dual-VCO-quantizer-loop structure. J Circuits Syst Comput 22(9) 1340–1349
7. Sahu AK, Chandra VK, Sinha GR (2015) System level behavioral modeling of CORDIC based ORA of built-in-self-test for sigma-delta analog-to-digital converter. Int J Signal Image Process Issues (2)
8. Reddy K (2012) A 16-mW 78-dB SNDR 10-MHz BW CT $\sum\Delta$ ADC using residue-cancelling VCO-based quantizer. IEEE J Solid-State Circuits 47(12):2916–2927
9. Sahu AK, Chandra VK, Sinha GR (2014) Improved SNR and ENOB of sigma-delta modulator for post simulation and high level modelling of built-in-self-test scheme. Int J Comput Appl 11–14
10. Pavan S, Krishnapura N, Pandarinathan R, Sankar P (2008) A power optimized continuous time $\Sigma\Delta$ ADC for audio applications. IEEE Trans Circuits Syst I 43(2):351–360
11. Straayer MZ, Perrott MH (2008) A 12-bit, 10-MHz bandwidth, continuous-time $\sum\Delta$ ADC with a 5-bit, 950MS/s VCO-based quantizer. IEEE J Solid-State Circuits 43(4):805–814
12. Zeller S, Muenker C, Weigel R, Ussmueller T (2014) A 0.039mm2 inverter-based 1.82mW68.6 dB-SNDR 10 MHz-BW CT-$\sum\Delta$-ADC in 65 nm CMOS using power- and area-efficient design techniques. IEEE J Solid-State Circuits 49(7):1–13
13. http://www.eeweb.com/blog/carmen_parisi/balancing-the-tradeoffs-in-new-generation-adcs

A Survey on Coverage Problems in Wireless Sensor Network Based on Monitored Region

Smita Das and Mrinal Kanti Debbarma

Abstract These days, Wireless Sensor Networks (WSNs) have enormous application in both research and commercial fields such as environmental observation to endangered species recovery, habitat monitoring to home automation, waste management to wine production, and medical science to military applications. For attaining network coverage in a WSN, normally a few to thousands of small, limited-power sensor nodes are required to be deployed in an interconnected fashion. While arranging the sensor nodes in a WSN, covering the region to be monitored is a difficult task and became a hot field of research for last few years. Coverage verifies the quality of monitoring some event in a meticulous environment by the sensor nodes. The idea behind this paper is to study thoroughly about the most recent literature of WSN coverage based on monitored region of the network. While doing so, we have mainly studied three different types: coverage area, point, and barrier coverage.

Keywords Coverage · Wireless sensor networks · Sensor nodes
Connectivity · Monitored region · Deployment strategy

1 Introduction

Micro-Electro-Mechanical Systems (MEMS) technology, allied with wireless communication technology, made it trouble-free for WSNs to achieve global attention in recent years. With respect to the conventional sensor nodes, the wireless sensor nodes are very small in size and are economical. Though these sensor nodes have limited amount of power, processing speed, and memory, still they are proficient enough to sense, evaluate, and collect data from an environment. There are enormous applications of WSNs which can be utilized on academics, avalanche rescue, biomedical, disaster management, environmental investigation, habitat monitoring, intruder detection, ocean water monitoring, weapon tracking, and many more. A WSN consists of a few to several thousands of sensor nodes. As the sensors can be

S. Das (✉) · M. K. Debbarma
National Institute of Technology Agartala, Agartala 799046, Tripura, India
e-mail: smitadas.nita@gmail.com

© Springer Nature Singapore Pte Ltd. 2019
M. L. Kolhe et al. (eds.), *Advances in Data and Information Sciences*, Lecture Notes
in Networks and Systems 39, https://doi.org/10.1007/978-981-13-0277-0_29

mounted very close to the monitored event, chances of getting excellent quantity of accurately sensed data become higher. From mid-1990s exploration in the field of WSN started and among all the research topics, coverage has got the highest preference since last few years. Coverage quantifies how well an area is sensed by the nodes they are deployed into and it is broadly associated with the deployment strategy of the sensor nodes. As sensor nodes have very limited energy, deployment of the nodes should be made in such a way that maximum coverage should be attained with the minimum energy uses.

In this paper, we have followed the survey [1] to get a rough structure of the proposed work, but our paper has surveyed the most recent literature of present time which may be helpful for the research community. The residue of this paper is structured as follows: Sect. 2 is all about the coverage classification. Section 3 discusses special approaches on monitored region in WSNs. Section 4 has a brief summary of all three types of coverage based on monitored region. Finally, we conclude the paper by identifying the research challenges in network coverage.

2 Coverage and its Classification

Coverage determines how well each point inside a WSN is sensed by a sensor node. To sense each point, number of nodes should be higher in order to achieve optimal accuracy. But again higher number of node advocates higher expenditure. Therefore, optimal coverage with minimum number of nodes is a challenge when the sensed region is unknown as well as adverse. Categorization of coverage can be done on the basis of node sensing models, target attributes, optimization schemes, monitored region, deployment domain, and node deployment strategy.

In this paper, we are studying the coverage issues based on the monitored region. In Fig. 1, classification of coverage problem is shown. Nodes in a WSN are supposed

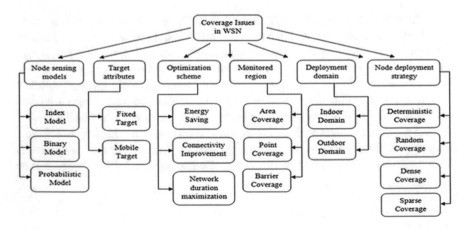

Fig. 1 Categorization of coverage

to be exposed to inconsiderate conditions. Even WSN faces great challenge with respect to network lifetime, nodes are fully dependent on battery power. Under such circumstance, deployment of nodes in a predetermined way might not be successful. To get effective coverage of the sensing field, generally, a large number of sensor nodes are deployed. If one sensor is unable to sense data or is inactive due to power failure, rest of the sensors can share its burden, thus providing better coverage.

3   Coverage Based on Monitored Region

Let us assume that, in a certain WSN, there is a set of sensors, $S = [s_1, s_2, \ldots s_n]$ in a 2D area X. For any specific sensor s_i, where i may vary from 1 to n, that is placed at the coordinate (x_i, y_i) inside X, let us again assume that sensor s_i has a sensing range of r_i which is termed as the sensing radius. Also, $X = x_i = \pi * r_i^2$, where i may vary from 1 to n. Any particular point inside X is said to be covered by sensor s_i, if that point is within r_i. In other words, we can say that coverage, C, may be denoted as a mapping from set S, which is a set of sensors, toward the set of area X, i.e., $C : S \rightarrow X$ such that $\forall s_i \in S, \forall (x_i, y_i) \in X$.

In order to reduce the cost of deployment as well as getting maximum coverage, it is very essential to decrease the number of nodes which is highly challenging, particularly when the sensing area is unfavorable. Therefore, in this section, a study is done on coverage issues based on the region that has to be monitored. On the basis of monitored region, coverage can be divided in the following way:

3.1   Area Coverage

The idea behind area coverage is to monitor a target region in such a way that each and every point inside that region is monitored. In 2003, area coverage problem was first conferred in [2] by Kar and Banarjee. Here, placement of the sensor nodes in a construct called an r-strip was proposed. Here, r is the sensing radius and each sensor is placed r distance away from the neighboring node. To have area coverage, strips were then placed in an overlapping fashion. In 2005, Zhang and Hou [3] have proposed the optimal geographical density control (OGDC) protocol which minimizes the overlapping of the sensing areas for $R_C \geq 2R_S$, where R_C is the communication range and R_S is the sensing range of the sensor. Gu et al. [4] in 2007 referred a unified sensing coverage design, which can support various protocols and also a two-level coverage protocol, uScan, under their coverage architecture. In 2009, Xing et al. [5] exhibited coverage problem under the data fusion model and calculated its performance limits to prove that the data fusion model can dramatically improve the sensing coverage compared with the traditional disk-based models. In 2012, Li et al. [6] proposed load balancing K-area coverage through an approach called Autonomous Deployment (LAACAD), for balancing the sensing workload to increase the network lifetime and provide k-coverage by controlling the convex optimization techniques.

In the most recent literature, Dash et al. [7] in 2014 proposed two novel coverage metrics—smallest k-covered segment and longest k-uncovered segment for measuring the area coverage quality while deploying sensors. In 2015, Vecchio and López-Valcarce in [8] proposed a distributed path planning and coordination technique to increase the sensing area coverage of deployed sensor nodes in a WSN. A set of static, immovable sensors supported by mobile nodes are used to design a framework for controlling their trajectories to improve coverage. In 2016, Gupta et al. in [9] proposed a meta-heuristic approach on the basis of Genetic Algorithm (GA) to identify the position of the sensor nodes to be placed so that all target points are k-covered and nodes are m-connected. They have presented a proficient design for chromosome representation and derived a fitness function with three objectives as use of minimum number of sensor nodes, coverage, and connectivity.

3.2 Point Coverage

The purpose of point coverage is to wrap up a set of point or targets, with known locations in a given sensing field, which needs to be monitored. In case of point coverage, it is very important to determining sensor nodes exact position and it ensures efficient coverage for a restricted number of static targets. In 2005, Cardei et al. [10] suggested one technique of point coverage to improve the sensor network lifetime. Each set cover is competent in providing coverage of all target points in the sensor field while each set cover is activated in different rounds. In 2008, Cheng et al. [11], first mathematically defined that sweep coverage in WSN is an NP-hard problem. They have suggested that the min-sensor sweep coverage problem cannot be solved other than the 2-approximation, including the case of 2. In 2010, Jianxin et al. [12] worked on target coverage problem with multiple sensing ranges for maximizing network lifetime for target coverage. They have introduced a global information-based Energy-efficient Centralized Greedy and Heuristic algorithm (ECGH) with a concept of 'remote target.' Medagliani et al. [13] in 2012 proposed an energy-efficient target detection application in WSNs with random node deployment and partial coverage based on duty cycling of the nodes. They have calculated the probability of detection of the missed target and the delay in target detection.

In the most recent literature, Gorain and Mandal [14] in the year 2014 proposed a 2-approximation algorithm for point coverage problem, where each mobile sensor node traverses every point of interest. The approximation factor of the proposed algorithm is 3-factor, if there are obstacles in the plane. Islam et al. [15] in 2015 proposed a novel cluster head based approach to provide maximum point coverage using minimum number of sensor nodes. TCDC (target coverage based on distributed clustering) minimizes the sensing redundancy and maximizes the number of sleeping node in the WSN and thus reducing power consumption. In 2016, Gorain and Mandal [16] again proposed an extension of their previous 2-approx algorithm for point coverage problem and developed a special coverage problem called energy-efficient sweep coverage or E-Sweep coverage problem. The problem is considered as NP-hard. They have also recommended an 8-approximation algorithm to solve this problem. Additionally, they have introduced another variation of sweep coverage problem named Energy Restricted sweep coverage problem or ER Sweep coverage problem.

3.3 Barrier Coverage

Barrier coverage refers to the detection of mobility across a boundary of sensors. Barrier coverage is specially used in intruder detection and whenever they cross a certain boundary, sensors detect the movement and send alert signal to the base station. In 2005, Kumar et al. [17] verified that a region is k-barrier covered on the basis of their proposed algorithm. They have shown the achievement of k-barrier coverage with the optimal deployment pattern and deterministic deployment of sensors. Balister et al. [18] in 2007 have suggested new technique to locate reliable density in a specified region of fixed size where sensor nodes are deployed as boundaries to identify the change in position of target. In 2009, Saipulla et al. [19] discussed that the linear deployment of nodes would represent a more practical sensor placement in different applications. Also, the authors proved that sensor deployment strategies are directly connected with the impact on the barrier coverage of a WSN.

In the most recent literature, Mostafaei and Meybodi [20] in 2014 proposed an algorithm based on learning automaton to assure strong barrier coverage of wireless sensor networks. With this algorithm, much network lifetime can be obtained and they have used a coverage graph of deployed sensor network as a base model to find near-optimal solution for barrier coverage problem. DeWitt and Shi [21] in 2016 proposed a unique solution to barrier coverage problem using energy harvesting algorithm and claimed optimal solutions in maintaining perpetual barrier coverage in energy harvesting sensor networks. They have also calculated the upper bounds of the metric and expressed that those bounds could be achieved in polynomial time with their algorithms. In 2017, Wang et al. [22] proposed an approach to analyze the effects of location errors on the minimum number of mobile nodes required to fill the gap in network when both stationary and mobile nodes are present. They have modeled the barrier coverage formation problem using a fault-tolerant weighted barrier graph. In the same year, Silvestri and Goss [23] had proposed an autonomous algorithm named MobiBar for k-barrier coverage formation with mobile sensors. In a finite time, all sensors stop moving and the final deployment provides the maximum level of k-barrier coverage attainable with the available sensors.

4 Summarization

Based upon the above discussions on area coverage, the summarization points are tabulated in Table 1. While we studied the various types of area coverage since the inception of the concept, we have cultured that most of the algorithms are distributed in nature. We have found [2, 5] centralized algorithms. Therefore, distributed algorithms are very much essential for WSN area coverage determination. In maximum cases, location information about the sensor is a prerequisite to resolve the coverage issues. In most of the papers, we have studied nature of deployment that is either random or deterministic/predefined. Dense deployment of sensor may raise the expenditure and sparse deployment is application specific. When we are considering area coverage, our target is to cover each and every point in the sensing field. Keeping this idea in mind if we deploy maximum number of sensors, deployment and maintenance cost will become very high. Therefore, the main research challenge for area coverage is to get maximum coverage with minimum number of node deployment. In the optimization scheme for coverage detection, energy saving and connectivity improvement play key role. While selecting the simulator, maximum authors either rely on NS2 due to open-source clause or Matlab due to its wide range of acceptability. Based upon the previous discussions on point coverage, the summarization points are charted in Table 2. On the basis of the types of algorithms, maximum authors have chosen distributed type of algorithm due to their effectiveness in WSN. Authors have selected different types of approaches such as heuristics/approximation based, cluster head based, or Point of Interest (POI) based, and the deployment strategy in almost all the cases is random. Also, in majority of the papers, network duration maximization is the optimization strategy to be obtained. In case of point coverage, POIs decide which points in the sensing field are required to be covered, and thus POI-based approaches are so much important. But, in the case of real-time systems, it is very difficult to guess the POIs prior to event happening. For that reason, it is assumed that POIs can occur at any place in the sensing field. Therefore, the main research challenge for point coverage is to identify the time delay to detect the POIs with respect to certain number of nodes present in the network and try to figure out how to optimize this delay in a moderate level. Finally, the main points on barrier coverage discussion have been tabulated in Table 3. With node deployment as random and connectivity improvement as prime optimization strategy, here the type of algorithm is mainly distributed in nature. As barrier coverage is principally related to the security applications in a WSN, hence, node location information is so much important. For obtaining prior node location, sophisticated hardware or GPS might be required which in turn increases the cost of deployment. Thus, the main research challenge for barrier coverage is to use computational geometry-based models in such a way that location errors of the nodes can be identified. Once the location errors can be identified properly, it will be easier to deploy the sensor nodes to improve the barrier coverage.

Table 1 Summary of area coverage

Year/Ref.	Optimization scheme	Deployment strategy	Features	Pros/Cons	Simulator
2003/[2]	Energy saving	Predefined	Sensor node minimization	Simple/Not feasible	–
2005/[3]	Energy saving, network duration maximization	Predefined	OGDC protocol, overlapping area minimization	Density control/Undetected upper bound of network lifetime	NS-2
2007/[4]	Energy saving, connectivity improvement	Random	Two-level coverage, generic switching global scheduling	Optimal coverage in linear time/Time complexity	TinyOS, MicaZ motes
2009/[5]	Connectivity improvement	Predefined	Coverage performance of large-scale WSN, stochastic fusion algorithm	Relation of coverage and network density/Application dependent	Simulation based on real and synthetic data traces
2012/[6]	Connectivity improvement	Random, sparse	LAACAD algorithm, k-area coverage	Sensing range minimization/High computational complexity	NS-2
2014/[7]	Connectivity improvement	Predefined	Smallest k-covered segment line for intruder	Longest-uncovered segment for defender/-	Matlab
2015/[8]	Network duration maximization	Random	Static infrastructure, trajectory control	Increased sensing area/ Undefined performance	Matlab
2016/[9]	Connectivity improvement, network duration maximization	Predefined	Position of sensor node, K-coverage, and m-connected	Efficient fitness function/Node failure	Matlab

Table 2 Summary of point coverage

Year/Ref.	Optimization scheme	Deployment strategy	Features	Pros/Cons	Simulator
2005/[10]	Network duration maximization	Random	Maximum set cover identification, node scheduling mechanism	Sleep mode for redundant nodes/Single coverage	Matlab
2008/[11]	Connectivity improvement	Random	CSWEEP centralized DSWEEP distributed	Increased coverage efficiency/Communication overhead	3D robot simulator Simbad
2010/[12]	Network duration maximization	Random	Energy-efficient centralized greedy heuristic algorithm (ECGH), energy-efficient distributed localized Greedy heuristic algorithm (EDLGH)	Eliminating redundant sensor nodes/Single coverage	C++
2012/[13]	Energy saving	Random	Probability of detection of missed target, delay in detection of an incoming target	Energy-efficient target detection/Partial coverage	Matlab, C
2014/[14]	Network duration maximization	Random	2-approximation algorithm for static POI and 3-approximation algorithm for mobile POI	Mobile nodes visit each POI/High communication overhead	C++
2015/[15]	Energy saving	Random	Decreases communication and computation overhead	Minimizes sensing redundancy/Static weight factors	NS-3
2016/[16]	Network duration maximization	Random	Change in approximation factor as 8	Energy-efficient coverage/High computational complexity	–

Table 3 Summary of barrier coverage

Year/Ref.	Optimization scheme	Deployment strategy	Features	Pros/Cons	Simulator
2005/[17]	Connectivity improvement	Predefined	Weak k-barrier coverage	Optimal deployment pattern/Local algorithm	Multi-agent simulations
2007/[18]	Connectivity improvement	Random	High connectivity	Strong barrier coverage/1-coverage	–
2009/[19]	Connectivity improvement	Predefined, random	Line-based normal random offset distribution	Strong barrier coverage/depends on variance of LNRO	Crossbow
2014/[20]	Network duration maximization	Random	Identify best node irrespective of no. of nodes	Strong barrier coverage/High computational complexity	WSN simulator
2016/[21]	Network duration maximization	Random	Energy harvesting	Perpetual barrier coverage/Works on small no. of nodes	Python, graph tool library
2017/[22]	Connectivity improvement	Random	Fault-tolerant weighted barrier graph	Removes unnecessary edges/Curve based barrier coverage	Matlab
2017/[23]	Connectivity improvement	Random	Self-reconfigurable	Dynamic/–	Opnet

5 Conclusion

Coverage of sensors with respect to monitored area is an important field for research activities. Coverage is further associated with different optimization schemes like energy efficiency, network duration maximization, and connectivity improvement. We have studied various papers on area, point, and barrier coverage to review the existing work, summarize their important qualities, and formulate some research challenges from the recent literature. In case of area coverage, getting maximum coverage with minimum number of nodes was the prime challenge and further works may be done in this aspect. In point coverage, additional works may be carried out in identifying the POIs with respect to nodes in the network. Finally, recognizing the location errors and developing a fault-tolerant system in case of barrier coverage are the prime focus. The critical points in all these types of coverage are computational geometry-based approach, random deployment, distributed algorithm, and prior node location information.

References

1. Zhu C, Zheng C, Shu L, Han G (2012) A survey on coverage and connectivity issues in wireless sensor networks. J Netw Comput Appl 35(2):619–632
2. Kar K, Banerjee S (2003) Node placement for connected coverage in sensor networks. In: WiOpt '03: modeling and optimization in mobile, ad hoc and wireless networks, 2 pp
3. Zhang H, Hou JC (2005) Maintaining sensing coverage and connectivity in large sensor networks. Ad Hoc Sensor Wireless Netw 1(1–2):89–124
4. Gu Y, Hwang J, He T, Du DHC (2007) Usense: a unified asymmetric sensing coverage architecture for wireless sensor networks. In: 27th international conference on distributed computing systems, 2007. ICDCS '07, IEEE, pp 8–8
5. Xing G, Tan R, Liu B, Wang J, Jia X, Yi CW (2009) Data fusion improves the coverage of wireless sensor networks. In: Proceedings of the 15th annual international conference on mobile computing and networking, ACM, pp 157–168
6. Li F, Luo J, Xin SQ, Wang WP, He Y (2012) Laacad: load balancing k-area coverage through autonomous deployment in wireless sensor networks. In: 2012 IEEE 32nd international conference on distributed computing systems (ICDCS), IEEE, pp 566–575
7. Dash D, Gupta A, Bishnu A, Nandy SC (2014) Line coverage measures in wireless sensor networks. J Parallel Distrib Comput 74(7):2596–2614
8. Vecchio M, López-Valcarce R (2015) Improving area coverage of wireless sensor networks via controllable mobile nodes: a greedy approach. J Netw Comput Appl 48:1–13
9. Gupta SK, Kuila P, Jana PK (2016) Genetic algorithm approach for k-coverage and m-connected node placement in target based wireless sensor networks. Comput Electr Eng 56:544–556
10. Cardei M, Thai MT, Li Y, Wu W (2005) Energy-efficient target coverage in wireless sensor networks. In: 24th annual joint conference of the IEEE computer and communications societies, INFOCOM 2005. Proceedings IEEE, vol 3. IEEE, pp 1976–1984
11. Cheng W, Li M, Liu K, Liu Y, Li X, Liao X (2008) Sweep coverage with mobile sensors. In: IEEE international symposium on parallel and distributed processing, 2008, IPDPS 2008, IEEE, pp 1–9
12. Wang J, Liu M, Lu M, Zhang X (2010) Target coverage algorithms with multiple sensing ranges in wireless sensor networks. In: Military communications conference, 2010-MILCOM, IEEE, pp 130–135
13. Medagliani P, Leguay J, Ferrari G, Gay V, Lopez-Ramos M (2012) Energy-efficient mobile target detection in wireless sensor networks with random node deployment and partial coverage. Pervasive Mob Comput 8(3):429–447
14. Gorain B, Mandal PS (2014) Approximation algorithms for sweep coverage in wireless sensor networks. J Parallel Distrib Comput 74(8):2699–2707
15. Islam MM, Ahasanuzzaman M, Razzaque MA, Hassan MM, Alelaiwi A, Xiang Y (2015) Target coverage through distributed clustering in directional sensor networks. EURASIP J Wireless Commun Netw 2015(1):167
16. Gorain B, Mandal PS (2014) Line sweep coverage in wireless sensor networks. In: 2014 sixth international conference on communication systems and networks (COMSNETS), IEEE, pp 1–6
17. Kumar S, Lai TH, Arora A (2005) Barrier coverage with wireless sensors. In: Proceedings of the 11th annual international conference on mobile computing and networking, ACM, pp 284–298

18. Balister P, Bollobas B, Sarkar A, Kumar S (2007) Reliable density estimates for coverage and connectivity in thin strips of finite length. In: Proceedings of the 13th annual ACM international conference on mobile computing and networking, ACM, pp 75–86
19. Saipulla A, Westphal C, Liu B, Wang J (2009) Barrier coverage of line-based deployed wireless sensor networks. In: INFOCOM 2009, IEEE, pp 127–135
20. Mostafaei H, Meybodi MR (2014) An energy efficient barrier coverage algorithm for wireless sensor networks. Wireless Pers Commun 77(3):2099–2115
21. DeWitt J, Shi H (2017) Barrier coverage in energy harvesting sensor networks. Ad Hoc Netw 56:72–83
22. Wang Z, Chen H, Cao Q, Qi H, Wang Z, Wang Q (2017) Achieving location error tolerant barrier coverage for wireless sensor networks. Comput Netw 112:314–328
23. Silvestri S, Goss K (2017) Mobibar: an autonomous deployment algorithm for barrier coverage with mobile sensors. Ad Hoc Netw 54:111–129

Flood-Prediction Techniques Based on Geographical Information System Using Wireless Sensor Networks

Naveed Ahmed, Atta-ur-Rahman, Sujata Dash and Maqsood Mahmud

Abstract This paper presents a comprehensive study of flood forecasting, analysis, and prediction using geographical information system (GIS) and wireless ad hoc sensor networks. The role of science and technology has moved the research towards new horizon, where scientists and engineers from all over the world are using GIS-based techniques for flood prediction and hydrological risk analysis. The radar satellite images are also most frequently used for identifying the flood catchment areas in specific disaster zone. The GIS domain also proves to be very helpful for us in geographical survey and to identify the tsunamis causing vast potential and economical damage. The input parameters for flood forecasting are also used for modeling GIS, depending on the environmental conditions and climatic parameters such as soil moisture, air pressure, direction of wind, humidity, and rain fall. The core objective of this research is to study various GIS based flood forecasting techniques. In this research study we have proposed a GIS-based flood forecasting model using neural network-based approach. Our proposed model is helpful for the researchers in predicting the upcoming disasters and to take necessary actions by the rescue authorities to save the life of thousands of people to be suffered from this critical circumstance.

N. Ahmed
Faculty of Engineering and Computer Science, National University of Modern Languages (NUML), Islamabad, Pakistan
e-mail: project_naveed@yahoo.com

Atta-ur-Rahman
Department of Computer Science, College of Computer Science and Information Technology, Imam Abdulrahman Bin Faisal University (IAU), Dammam, KSA, Saudi Arabia
e-mail: aaurrahman@iau.edu.sa

S. Dash (✉)
Department of Computer Science & Application, North Orissa University, Odisha, India
e-mail: sujata238dash@gmail.com

M. Mahmud
Department of Management Information System, College of Business Administration, Imam Abdulrahman Bin Faisal University (IAU), Dammam, KSA, Saudi Arabia
e-mail: mmahmud@iau.edu.sa

© Springer Nature Singapore Pte Ltd. 2019
M. L. Kolhe et al. (eds.), *Advances in Data and Information Sciences*, Lecture Notes in Networks and Systems 39, https://doi.org/10.1007/978-981-13-0277-0_30

Keywords GIS · Flood forecasting techniques · Wireless sensor networks
Particle swarm optimization · Artificial neural fuzzy inference systems

1 Introduction

Flood prediction and river flow modeling is one of the important problems that have attracted the large number of scientists from all over the world. Nowadays availability of accurate flood forecasting technique and methods helps in the reduction of Floods and droughts. Ad hoc wireless Sensor Networks are significantly used in Real-Life applications especially the major application areas includes telemedicines, wireless body area networks vehicle ad hoc networks, underwater wireless sensor networks and Disaster Management. The domain of the Disaster Management includes river floods, hurricanes, fires, and earth quakes which provides a great danger and risk factor for the existing human population. Wireless Sensor Networks can be classified as low-power multi-hopping system capable to transfer the information from one node to another via using Ad hoc Relay Stations. Different researchers and scientists had proposed the diversified architecture for Disaster survivor detection in critical circumstances using Ad hoc wireless sensor Network Architecture. Our major objective in this research is to design a comprehensive architecture for flood risk assessment using GIS-based approach along with its integration with ad hoc wireless sensor network. The core issue for this research mainly depends upon classification of determining flood level duration and its intensity before this extreme event occurs. The data sets collected from the sensors nodes are used for flood forecasting using neural network based approach [1]. The complex problem of flood forecasting can easily be solved efficiently using GIS, WSN, and Artificial Neural networks based technique. The research work also focuses on the design of hydrographs that indicates the location of possible discharges.

Remainder of paper is structured as follow. Section 2 discusses Flood Prediction techniques, Sect. 3 presents Proposed Model for Flood forecasting using ANNFIS, Sect. 4 presents proposed methodology based on input parameters for flood forecasting and finally conclusion and future research direction are presented in Sect. 5.

2 Flood Prediction Techniques

This section provides the review of existing flood forecasting techniques in the literature.

2.1 Existing Flood Forecasting Techniques and Models

Rao et al. [2] had developed the flood forecasting Model for Godavari Basin. The Author had also proposed the distributed Modeling approach for topographic and metrological parameters which are used to calculate the extreme Runoff process. The kinematic wave mathematical model is designed to achieve runoff model which helps in predicting the terrains. The soil conservation service (SCS) Unit hydrograph technique has been adopted to derive graphs from gauge rainfall data and runoff process from large number of water sheds. According to Biondi and De Luca [3] stochastic and distributed modeling technique has been used for numerical weather prediction in case of river discharge. The mechanism for using ensemble flood forecasting in case of heavy rainfall arises, particularly to predict flood warnings earlier before the critical condition occurs. The solution for flood forecasting is determined in simplest form, based on probabilistic flood prediction which is very useful factor in obtaining the estimation of flood Risk [4]. Fiorentio et al. [5] had performed the detailed analysis of flood frequency distribution and compared the simulation results with distributed hydrological modeling (DREAM) along with rainfall generator scheme (IRP). The Methodological scheme has been adopted which consists of occurrence of dry and rainy intervals on the basis of exponential (wet) and Weibull (dry) distribution. According to Rozalis et al. [6] Flood events are generated by extreme rainfall events with relatively high rain fall intensity due to thunder storms. One of the main objectives of the current research is to use relatively simple and flexible model that can be applied over gauged and ungauged water sheds. The authors has also proposed the kinematic wave method for determining flood flow routing within specified zone. The model input parameters for rainfall is obtained directly from Radar satellite images, along with the rain fall gauges. The model performance was calculated by comparing different hydrographs over the study area of 20 selected locations for identifying peak Flow discharge, runoff depth rain depth along with maximum rain rate. Ren et al. [7] had presented a new classified real time flood forecasting frame work by coupling fuzzy clustering and neural networks with hydrological Modeling. The fuzzy clustering Model is used to classify the historical floods from the available flood records in different categories which is used to calculate flood peak and runoff depth. The conceptual hydrological model used for generating optimal set of parameters for flood prediction using genetic algorithm. The artificial neural network model is trained to predict the real-time flood events using the real-time rainfall data. According to Nie et al. [8] rapid climatic change is an important factor for the occurrence of flash flood events. The variation in the comprehensive flood index in certain regions is due to functional decline of forests during the last few years. The changes in the flood disaster are analyzed by using SPSS and Arc GIS simulation tool at temporal and spatial scale. The FFT (fast Fourier Transform) is used to identify the Flood Trends in different regions of China from 1980 to 2009. Along with the Advancement in technology Alferi et al. [9] had focused his research towards numerical weather forecasting based early Flood warning system which is based on Ensemble Flood Predictions. The use of metrological data sets such as (COSMO) for the 30 years

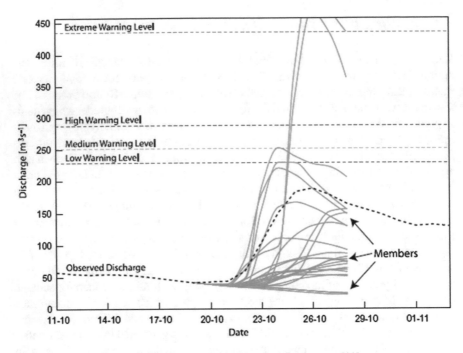

Fig. 1 Example of an ensemble hydrographs for historical flood events [11]

entire period is used to drive discharge climatology and also results in early warning threshold. According to Alferi et al. [9] the gamma probability distribution is the best method for the quantitative analysis of water flow, which is an optimal method in Flood warning system. The Future Analysis in Flood Prediction for an event based approach using time window of appropriate duration improves the threshold analysis, also in terms of False alarm rates and hit rates. Ahmad et al. [10] had proposed the integrated Model of Flood prediction using GIS-based wireless sensor networks for calculating the impact of Flood damage during the monsoon regions especially in the areas of Sind Pakistan. The rain fall data during the last few years had been collected from Pakistan Metrological Department (web source). Relevant data and information had been used to predict warning relevant to Floods, Thunder storms and rainfall using ARC GIS Simulation tool. Ahmad et al. [10] had also proposed the mathematical Model for Flood prediction which also provides the impact of Flood disaster in the selected region. The hydrograph in Fig. 1 provides in deep detail about the observed discharge peak level within different warning levels in the Flood-affected areas. The Operational methods of Research focuses on Numerical weather prediction also known as ensemble Prediction Systems. One of the major short falls of using EPS for Flood forecasting system designed for hydrological application is limited to low frequency of Floods in disaster effected region.

According to Neal et al. [12] space-time sampling strategy is used in Flood forecasting using wireless sensor nodes. The proposed Flood forecasting Model forecasts

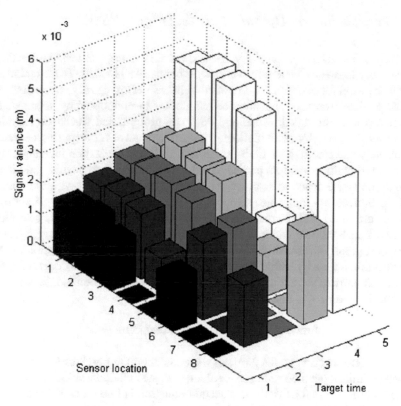

Fig. 2 Example of deployment of Sensor location w.r.t. sensor location within a specified time [12]

using Web-based GIS. Flood forecasting Model takes the real-time data from Gateway node deployed in the region of Flood effected area. Gateway node is connected to several sensor nodes working in peer to peer architecture. Neal et al. [12] had also developed a technique named as Ensemble Transform Kalman Filter which is used to estimate the potential value of uncertainties in upcoming Flood forecast.

The flood forecasting Model takes the real-time data from Gateway node deployed in the region of Flood-effected area. Gateway node is connected to several sensor nodes working in peer to peer architecture. Neal et al. [12] had also developed a technique named as Ensemble Transform Kalman Filter which is used to estimate the potential value of uncertainties in upcoming Flood forecast. Figure 2 provides the three-dimensional view related to the deployment of sensor nodes with Q possible measurements at eight different locations with respect to the signal variance intensity measured in meters over the specified coverage range. The adaptive sampling results are based on five different targeted time events which strongly represents the working of sensors within specified time interval along with its signal variance intensity at discrete events. Hugs et al. [13] had designed an embedded computing platform named as grid stix-based WSN.

2.2   Particle Swarm Optimization Technique (PSO)

The most latest technique used by researchers now a days is PSO and artificial neural fuzzy inference system for various problems like [14–19]. These techniques have being used for extreme flood prediction in case of emergency circumstances.

The particle swarm optimization technique has been adopted by scientists and researchers in hydrological Modeling. PSO is a group-based stochastic technique falls under the category of Evolutionary algorithms and soft computing developed by Kennedy and Eberhart in 1995. In particle swarm optimization technique there is a group of multiple random particles located within specified position and moves through the entire space to search for potential solution. Along with the solution identification these particles also learns within a group using neural network approach [20]. According to [21] the emphasis of PSO algorithm focuses on the best fitness position of each particle within the entire space named as personal best (pbest) and global best (gbest) collectively depending upon the movement acceleration towards the next particle velocity within the hyper space. There are two major equations which are used in particle swarm optimization technique. The first equation is known as movement equation which is described as follows:

$$Presentlocation = Previouslocation + V_i \Delta t \tag{1}$$

The present location of the particles depends upon the previous location of the particles within a specified vector space along with the individual velocity of particles in the specific interval of time. The second equation is known as velocity update equation which can be described as follows:

$$V_i = wV_{i-1} + C_1 * rand() * (Pbest - preslocation)$$
$$+ C_2 * rand() * (gbest - preslocation) \tag{2}$$

The velocity update equation describes the change in velocity for the particles within the entire movement space known as gbest and pbest. In the Eq. (2) V_i is the initial velocity of the particles, Δt is the time interval for the movement of particles within a hyper space, V_{i-1} is the previous velocity, random() is the random number value for example (0, 1, 2, 3…) and C_1 and C_2 are the acceleration of the particles within the entire space. The strengths of the research work are based on application of PSO algorithms specially in artificial neural networks for enhancing the learning process. Along with this the authors had also proposed a unique technique for calibrating the daily rainfall runoff model named as PSONN (Particle swarm optimization neural networks). The input parameters for the PSONN model are temperature, moisture content, and evaporation.

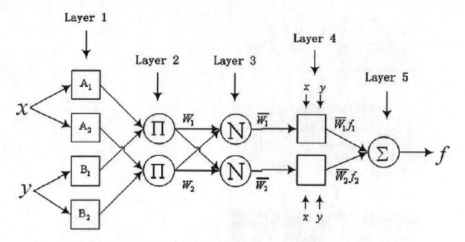

Fig. 3 Graphical representation of five layer feed forward ANFIS system [20]

2.3 Artificial Neural Based Fuzzy Inference System (ANFIS)

ANFIS technique is a multilayer feed forward back propagation network which is capable to forward the weighted connections from input to output layers [22]. The ANN Model identifies a set of parameters which serve as a basis of IF–then fuzzy rules based on appropriate member functions. The Sugeno inference system has been adopted which provides efficient mathematical modeling and optimization technique. The first-order sugeno fuzzy model can be mathematically expressed as follows:

$$Rule1 : If\ X\ is A_1\ and\ Y\ is B_1,\ then f_1 = p_1 x + q_1 y + r_1 \qquad (3)$$

$$Rule2 : If\ X\ is A_2\ and\ Y\ is B_2,\ then f_2 = p2x + q_2 y + r_2 \qquad (4)$$

The above-mentioned rule-based equations depends on the output function f corresponds to the input vector value x and y. The values p, q, and r represent the constant quantities.

Figure 3 demonstrates the graphical representation of ANFIS system which consists of input layer mentioned as Layer 1, hidden layer mentioned as Layer 2, 3, and 4 and finally the output layer mentioned as Layer 5. The most common Neural Network Model is classified as MLP (Multi-layer perception) neural network. The major goal of this supervised network is to map the input into output.

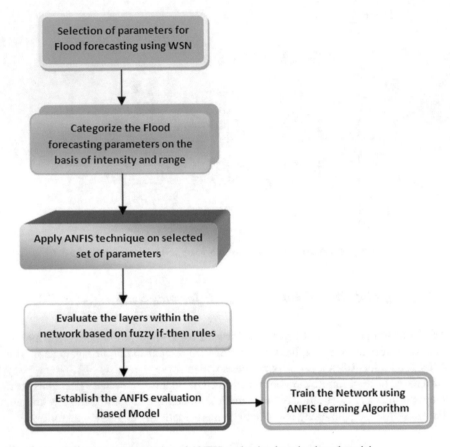

Fig. 4 Block diagram representation of ANFIS evaluation learning based model

3 Proposed Mathematical Model for Flood Forecasting

The following section describes the proposed model for Flood forecasting based on artificial neural fuzzy inference system.

3.1 Adaptive Neuro-Fuzzy Inference System Algorithm

The selection of parameters for flood forecasting using the wireless sensor networks plays a vital role in training the neural network using ANFIS learning algorithm (Fig. 4).

The selection of parameters for flood forecasting is very important factor in flood forecasting and disaster risk analysis. The key parameters for flood forecasting are humidity, rainfall, temperature, pressure, and wind speed. The parameters are selected using wireless sensor network based architecture. The next step is to categorize the flood forecasting parameters based on flood intensity and range. The heavy intensity rainfall in millimeters for the long period of time, for example, one week creates a potential impact of flood in rivers, streams and water basins especially in rural areas. The next step is to apply ANFIS technique on the selected set of parameters. The ANFIS system in this research scenario emphasizes on rainfall parameter which results in increased water level intensity within the region of interest. The next step is to apply fuzzy if-then rules based on resulting mathematical equations as mentioned in Eqs. 1 and 2. In this step, we have also design the mathematical representation of the function f1 and f2 based on the fuzzy sets. Finally the equation for the overall output is represented which is the product of the resulting function and the overall output of the system. It is also possible to train the neural network learning algorithm based on adaptive neuro-fuzzy inference system.

3.1.1 Mathematical Representation of Adaptive Neuro-Fuzzy Inference System Algorithm

The following equations describe in deep detail about the mathematical representation of artificial neuro-fuzzy inference system [23].

$R_f = rainfall$

$W_l = waterlevel$

If R_f is A_1 and W_L is B_1, then

$W_l = waterlevel$

If R_f is A_1 and W_L is B_1, then

$$f_1 = p(R_F) + q(W_L) + r_1 \tag{5}$$

If R_f is A_2 and W_L is B_2, then

$$f_2 = p(R_F) + q(W_L) + r_2 \tag{6}$$

Where A_1, A_2 and B_1, B_2 are fuzzy sets

Layer 1: $O_i^1 = \mu_{A_i}(R_F)$

Layer 2: $W_i = \mu_{A_i}(R_F) * \mu_{B_i}(W_L)$ $i = 1, 2, \ldots$

Layer 3: $\overline{W}_i = \frac{w_1}{w_1 + w_2} i = 1, 2, \ldots$

Layer 4: $O_i^4 = \overline{W}_i f_i = \overline{W}_i (P_i(R_F) + Q_i(W_L) + r_i)$

Layer 5: Overall Output $= \sum_i \overline{W}_i f_i$

The output functions for the equations depend on rainfall and water fall parameters multiplied by the constant values. The Layer 1 focusses on the output function, layer 2 and layer 3 describes about the weighted functions and finally layer 4 describes about the output function for the product of weight and input function. The input parameters for the proposed mathematical model are rain fall intensity will be measured in millimeters and water level will be calculated in cubic feet per meters.

3.2 Calculating Flood Disaster Risk Ratio by Using Probability Density Function

The probability of flood disaster risk ratio is calculated by using Bayesian decision theory which is based on the following set of parameters.

$T_H \rightarrow Temperature\,(High)$

$T_L \rightarrow Temperature\,(Low)$

$T_C = T_H - T_L\,(Temperature\,change)$

$R \rightarrow Rainfall\,Intensity\,(mm)$

$H \rightarrow Humidity\,(\%)$

$e \rightarrow prediction\,error$

$\Phi \rightarrow prediction\,cons\tan t$

According to Bayesian decision theory probability density model can be applied to calculate the humidity and rainfall. The equations can be derived on the basis of generic formula derived from [17] mentioned as follows:

$$P(T_C|R) = \frac{P(R|T_C)P(T_C)}{P(R)} \tag{9}$$

$$P(T_C|H) = \frac{P(H|T_C)P(T_C)}{P(H)} \tag{10}$$

The linear regression function based on probability density model can be described as follows:

$$g(\varphi^t|R, H) = R\phi^t + H \tag{11}$$

Finally the regression function r can be defined as follows:

$$r = g(x) + \varepsilon, \tag{12}$$

where epsilon is the prediction error.

4 Proposed Methodology for Flood Prediction

The proposed methodology is designed on the basis of following set of parameters. The selected parameters for flood forecasting are rainfall intensity which is measured

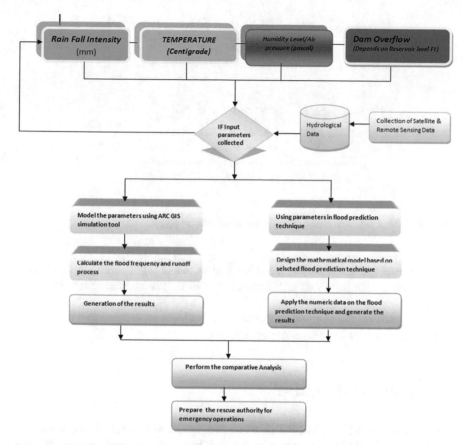

Fig. 5 Proposed methodology based on input parameters for flood prediction

in millimeters, temperature which is calculated in centigrade, humidity level along with air pressure is calculated in Pascals. Finally the dam overflow factor which is dependent on rainfall intensity is calculated in water reservoir level in feet's. The decision is represented by diamond. The decision step is based on the selection of input parameters (Fig. 5).

If the input parameters are successfully determined there are multiple possible ways which are mentioned as follows.

- The first option is to select the numerical values of input parameters and use the parameters an input to ARC GIS simulation tool.
- The next step is to calculate the flood frequency and runoff process using ARC GIS simulation tool.
- Finally the results are generated on the basic of graphs.

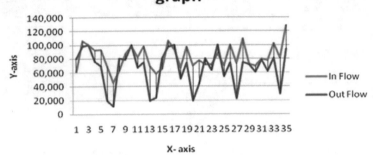

Fig. 6 Graphical representation of actual inflow and outflow graph

- The second option is to select the numerical values of input parameters and apply the values on the proposed mathematical model to generate the results including graphs.
- Finally perform the comparative analysis of both the techniques and prepare the rescue authorities for emergency operations.

5 Results and Discussion

The data has been obtained from Pakistan metrological department during the month of September 2014 [24]. The structure of data is based on designed capacity of water storage level of River Indus. It also depends upon the actual in Flow and outflow of water level based on reservoir elevation. The comparative analysis had been obtained on the basis of numerically weather prediction normally classified as rainfall intensity.

Figure 6 provides the graphical information related to actual Inflow and outflow depending on water reservoir level. The redline indicated the maximum outflow of water depending upon its storage capacity and blue line indicates the moderate level of actual inflow depending on reservoir storage designed capacity.

Figure 7 provides the graphical representation of very high intensity of actual inflow and outflow depending on multiple numerical ranges. The numerical ranges along x-axis and y-axis provide the information related to the water level measured in cubic centimeters. It has been observed from the simulation results that the outflow of water as mentioned in red colour increases depending on the storage reservoir level elevation and capacity to store the water in maximum cubic centimeters.

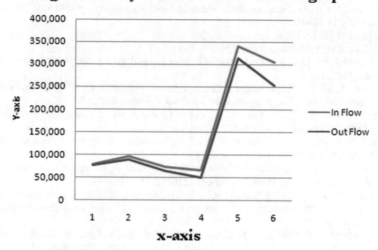

Fig. 7 Graphical representation of very high intensity range inflow and outflow

6 Conclusion and Future Work

This research paper investigates how to build standard rules and regulations to be followed in case of any flood emergency circumstances. Majority of authors have discussed about the applications of GIS in flood circumstance. The proposed flood hazard model focuses on the comparative analysis of the calculated and estimated results. On the basis of these results we can predict the intensity of flood disaster in the specific region of interest. The proposal is based on a combination of GIS, Wireless Sensor Network and Artificial Neuro-Fuzzy Inference System.

References

1. Seal V, Raha A, Maity S, Mitra SK, Mukherjee A, Naskar MK (2012) A real time multi-variate robust regression based flood prediction model using polynomial approximation for wireless sensor network based flood forecasting systems. In: Advances in computer science and information technology. Springer, Berlin, Heidelberg, pp 432–441
2. Rao KHVD, Rao VV, Dadhwal VK, Behera G, Sharma JR (2011) A distributed model for real-time flood forecasting in the Godavari basin using space inputs. Int J Disaster Risk Sci 2(3):31–40
3. Biondi D, De Luca DL (2013) Performance assessment of a Bayesian forecasting system (BFS) for real-time flood forecasting. J Hydrol 479:51–63
4. Cloke HL, Pappenberger F (2009) Ensemble flood forecasting: a review. J Hydrol 375(3): 613–626
5. Fiorentino M, Manfreda S, Iacobellis V (2007) Peak runoff contributing area as hydrological signature of the probability distribution of floods. Adv Water Resour 30(10):2123–2134

6. Rozalis S et al (2010) Flash flood prediction using an uncalibrated hydrological model and radar rainfall data in a mediterranean watershed under changing hydrological conditions. J Hydrol 394(1):245–255
7. Ren M et al (2010) Classified real-time flood forecasting by coupling fuzzy clustering and neural network. Int J Sedim Res 25(2):134–148
8. Nie C et al Spatial and temporal changes in flooding and the affecting factors in China. Nat Hazards 61(2):425–439
9. Alfieri L, Thielen J, Pappenberger F (2012) Ensemble hydro-meteorological simulation for flash flood early detection in southern Switzerland. J Hydrol 424:143–153
10. Ahmad N et al (2013) Flood prediction and disaster risk analysis using GIS based wireless sensor networks, a review. Int J Adv Sci Technol
11. Chen J, Hill AA, Urbano Lensyl D (2009) A GIS-based model for urban flood inundation. J Hydrol 373(1):184–192
12. Neal Jeffrey C, Atkinson Peter M, Hutton Craig W (2012) Adaptive space–time sampling with wireless sensor nodes for flood forecasting. J Hydrol 414:136–147
13. Hughes D et al (2006) Using grid technologies to optimise a wireless sensor network for flood management. In: Proceedings of the 4th international conference on embedded networked sensor systems. ACM
14. Atta-ur-Rahman, Qureshi IM, Malik AN, Naseem MT (2016) QoS and rate enhancement in DVB-S2 using fuzzy rule base system. J Intell Fuzzy Syst (JIFS) 30(1):801–810
15. Atta-ur-Rahman, Qureshi IM, Malik AN, Naseem MT (2014) Dynamic resource allocation for OFDM systems using differential evolution and fuzzy rule base system. J Intell Fuzzy Syst (JIFS) 26(4):2035–2046. https://doi.org/10.3233/ifs-130880
16. Atta-ur-Rahman, Qureshi IM, Malik AN, Naseem MT (2014) A real time adaptive resource allocation scheme for OFDM systems using GRBF-neural networks and fuzzy rule base system. Int Arab J Inf Technol (IAJIT) 11(6):593–601
17. Atta-ur-Rahman (2017) Applications of softcomputing in adaptive communication. Int J Control Theory Appl 10(18):81–93
18. Atta-ur-Rahman, Qureshi IM, Malik AN (2012) Adaptive resource allocation in OFDM systems using GA and fuzzy rule base system. World Appl Sci J (WASJ) 18(6):836–844
19. Atta-ur-Rahman (2017) Applications of evolutionary and neuro-fuzzy techniques in adaptive communications. In: Modeling, analysis and applications of nature-inspired metaheuristic algorithms, 1st edn, chapter: 10. IGI Global, pp 183–217
20. Kuok Kuok King, Harun S, Shamsuddin SM (2010) Particle swarm optimization feedforward neural network for modeling runoff. Int J Environ Sci Technol 7(1):67–78
21. Eberhart RC, Shi Y (2001) Particle swarm optimization: developments, applications and resources, Proc. Proceedings of the IEEE Conference on Evolutionary Computation, ICEC, pp 81–86
22. Wang W-C et al (2009) A comparison of performance of several artificial intelligence methods for forecasting monthly discharge time series. J Hydrol 374(3):294–306
23. http://www.cmpe.boun.edu.tr/ethem/i2ml2e
24. http://www.pmd.gov.pk/FFD/cp/floodpage.htm

An EHSA for RSA Cryptosystem

Manisha Kumari, Deeksha Ekka and Nishi Yadav

Abstract This paper presents a new Enhanced Hybrid Security Algorithm (EHSA) for RSA cryptosystem, which is compared with an encryption algorithm using dual modulus in various aspects such as key generation time, encryption time, decryption time, and data security. The proposed scheme uses eight different prime integers for the computation of two modulus values, which generate the two different public keys and private keys. Thus the complexity involved in factorizing the modulus value increases. Also, the proposed algorithm uses double encryption and decryption using double private and public keys to provide security against Brute-force attack. Therefore, if an intruder detects a single key of cryptosystem even then it is not possible to decrypt the message. Also, it is difficult to factorize the modulus value into its four prime factors. Thus, it enhances the security of encrypted data two times.

Keywords EHSA · RSA cryptosystem · Dual modulus · Prime number
Encryption · Decryption

1 Introduction

The degree of protection to harm the data and resistance to harm the data is known as security. Data security is ensured when it includes confidentiality, integrity, and availability of data. Confidentiality ensures that given information can only be accessed by authorized person. Integrity specifies the originality of data, and ensures that data is not being modified [1, 2]. Availability refers to assurance that user has access to information anytime and to anywhere in the network [3].

M. Kumari · D. Ekka · N. Yadav (✉)
School of Studies in Engineering and Technology, Guru Ghasidas University,
Bilaspur, CG, India
e-mail: nishidv@gmail.com

M. Kumari
e-mail: manisha1921995@gmail.com

D. Ekka
e-mail: deekshaekka88@gmail.com

© Springer Nature Singapore Pte Ltd. 2019
M. L. Kolhe et al. (eds.), *Advances in Data and Information Sciences*, Lecture Notes in Networks and Systems 39, https://doi.org/10.1007/978-981-13-0277-0_31

As the communication over Internet is increasing day by day, intruder would use more advanced techniques and thus information security over the Internet becomes more vital. Nowadays, it is common to exchange personal data on the Internet. So data security is crucial. In order to communicate over Internet the sender has to encrypt the message or plain text with receiver's public key and then the receiver has to decrypt the encrypted text or cipher text using private key [4].

Cryptography for data security is a very powerful method for protection of data from being stolen. Cryptography is a method to encode the information to keep the information being hacked by a third party [3]. The most popular method of encryption is symmetric-key encryption. In this method, the same key is used for both encryption and decryption process. Symmetric-key encryption takes place either in block cipher or in a stream cipher. As we are using the same key for both the encryption and decryption process, thus the computational power of this encryption technique is small.

While in asymmetric-key encryption technique, different keys are used for encryption and decryption process. It is also known as public key encryption. This encryption technique is slow and impractical in case of large amount data [5].

The hash function involves a mathematical function to irreversibly encrypt the data. It consists of algorithms like message digest, hash function algorithm [6]. Cloud computing is widely accepted around the world. But the security of data on cloud server is a challenging issue. The best way to secure the information on cloud server is a security algorithm [7–9].

2 Background Details and Related Work

2.1 Rivest, Shamir, and Adleman (RSA) Algorithm

In conventional RSA scheme, two large prime numbers say "p" and "q" are used for the computation of variable n. The security of RSA algorithm depends on the difficulty of factorization of large composite integer "n" [10–12]. The RSA scheme is as follows [13]:

Key generation:

1. Select two large random prime numbers, p and q of approximately equal size that their product $n = p * q$ is desired bit length.
2. Compute $n = p * q$ and $\varnothing(n) = (p-1) * (q-1)$.
3. Choose a positive integer e, such that $1 < e < \varnothing(n)$, such that GCD $(e, \varnothing(n)) = 1$.
4. Compute the value of secret Exponent d, $1 < d < \varnothing(n)$, such that $e * d = 1 (\bmod \varnothing(n))$.
5. The pair (e, n) is a public key and (d, n) is a private key. The values d, p, q and $\varnothing(n)$ should be kept as a secret.

Where,

- n is a modulus.
- e is the public exponent or encryption exponent or simply the exponent.
- d is the secret exponent or decryption exponent.

Encryption:

Suppose user A wants to send message 'm' to user B

1. Obtain the public key (e, n) of user B.
2. Represent the plaintext as positive integer m.
3. Compute the ciphertext $c = m^e$ mod n, using user B's public key.
4. Send the cipher text c to user B.

Decryption:

User B will retrieve the original message from ciphertext sent by the user A.

1. Use private key (d, n) to compute $m = c^d$ mod n.
2. Extract the plaintext m from c.

2.2 Encryption Algorithm Using Dual Modulus

In this algorithm, double-mod operation is applied in encryption as well as decryption process using two private keys "e1" and "e2" and public keys "d1", and "d2" respectively. More than two large prime numbers say "p1", "p2", "q1", and "q2" are used for generation of modulus values say "n1" and "n2". Dual modulus encryption technique seems to be impractical, as it takes a large amount of computational time for the generation of public key and private key. But it also enhances the security two times. Dual modulus algorithm is as follows [14]:

Key Generation:

1. Select four large random primes, p1, p2 and q1, q2 of an approximately equal size.
2. Compute $n1 = p1 * q1$ and $n2 = q1 * q2$.
3. Compute Euler's totient $\emptyset(n1) = (p1-1) * (p2-1)$ and $\emptyset(n2) = (q1-1) * (q2-1)$.
4. Choose two positive integer e1, e2, such that $1 < e1 < \emptyset(n1)$, GCD $(e1, \emptyset(n1)) = 1$ and $1 < e2 < \emptyset(n2)$, GCD $(e2, \emptyset(n2)) = 1$.
5. Compute the secret Exponent d1, d2, such that $1 < d1 < \emptyset(n1)$, $e1 * d1 = 1$ (mod $\emptyset(n1)$) and $1 < d2 < \emptyset(n2)$, $e2 * d2 = 1$ (mod $\emptyset(n2)$).
6. The public key is (e1, e2, n1, n2) and the private key is (d1, d2, n1, n2). Keep all the values d1, d2, p1, p2, q1, q2, $\emptyset(n1)$ and $\emptyset(n2)$ secret.

Where,

- n1 and n2 are known as modulus.
- e1 and e2 are known as the public exponent or encryption exponent or simply the exponent.
- d1 and d2 are known as the secret exponent or decryption exponent.

Encryption:
Suppose, user A wants to send a message to user B.

1. Obtain the public key (e1, e2, n1, n2).
2. Represent the plaintext as positive integer m.
3. Compute the ciphertext $c = ((m^{e1} \bmod n1)^{e2} \bmod n2)$.
4. Send the cipher text c to user B.

Decryption:
Now, user B will retrieve the original message.

1. Use private key (d1, d2, n1, n2) to compute $m = ((c^{d2} \bmod n2)^{d1} \bmod n1)$.
2. Extract the plaintext m from c.

2.3 Enhanced Method for RSA Cryptosystem Algorithm

In this algorithm, we are using three large prime numbers say "p", "q", and "r" instead of two prime numbers to generate public and private key. Here in ERSA, it is more difficult to factorize the modulus "n" into its three prime factors [15]. Thus, it enhances the security of conventional RSA. Steps involved in this algorithm is as follows:

Key Generation:

1. Generate three large random primes, p, q, and r of an approximately equal size that their product $n = p * q * r$ is desired bit length.
2. Compute $n = p * q * r$ and $\emptyset(n) = (p - 1) * (q - 1) * (r - 1)$.
3. Choose a positive integer e, such that $1 < e < \emptyset(n)$, such that GCD (e, $\emptyset(n)$) = 1.
4. Compute the secret Exponent d, $1 < d < \emptyset(n)$, such that $e * d = 1 \pmod{\emptyset(n)}$.
5. The public key is (e, n) and the private key is (d, n). Keep all the values d, p, q and $\emptyset$ secret.
    Where,

    - n is known as the modulus.
    - e is known as the public exponent or encryption exponent or simply the exponent.
    - d is known as the secret exponent or decryption exponent.

Encryption:
User A wants to send message "m" to user B.

1. Obtain the public key (e, n).
2. Represent the plaintext as positive integer m.
3. Compute the ciphertext $c = m^e$ mod n.
4. Send the cipher text c to user B.

Decryption:
User B will retrieve the message from the ciphertext.

1. Use private key (d, n) to compute $m = c^d$ mod n.
2. Extract the plaintext m from c.

3   Proposed Approach

The basic idea of the proposed approach is based on encryption algorithm using dual modulus and enhanced method for RSA cryptosystem algorithm. Using dual modulus in proposed algorithm, we introduce double-mod operation-based encryption and decryption using two public keys say "e1" and "e2" and two private keys say "d1" and "d2", respectively. Using dual modulus operation in proposed algorithm improves the security of data as compared with RSA cryptosystem to a very large extent. In the proposed approach, even if an intruder gets to succeed in detection of private key, still our data is secured as we have encrypted our data with two different public keys. Thus, the proposed algorithm is more secure than conventional RSA cryptosystem. Enhanced method for RSA cryptosystem algorithm is being used in the proposed algorithm provides an idea of using four prime numbers for the calculation of each modulus value such as "p1", "p2", "p3", and "p4" are used for the generation of "n1" and "q1", "q2", "q3", and "q4" are used for the generation of "n2" [14]. It is more difficult to factorize the modulus value into its four composite prime factors, it might increase the complexity of computation but it also enhances the security of data in the proposed approach. Here, in the proposed scheme, totally eight prime numbers has been used to generate the public key and private key.

Features of the proposed algorithm for RSA cryptosystem are as follows:

1. Four prime numbers are used to generate modulus say n1 and n2 [14].
2. Double-mod operation-based encryption using two public keys [4].
3. Double-mod operation-based decryption using two private keys [4].

Steps involved in the proposed scheme are as follows:

Key Generation:

1. Generate large, random primes p1, p2, p3, p4, p5, p6, p7, p8.
2. Compute n1 = p1 * p2 * p3 * p4.
3. n2 = q1 * q2 * q3 * q4.

4. Compute Euler's totient function.
5. $\emptyset(n1) = (p1 - 1) * (p2 - 1) * (p3 - 1) * (p4 - 1)$.
6. $\emptyset(n2) = (q1 - 1) * (q2 - 1) * (q3 - 1) * (q4 - 1)$.
7. Generate two integers e1 and e2 such that $1 < e1 < \emptyset(n1)$, $GCD(e1, \emptyset(n1)) = 1$ and $1 < e2 < \emptyset(n2)$, $GCD(e2, \emptyset(n2)) = 1$.
8. Compute the private key d1 and d2 such that $e1 * d1 \bmod \emptyset(n1) = 1$ and $e2 * d2 \bmod \emptyset(n2) = 1$.
9. (e1, e2, n1, n2) is public key and (d1, d2, n1, n2) is private key.

   - n1 and n2 are modulus.
   - e1 and e2 are public exponent or encryption exponent.
   - d1 and d2 are private exponent or decryption exponent.

Encryption:
User A wants to send message "m" to user B.

1. Obtain the private key of user B, i.e., (e1, n1) and (e2, n2).
2. Represent the message as positive integer m.
3. Encrypt the plaintext m by $c = ((m^{e1} \bmod n1)^{e2} \bmod n2)$.
4. Send the cipher text c to user B.

Decryption:
User B will retrieve the original message as

1. Decrypt the ciphertext c by $m = ((c^{d2} \bmod n2)^{d1} \bmod n1)$.
2. Extract the plaintext m from the cipher c.

The public keys "e1" and "e2" and private keys "d1" and "d2" are dependent on "n1" and "n2", respectively. In this scheme, plain text "m" is converted into final cipher text "c2" by applying encryption process twice and the original message "m" can be retrieved by the receiver by applying decryption process twice.

Proof cipher text generated by sender by using message "m" is initially encrypted using "e1" to generate intermediate cipher text "c1" and then "c1" is encrypted again using "e2" to generate final cipher text c2. Encryption process is done twice as $c1 = m^{e1} \bmod n$ and $c2 = c1^{e2} \bmod n$. The same process is followed for decryption, "c2" is decrypted into "c1" using "d2" and "c1" is decrypted using "d1" to retrieve the original message "m". Decryption process is done twice as $c1 = c2^{d2} \bmod n$ and $m = c1^{d1} \bmod n$.

Example:

1. Select prime numbers $p1 = 31$, $p2 = 37$, $p3 = 41$, $p4 = 43$, $q1 = 47$, $q2 = 53$, $q3 = 59$, and $q4 = 61$.
2. Calculate modulus value $n1 = p1 * p2 * p3 * p4 = 2,022,161$ and $n2 = q1 * q2 * q3 * q4 = 8,965,109$.
3. Calculate totient function phi, $\Phi(n1) = (p1 - 1) * (p2 - 1) * (p3 - 1) * (p4 - 1) = 1,814,400$ and $\Phi(n2) = (q1 - 1) * (q2 - 1) * (q3 - 1) * (q4 - 1) = 8,324,160$.

4. Choose public exponent e1 and e2 such that $1 < e1 < \Phi(n1)$ and gcd(a1, $\Phi(n1)) = 1$ and $1 < e2 < \Phi(n2)$ and gcd(e2, $\Phi(n2)) = 1$. Then, $e1 = 1,814,383$ and $e2 = 8,324,143$.

5. Compute private exponent d1 and d2 such that $e1 \times d1 \bmod(\Phi(n1)) = 1$ and $e2 \times d2 \bmod(\Phi(n2)) = 1$. Then $d1 = 533,647$ and $d2 = 7,344,847$.

6. Public key: $(e1, n1) = (1,814,383, 2,022,161)$ and $(e2, n2) = (8,324,143, 8,965,109)$. Private Key: $(d1, n1) = (533,647, 2,022,161)$ and $(d2, n2) = (7,344,847, 8,965,109)$.

Say A wants to encrypt the message $m = 21$

- Encryption:

$$c = \left((m)^{e1} \bmod n1\right)^{e2} \bmod n2)$$

$$= \left((21)^{1,814,383} \bmod 2,011,161\right)^{8,324,143} \bmod 8,965,109) = 6\tilde{n}$$

Say B wants to decrypt the cipher text c, then

- Decryption:

$$m = \left((c)^{d2} \bmod n2\right)^{d1} \bmod n1)$$

$$= \left((6\tilde{n})^{7,344,847} \bmod 8,965,109\right)^{533,647} \bmod 2,022,161) = 21$$

4 Experimental Setup and Results

The proposed algorithm is being compared with RSA using dual modulus on the basis of following parameters such as

1. Key generation time
2. Encryption time
3. Decryption time
4. Security of data.

Performance of EHSA by taking various combinations of prime numbers for desired key length is given in Table 1.

Results:
1. Key Generation Time
The proposed algorithm takes less time than RSA using dual modulus in key generation. Using both dual modulus operation and eight prime numbers might increase the complexity but it also enhances the security. Comparison of key generation time in the proposed algorithm and RSA with dual modulus for various combinations of prime number is shown in Fig. 1.

Table 1 Performance of EHSA

Combinations	Key generation time (ms)	Encryption time (ms)	Decryption time (ms)
1	915	115	190
2	928	131	172
3	1126	124	169
4	1262	128	199
5	1276	117	186
6	1312	121	178
7	1390	127	193
8	1487	120	185
9	1487	131	199
10	1677	131	201

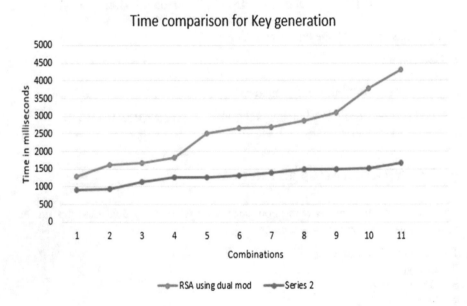

Fig. 1 Analysis of key generation time

2. Encryption Time

Proposed algorithm takes comparatively equal encryption time than RSA using dual modulus but it also improves the security of data. The comparison of encryption time is shown in graph in Fig. 2.

3. Decryption Time

There is a negligible difference in decryption time of EHSA and RSA using dual modulus. Therefore, decryption time of EHSA and RSA using dual modulus are comparatively equal. This comparison is shown in graph in Fig. 3.

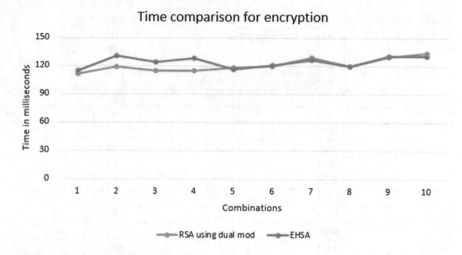

Fig. 2 Analysis of encryption process time

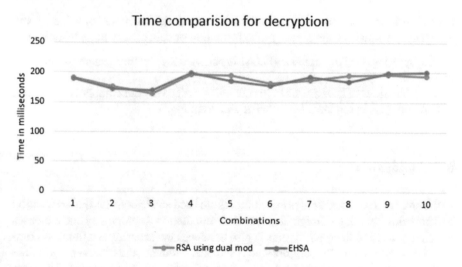

Fig. 3 Analysis of decryption process time

4. Security of Data

Proposed scheme is more secure as compared to conventional RSA cryptosystem, ERSA (Enhanced method for RSA cryptosystem) and RSA with dual modulus. Proposed scheme is using double encryption and decryption with two different public keys and two different private keys along with eight prime integer has been used

Table 2 Comparison of EHSA with dual-mod RSA in terms of key generation time

Combinations	Key generation time (ms)		EHSA versus Dual-mod RSA (Speed up (%))
	Dual-mod RSA	EHSA	
1	1292	915	29.18
2	1611	928	42.40
3	1676	1126	32.82
4	1818	1262	30.58
5	2664	1276	52.10
6	2698	1312	51.37
7	2868	1390	51.53
8	3084	1487	51.78
9	3789	1487	60.75
10	4319	1677	61.17

instead to four or six prime integers for the computation of modulus value n1 and n2. Hence, it enhances the security of information or data at very large extent.

The speed up (%) of EHSA and Dual Modulus RSA for Key generation is given in Table 2.

Ten combinations of keys are taken in account to deduce the fair comparison between EHSA and RSA using Dual Modulus as shown in Fig. 1.

5 Conclusions

In the proposed scheme, eight prime numbers are used and encryption and decryption had been done twice, so it might increase the computational complexity and increases the encryption and decryption time. It also enhances the security two times as compared to conventional RSA cryptosystem and also it reduces the key generation time in the proposed scheme than RSA using dual modulus. Also, the proposed scheme is more secure against the Brutal-Force Attack than conventional RSA. Even if an intruder detects the private key still it is not possible to retrieve the information from the ciphertext as sender has encrypted the information twice. Therefore, the proposed algorithm improves the performance and security of RSA using dual modulus.

References

1. Suja GJ (2016) New approach for highly secured I/O transfer with data on timer streaming. IEEE

2. Aboud SJ, AL-Fayoumi MA, Al-Fayoumi M, Jabbar HS An efficient RSA public key encryption scheme. In: Fifth international conference on information technology: new generations
3. Ramaporkalai T (2017) Security algorithm in cloud computing. Int J Comput Sci Trends Technol (IJCST-2017)
4. Goel A, Manu (2017) Encryption algorithm using dual modulus. In: 3rd IEEE international conference on computational intelligence and communication technology. IEEE—CICT
5. Jeeva AL, Dr Palanisamy V, Kanagaram K (2012) Comparative analysis of performance efficiency and security measures of some encryption. Int J Eng Res Appl (IJERA)
6. Pancholi VR, Dr Patel BP (2013) Enhancement of cloud computing security with secure data storage using AES. Int J Innov Res Sci Technol (IJIRST-2016)
7. Madaan S, Agarwal RK Implementation of identity based distributed cloud storage encryption scheme using PHP and C for Hadoop file system
8. Prabu Kanna G, Vasudevan V (2016) Enhancing the security of user data using the keyword encryption and hybrid cryptographic algorithm in cloud. In: International conference on electrical, electronics and optimization techniques (ICEEOT)
9. Osseily HA, Haidar AM, Kassem A Implementation of RSA encryption using identical modulus algorithm
10. Verma S, Garg D (2014) An improved RSA variant. Int J Adv Technol
11. Bhandari A, Gupta A, Das D (2016) Secure algorithm for cloud computing and its applications. IEEE
12. Dongjiang L, Yandan W, Hong C The research on key generation in RSA public-key cryptosystem. In: 2012 Fourth international conference on computational and information sciences
13. Ambedkar BR, Gupta S, Goutam P, Bedi SS (2011) An efficient to factorize the RSA public key encryption. In: 2011 international conference on communication system and network technologies
14. Panda PK, Chhattopadhyay S (2017) A hybrid security algorithm for RSA cryptosystem. In: International conference on advanced computing and communication system (ICACCS)
15. Prof Dr Al-Hamami AH, Aldariseh IA (2012) Enhanced method for RSA cryptosystem algorithm. In: International conference on advance computer science applications and technologies

Securing ZRP Routing Protocol Against DDoS Attack in Mobile Ad Hoc Network

Munesh C. Trivedi, Sakshi Yadav and Vikash Kumar Singh

Abstract Hybrid routing protocols especially zone routing protocol (ZRP) is getting much attention in recent years due to its superior characteristics over pro-active and reactive routing protocols. Because of its wide applicability in ad hoc networks, the demand for improving its quality of services (QoS) is also increased day by day. In most of the applications the QoS of any system is measured by their availability. It means system availability is the critical parameter of QoS. However, to degrade the QoS of any system attacker performs the denial of service (DoS) attack. Moreover, the DoS attack is strengthened by making it distributed in nature called the Distributed DoS (or DDoS) attack. Securing ZRP routing protocol in ad hoc environment is challenging due to fundamental characteristics of ad hoc networks. In this paper a new model is introduced to make routing protocol secure against DDoS attack in MANET. The proposed model uses intrusion prevention scheme in which some nodes are specified as IPN node to protect the network from DDoS attack. These specified IPS nodes scan the neighbor nodes regularly which comes in its zone and radio range for identifying the misbehavior nodes. On the identification of these misbehavior (i.e., frequent packet passing other than TCP and UDP packets) nodes, these node are blocks by IPN nodes. IPN nodes also send the blocked nodes information to all of the legitimate sender nodes in the network to change their routes. To testify the effectiveness of the proposed model against DDoS attack experiments performed on network simulator tool (NS2.35) on Ubuntu 12.04 LTS platform. Simulation results show that proposed model makes ZRP protocol efficient to handle DDoS attack.

Keywords Distributed denial of services attack · Intrusion prevention · ZRP routing protocol · MANET

M. C. Trivedi (✉) · S. Yadav
Department of Computer Science & Engineering, ABES Engineering College, Ghaziabad, India
e-mail: munesh.trivedi@gmail.com

S. Yadav
e-mail: sakshi91193@gmail.com

V. K. Singh
Department of Computer Science & Engineering, IGNTU, Amarkantak, India
e-mail: drvksingh76@gmail.com

© Springer Nature Singapore Pte Ltd. 2019
M. L. Kolhe et al. (eds.), *Advances in Data and Information Sciences*, Lecture Notes in Networks and Systems 39, https://doi.org/10.1007/978-981-13-0277-0_32

1 Introduction

MANET is enthralling and emerged as a new field of research to resolve various real-world challenges. MANET is the group of communicating devices that have small-size batteries, simple circuit, cheap, limited memory, and small processing capability. [1–3]. In future, MANET is supposed to be consist of thousands of cheaper devices, each device having more networking power with limited computational and communication power [4–9] which provides the freedom to establish a scalable mobile ad hoc network. A wireless mobile ad hoc network contained of numerous small nodes that are capable to send and receive the messages. Due to the fundamental properties and working style (like the absence of central controller, wireless communication channel, shared medium, limited computing capability, small storage, short communication range, limited bandwidth, etc.) [10–12] of ad hoc networks, imposing the security scheme is very difficult. Specially for protecting the networks against DDoS attacks. To solve this problem in past number of methods, models, frameworks and algorithms has been proposed. Some of these works are general and they can handle any type of attacks while other are attack specific and handle only the specific attack (such as black hole, gray hole, denial of services attack, etc.) for which they have designed. Some of the good works in this field are discussed below.

Qian et al. [13] proposed a framework which provides two features first is security and second is survivability these both features are very important for several applications in wireless sensor network. They have proposed the security and survivability in one architecture with composite sensor nodes. To explain the co-operation between survivability and security, they also design a key management policy for security in this architecture and analyses this scheme for the proposed framework to provide the security and survivability both together. The results of this scheme shows that how a good design can give better improvement for both the security and survivability in wireless sensor networks, but in some situations the framework is not able to balance in survivability and security, so this is the small drawback of this scheme. To overcome this shortcoming Thein et al. [14] proposed framework that is able to provide all services when, attacks and failure of network occurs. All of the above methods are computationally inefficient and may not be suitable for practical real-time applications. To solve the limitations of above methods Sachan and Khilar [15] proposed the method A-AODV. In this hashed message authentication code (HMAC) is utilized for end-to-end authentication. Moreover, to verifying the integrity of the control messages hash chain is utilized. Since this method used HMAC that is verified only by the intended receiver, therefore, we cannot apply this technique to verify and authenticate broadcast message such as RRER message. In 2014, Arya and Rajput [16] have proposed the framework to impose the security scheme in AODV routing protocol using nested message authentication code. In this work author has also utilized the beauty of key pre-distribution technique in which symmetric keys are distributed at the network deployment time to minimize the limitations of work [15] that distribute the security keys at the time when communication establishing between the source node and destination node. The work of [16] efficiently protects the ad hoc networks from many routing attacks. The drawback of this framework is that it is inefficient to handle the insider attacker, i.e., our genuine node is compromised by the attacker. Rajput and Trivedi [17] to extend the work of [16] to make secure MANETs against frequently occurring attacks. Here, zone routing protocol (ZRP) is utilized to major the effectiveness of the proposed security scheme.

All of the methods discussed above efficiently handle the specific type of routing attacks such as black hole, gray hole, sniffing, etc., but handling of denial of services specifically in distributed nature still challenging and open filed for research.

The remaining part of this manuscript is organized as: Sect. 2 describes the proposed methodology. Section 3 explains simulation result and discussion, and finally conclusion given in Sect. 4.

2 Securing ZRP Protocol Against DDoS Attack

Inspired from the above problem, to make the network secure detection and prevention of routing attacks is tremendously difficult task in the case of topology and infrastructure independent network, e.g., MANET, WSN, etc. In most of the real-world applications DDoS attack [18, 19] is crucial. Our prime goal is to develop a security model, which can effectively handle DoS attack in mobile ad hoc network. To handle DDoS attack in this paper a novel security model is developed which is summarized as follows.

2.1 The Proposed Security Model

The proposed security model introduced a novel technique to secure AODV routing protocol in MANET. The proposed model consists of following modules:

1. Procedure to introduce DDoS attack in normal network.
2. Algorithm to protect the routing from attack.

In proposed model, a security module is created to protect the network against DDoS attack. For this proposed prevention algorithm is implemented on the already attack infected network. Here, some specific nodes are chosen and set as intrusion prevention nodes (IPN). The IPN are continuously scanning all the neighbor nodes which comes in its radio range inside the zone of the network for the purpose of identifying the devices which are participating in huge and frequent unwanted packets sending to the particular node of the network. On the detection of such nodes, IPN blocks all activities of these devices in the network and send the reply request to the authentic device, which was asking to the target device for routing their packets.

We proposed two algorithms in our methodology, first algorithm to launch the DDoS attack in the normal scenario and second algorithm to prevent the network from the effect of DDoS attack which are as follows:

1. *Launching the DDoS Attack*:

To launch DDoS attack, in this paper separate module has been designed. It is used to examine the consequence of DDoS attack under ZRP routing protocol. The malicious nodes spread harmful and unwanted packets into the network for the purpose of consuming the network bandwidth. Consequently, authentic nodes are unable to share the important information from the available routes. Detailed pseudocode for launching the DDoS attack is given in Algorithm 1.

2. *The proposed DDoS Attack prevention Algorithm*:

In this paper intrusion prevention system based DDoS attack prevention algorithm is proposed. Here, some specific nodes are set as IPN, which continuously monitor its entire neighbor nodes comes in its radio range for identifying their behavior. On the identification of malicious nodes, IPN keep their malicious activities in its routing table then specific IPN node (node which detect the malicious activity. Let for notation convenient we use notation IP for such node) scan every message, if IP find any abnormal modification then it block that identified malicious node. After this, IP informed the authentic node about this malicious activity. On the receiving of such alert from the IP node, authentic node alters its route computation scheme and search for other alternative route. After this it updates its routing table and sends the messages for destination node using updated secure route. Detailed pseudocode given in Algorithm 2.

ALGORITHM 1: The DDoS attack launching algorithm

Step1: Deploy Mn as DDoS spreading nodes

Step2: Deploy N as authentic nodes

Step3: Set Mn^{th} node as abnormal node // Last node as DDoS.

Step4: Broadcasting capability of malicious nodes = 2000 // each malicious node can broadcast the packets to 2000 nodes in the network

Step5: Fixed size of the packet = 512 bytes.

Step6: Packet speeding speed = 1packet/second

Step7: Set M_Port ; // for spreading malicious packets via Mn nodes.

Step8: Fixed **Scanrate** = 1.

Step9: SPmax = Mn* Packet size; //total packet transmission/sec.

Step10: Ψ= (**SPmax** * **Scanrate** / **abnormal node**); //malicious packets send to all normal nodes.

Step11: Incorporated routing protocol = ZRP;

ALGORITHM 2: Algorithm to Prevent N/W from DDoS Attack

Notations:

M: total number of mobile communicating devices;**S:** is the sender node;// S ∈M

D:is the destination node;   // D ∈M

t_0: is the start simulation time, **R:** is the radio range of each node

IP: intrusion prevention node //IP ∈IPN, IPN ∈M;

Max_Lim: allowed thresholding load for each node

Step1: Deploy M nodes with ZRP routing protocol

Step2:IP monitors all its neighbor nodes and storetheir behavior in routing table;

Pseudo code for this step:

If (Mnth Node load ÃMax_Lim load)

{IPmaitaing table for all Congestion Spreader node **Mn**;

 Send(minimize data rate to Mnth node) **{if** (Mnth node change its data rate)

    { Exist in Route}   **else**   { Send reply packet to**S node** about Mnth node;

IPnode blocksMnth node; } }

Recompute_path (); // initiate fresh rout establishment }

Else: establish route issecure;}

Step3: Recompute_Path (sender, destination, route-pkt)

Pseudo code for this step:

{If (Mn node in**R** ‖ neighbor   == true ‖ Mnth node = false)

{create route table (); Receive packet at destination;   }

Else: destination unreachable or node out of range; }

Step4: Send acknowledgement to S;

Step11: Ssendsinformation through secure path;

Step12: Session is terminated;

The given algorithm is efficient to handle and protect the network from DDoS attack. The proposed algorithm is very simple to understand and implements, and computationally efficient. It blocks all the DDoS infection responsible nodes in the network.

3   Simulation and Result Analysis

In this paper, ZRP routing protocol is used to test the effectiveness of the proposed security scheme. For simulating the proposed algorithm NS-2.35 simulator [20] is used. The required environmental and experimental parameters required to simulate the real-time ad hoc network are described in the below Table 1.

Here, four performance measuring metrics are used which are named as: Average and to End Delay (AE-ED), Packet Delivery Ratio (PDR) [16], Packet Loss Percentage (PLP), and Normalized Routing Load (NRL).

To verify the effectiveness of the proposed algorithm we have to compare the results of the proposed algorithm with original ZRP and ZRP with DDoS attack under various pause time and different number of malicious nodes.

First, the experiments are performed on 1000 nodes under ad hoc environment. Here, different pause times are used to show the impact of nodes mob-ability on

Table 1 Experimental parameters

Routing protocol	ZRP
Simulation tool	NS2(v-2.35)
Transmission range	100 m
Simulation time	150 s
Number of devices	1000
Networking area size	800 m × 600 m
Transmission range	250 m
Node moving speed	0–20 m/s
Application traffic	CBR
Number of malicious nodes	10, 20, 30, …, 100
Size of packets	512 bytes
Traffic rate	4 packet/s
Mobility model	Random way-point model
Mac model	802.15.4

Fig. 1 Pause time versus PDR

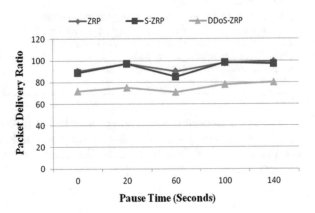

the performance. These results are shown in Figs. 1, 2, 3 and 4. Figure 1 show the packet delivery ratio for original ZRP, ZRP with DDoS attack and ZRP with our proposed secure scheme (S-ZRP). Here, approximately 5% of the total nodes are worked as malicious nodes. Figures 2, 3 and 4 show the AE-ED, PPL and NRL under 1%, 9%, 4% malicious nodes respectively. Form these figure we can conclude that the proposed algorithm effectively imposed the security mechanism with negligible overhead.

Additionally, to further verify the effectiveness of the proposed algorithm used different number of malicious nodes experiments are also performed for this and performances are reported on above four metrics in Figs. 5, 6, 7 and 8. Here, experiments are computed on the presence of 10, 20, 30, 40, 50, 60, 70, 80, 90, and 100 malicious nodes. Here, pause time is fixed to 140 s. From the results shown in Figs. 5,

Fig. 2 Pause time versus AE-ED

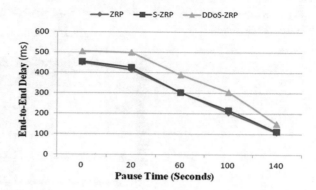

Fig. 3 Pause time versus PPL

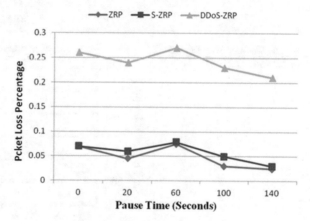

Fig. 4 Pause time versus NRL

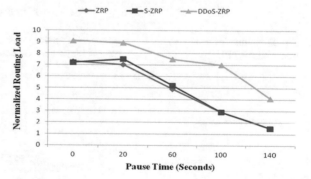

6, 7 and 8 we can conclude that the proposed algorithm can effectively handle the small as well as heavy malicious activities.

Fig. 5 Number of malicious
nodes versus PDR

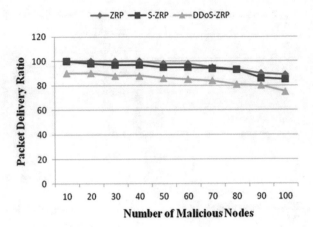

Fig. 6 Number of malicious
nodes versus AE-ED

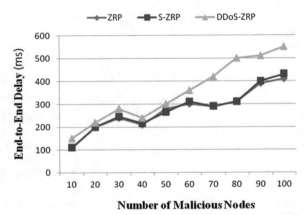

Fig. 7 Number of malicious
nodes versus PPL

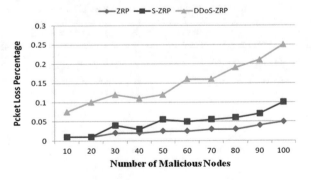

Fig. 8 Number of malicious nodes versus NRL

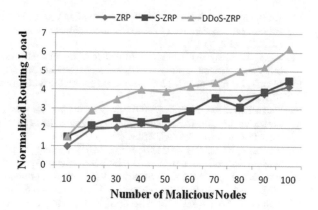

4 Conclusion

In this work, a novel security scheme for protecting the MANET against DDoS attacks has been proposed. Experiments are performed on three scenarios to analyze the performance. Here, ZRP routing protocol is utilized for experimenting the attack at network layer. The simulation results show that proposed security algorithm successfully protects the ad hoc network against DDoS attacks with negligible overhead increment.

References

1. Kalwar S (2010) Introduction to reactive protocol. IEEE Potentials 29(2):34–35
2. Song J-H, Wong VW, Leung VC (2004) Efficient on-demand routing for mobile ad hoc wireless access networks. IEEE Trans Sel Areas Commun 22(7):1374–1383
3. Zapata MG, Asokan N (2002) Securing ad hoc routing protocols. In: Proceedings of the 1st ACM workshop on wireless security. pp 1–10
4. Abusalah L, Khokhar A, Guizani M (2008) A survey of secure mobile ad hoc routing protocols. IEEE Commun Surv Tutor 10(4):78–93
5. Sharma AK, Trivedi MC (2016) Performance comparison of AODV, ZRP and AODVDR routing protocols in MANET. In: 2nd IEEE international conference CICT, Feb 2016. pp 12–13
6. Trivedi MC, Sharma AK (2015) QoS Improvement in MANET using particle swarm optimization algorithm. In: Proceedings of the 2nd Springer's international congress on information and communication technology (ICICT-2015). Udaipur, Rajasthan, India, pp 181–189
7. Rajput SS, Kumar V, Dubey K (2013) Comparative analysis of AODV and AODV-DOR routing protocol in MANET. Int J Comput Appl 63(22):19–24
8. D'Souza R, Varaprasad G et al (2012) Digital signature-based secure node disjoint multipath routing protocol for wireless sensor networks. Sens J IEEE 12(10):2941–2949
9. Sharma N, Gupta A, Rajput SS Yadav V (2016) Congestion control technique in MANET: a survey. In: 2nd IEEE international conference on CICT, Feb 2016. pp 280–282
10. Chakrabarti S, Mishra A (2001) Qos issues in ad hoc wireless network. IEEE Commun Mag 39(2):142–148

11. Xu B, Hischke S, Walke B (2003) The role of ad hoc networking in future wireless communications. In: Proceedings of the IEEE international conference on communication technology (ICCT), vol 2. pp 1353–1358
12. Zhang Y, Soong B-H (2006) Performance of mobile networks with wireless channel unreliability and resource inefficiency. IEEE Trans Commun 5(5):990–995
13. Qian Y, Lu K, Tipper D (2007) A design for secure and survivable wireless sensor networks. Wirel Commun IEEE 14(5):30–37
14. Thein T, Lee SM, Park JS (2009) Improved method for secure and survivable wireless sensor networks. In: 11th IEEE international conference on sensor networks proceedings computer modelling and simulation UKSIM'09. pp 605–610
15. Sachan P, Khilar PM (2011) Securing AODV routing protocol in MANET based on cryptographic authentication mechanism. Int J Netw Secur Appl (IJNSA) 3(5)
16. Arya KV, Rajput SS (2014) Securing AODV routing protocol in MANET using NMAC with HBKS technique. In: IEEE international conference on SPIN, Feb 2014. pp 281–285
17. Rajput SS, Trivedi MC (2014) Securing ZRP routing protocol in MANET using authentication technique. In: IEEE international conference on CICN, Nov 2014. pp 872–877
18. Stalling W (2006) Cryptography and network security, 4th edn. Pearson Education, India
19. Forouzan BA (2008) Cryptography and network security, 2nd edn. Tata McGraw-Hill Higher Education, India
20. Issariyakul EHT (2009) Introduction to network simulator NS2. Springer Science and Business Media, NY, USA

Printed in the United States
By Bookmasters